Strong Interactions at Low and Intermediate Energies

Proceedings of the 13th Annual HUGS at CEBAF

Strong Interactions at Low and Intermediate Energies

Jefferson Laboratory, USA 26 May-12 June 1998

Editor

J. L. Goity

Hampton University, USA

World Scientific

Singapore • New Jersey • London • Hong Kong

Published by

World Scientific Publishing Co. Pte. Ltd.
P O Box 128, Farrer Road, Singapore 912805
USA office: Suite 1B, 1060 Main Street, River Edge, NJ 07661
UK office: 57 Shelton Street, Covent Garden, London WC2H 9HE

British Library Cataloguing-in-Publication Data
A catalogue record for this book is available from the British Library.

STRONG INTERACTIONS AT LOW AND INTERMEDIATE ENERGIES

ISBN 981-02-4279-4

Printed in Singapore by Regal Press (S) Pte. Ltd.

CONTENTS

International Advisory Committee

O. K. Baker (Hampton University and Jefferson Lab),
W. W. Buck (Hampton University),
R. Carlini (Jefferson Lab),
K. deJager (Jefferson Lab),
F. L. Gross (College of William and Mary and Jefferson Lab),
N. Isgur (Jefferson Lab),
J-M. Laget (Saclay),
B. Mecking (Jefferson Lab),
M. P. Locher (Paul Scherrer Institute),
W. Roberts (Old Dominion University and Jefferson Lab),
J. D. Walecka (College of William and Mary).

Local Organizing Committee

J. L. Goity (Chair) (Hampton University and Jefferson Lab),
S. Ewing (Jefferson Lab),
P. Gueye (Hampton University),
C. Pittman (Hampton University),
R. Williams (Hampton University).

LECTURERS

Henk Blok Vrije Universiteit Amsterdam and NIKHEF.

Nora Brambilla University of Vienna.

Lawrence Cardman Jefferson Lab.

Thomas Cohen University of Maryland.

Charles Hyde-Wright Old Dominion University.

Peter Lepage Cornell University.

James Napolitano Rensselaer Polytechnic Institute.

Anatoly Radyushkin Old Dominion University, and Jefferson Lab.

Paul Souder Syracuse University.

LIST OF PARTICIPANTS

Ciciani, Luca	Institute de Physique Nucleaire, Orsay, France.
Horton, Theodore	North carolina State University, NC, USA.
Lebedev, Oleg	Virginia Polytechnic Institute, VA, USA.
Ivanov, Emil	University of Notre Dame, IN, USA.
Malov, Denis	Vanderbilt University, TN, USA.
Niyazov, Rustam	Old Dominion University, VA, USA.
Radu, Aurelian	Institute of Space Science, Bucharest, Romania.
Ramalho, Gilberto	University of Lisbon, Lisbon, Portugal.
Seal, Sudip	New Mexico State University, NM, USA.
Shinozaki Akihisa	University of Regina, SK, Canada.
Skabelin, Alexander	MIT, MA, USA.
Thaxton, Christopher	North Carolina State University, NC, USA.
Van der Meer, Robertus	Vrije University Amsterdam, The Netherlands.
Vansyoc, Kelley Gene	Old Dominion University, VA, USA.
Volmer, Jochen	Vrije University Amsterdam, The Netherlands.
McKee, David	New Mexico State University, NM, USA.
Puglia, Steven	University of Connecticut, CT, USA.
Llanes Estrada, Felipe	North Carolina State University, NC, USA.
Hafidi, Kawtar	DAPNIA/SPhN, Saclay, France.
Farah, Yusuf	North carolina A&M University, NC, USA.
De Vita, Raffaella	University of Genova, genova, Italy.

PREFACE

The 13th annual Hampton University Graduate Studies at the Continuous Electron Beam Accelerator Facility (HUGS at CEBAF) took place at Jefferson Lab and at Hampton University from May 26 to June 12, 1998. The three-week program provided theoretical and experimental lectures on several areas of strong interactions that are of high current interest, and attracted an attendance of more than twenty graduate students. This volume contains lectures that were offered in the program and represents a good cross section of the material that was presented. The main focus of the lectures is strong interaction physics at low and intermediate energies, and convey the current theoretical and experimental status of several key developments that are helping understand the non-perturbative regime of QCD.

The contents of this volume are as follows: Henk Blok's lectures address the foundations of electron scattering on nucleons and nuclei and give an ample view of this extended field; Peter Lepage's lectures give an introduction to Lattice QCD with its methods and recent developments such as improved actions, and also include examples of lattice simulations; Anatoly Radyushkin's lectures give an introduction to the important method of QCD sum rules and a number of applications that are currently very important; Nora Brambilla's lectures discuss quark confinement including approaches that are based on QCD and several applications; Jim Napolitano's lectures give an introduction to the experimental search and study of meson and baryon resonances, which is one of the important areas of focus at Jefferson Lab; Paul Souder's lectures address parity violation in electron scattering and its importance in the search of the strangeness content of the nucleon. Other lectures not included in this volume were given by Thomas Cohen on the Skyrme model, Charles Hyde-Wright on virtual Compton scattering and the Gerasimov-Drell-Hearn sum rule, and by Larry Cardman who gave an overview of the physics program of Jefferson Lab.

We would like to thank here all the lecturers for the excellent lectures, for the exciting atmosphere they created, and for giving so generously of their time in preparing and delivering the lectures as well as the contributions in this volume. We would also like to thank the International Advisory and Local Organizing Committees. Special thanks go to Susan Ewing, Carlane

Pittman, Paul Gueye, and Robert Williams for their great contributions that made the organization of the program possible. We also thank Rita Chambers, Montique McClary, Vevelyn Nazario, Mary Beth Stuart, Linda Ware, and many others who in different ways helped with the program.

The 13th HUGS at CEBAF is supported by the Department of Energy under grant number DE-FG05-87FR40380. Jefferson Lab provided the facilities such as lecture room, computer access and partial financial support.

(E,E′X) REACTIONS

H. P. BLOK
Department of Physics and Astronomy, Vrije Universiteit,
de Boelelaan 1081, 1081 HV Amsterdam, The Netherlands
and
NIKHEF, P.O. Box 41882, 1009 DB Amsterdam, The Netherlands

After an extensive introduction on (inclusive) electron scattering the important concepts (kinematical variables, invariants, cross section in terms of structure functions) of (e,e′x)-reactions are discussed. Then several examples are given how those concepts are used in practice to study major physical questions, like the single-particle structure of nuclei, nucleon-nucleon correlations, pion production, character of nucleon resonances and the form factor of the pion.

1 Introduction

Much information on the structure of hadronic systems has been obtained from elastic and inelastic scattering of high-energy electrons. Examples are the elastic form factors of nucleons and nuclei, from which charge and magnetization disctributions have been extracted, transition form factors to nucleon resonances and excited states of nuclei, which gave insight in the character of these excitations, and the quark-parton structure of the nucleon, as revealed by deep-inelastic scattering.

More detailed studies are possible with (e,e′x) coincidence reactions. For instance, the contributions of various orbitals for the proton to the overall charge distribution of nuclei have been investigated extensively with the (e,e′p) reaction. Nucleon-nucleon correlations were studied with the (e,e′d) reaction, and recently also with the (e,e′pp) reaction. And exclusive or semi-inclusive (e,e′π) and (e,e′K) reactions allow for a deeper investigation of the character of N* resonances and of quark distributions.

In these lecture notes we first discuss the formalism of inclusive electron scattering, since many of the concepts used in (e,e′x) reactions are used already there. This is done starting on a simple level so that the text is sort of self contained, also for students not familiar with this field. Then we present the general framework of (e,e′x) reactions: kinematical variables, Lorentz invariants that can be constructed from it, cross section in terms of structure functions, and how to determine the latter separately. After that we discuss several examples showing how one can study certain nuclear and nucleonic questions with the (e,e′x) reaction:

⋆ plane-wave and distorted-wave impulse approximation (PWIA/DWIA) de-

scription of the (e,e′p) reaction and the single-particle structure of nuclei,
⋆ study of nucleon-nucleon (NN) correlations with (e,e′d) and (e,e′pp) reactions,
⋆ the ^{1}H(e,e′π^0)p reaction near threshold as a test of Chiral Perturbation Theory,
⋆ the same reaction in the Δ-region to study the character of the $\Delta(1232)$ isobar,
⋆ use of the ^{1}H(e,e′π^+)n reaction above the resonance region to determine the charge form factor of the pion.

The selection made is a personal one, based on expertise and interest, and considering the physics programme at the Thomas Jefferson National Accelerator Laboratory. There exist many other examples. However, the ones chosen cover a wide range and thus may serve to clarify several issues relevant to the description of (e,e′x) reactions. Polarization degrees of freedom are completely left out of the discussion of the formalism, and are occasionally mentioned only in the examples. This does not mean that polarization is unimportant, to the contrary, but a fair treatment would go far beyond the scope of these notes.

Refs. [1,2,3] give some texts that may be useful for introductory, parallel, or further reading.

2 (e,e′) reactions

2.1 Introduction

Electrons as a probe of nuclear structure have some distinct advantages over other probes like hadrons or γ-rays:
a. The interaction between the electron and the nucleus is known; it is the electromagnetic interaction with the charge ρ and the current $\mathbf{J}$ of the nucleus: $V_{int} = \rho\phi + \mathbf{J}.\mathbf{A}$, where ϕ and $\mathbf{A}$ are the scalar and vector potentials generated by the electron.
b. The interaction is weak, so that in almost all cases it can be treated in the one-photon exchange approximation, i.e., two-step processes (two-photon exchange) are small [a].
c. The energy $\hbar\omega$ and linear momentum $\hbar c|\mathbf{q}|$ transferred to the nucleus in the scattering process can be varied independently from each other (see Fig. 1), in contrast to γ-decay, where $\omega = |\mathbf{q}|c$ [b]. This is very important, as for a certain

[a] This statement applies not to (charge) elastic scattering from nuclei. The interaction with the Coulomb field of the nucleus is so strong that it has to be treated to all orders. On top of that in very special cases one has to include two-step processes via inelastic states, called dispersion corrections, see Ref. [4]

[b] In the rest of the text we will use natural units $\hbar = c = 1$.

$|\mathbf{q}|$ one effectively measures a Fourier component of ρ or $\mathbf{J}$. By varying $|\mathbf{q}|$ all Fourier components can be determined and from these the radial dependence of ρ and $\mathbf{J}$ can be reconstructed. How fine details in ρ or $\mathbf{J}$ can be "seen" depends on $|\mathbf{q}|$, so on the de Broglie wavelength λ of the electron.
d. Because the photon has no charge, in γ-decay only the $\mathbf{J}.\mathbf{A}$ interaction plays a role, which results in *two* (not *three*, because the Lorentz condition $\nabla.\mathbf{A}$ restricts $\mathbf{A}$ to only two independent components) kinds of transitions, which after multipole decomposition yield the so-called electric (Eλ) and magnetic (Mλ) transitions. In electron scattering one also has the Coulomb interaction between the charges in the nucleus and the electron, resulting in charge (Cλ) transitions. These three types of transitions each obey their own selection rules (see Sec. 2.6).

Electron scattering has also some disadvantages:
a. The interaction is weak, so cross sections are small, but one can use high currents and thick targets.
b. Neutrons are less accessible than protons, since the generally dominant Coulomb scattering is almost absent for neutrons, as they do not have a net electric charge.
c. Because electrons are very light particles, they easily emit Bremsstrahlung. Together with some other processes this gives rise to radiative tails. Particularly in (e,e′x) reactions corrections for those radiative processes are not trivial [5].

2.2 Mott cross section and elastic form factor

Before discussing in general (e,e′) scattering we first treat elastic scattering from a charge distribution, thereby introducing the important concept of a form factor.

Both in classical mechanics and in non-relativistic quantum mechanics the differential cross section for elastic scattering of a point particle with charge ze from a point charge Ze via the Coulomb interaction is given by the Rutherford formula:

$$\frac{d\sigma}{d\Omega} = \left(k\frac{zZe^2}{2mv^2}\right)^2 \frac{1}{\sin^4(\theta/2)}. \tag{1}$$

(The value of k depends on the system of units. In the rational system k = 9 x 10^9). For electrons there are three changes:
a. a Lorentz contraction factor $1-\beta^2$ ($\beta = v/c$),
b. a factor $1-\beta^2\sin^2(\theta/2)$ due to the helicity of the electron (helicity means, that spin and momentum of the electron are correlated: $\sigma.\mathbf{p}/p = \pm v/c$); because the Coulomb interaction does not change the direction of the spin, but

does change the direction of $\mathbf{p}$, one gets this reduction factor; look at scattering over 180° to understand the process most easily),
c. the Rutherford formula is exact, independent of the values of z and Z. In the relativistic case there are higher order (in $zZ\alpha$) terms, where $\alpha = 1/137.06$ is the fine-structure constant.

If we neglect c. we get for electron scattering from a charge Ze:

$$\frac{d\sigma}{d\Omega} = \left(\frac{Ze^2}{2mc^2\beta^2}\right)^2 (1-\beta^2)(1-\beta^2\sin^2(\theta/2))\frac{1}{\sin^4(\theta/2)}, \tag{2}$$

which in the limit $\beta = 1$ gives (with $mc^2/\sqrt{1-\beta^2} = E$) the Mott cross section:

$$\sigma_M = \left(\frac{Ze^2}{2E}\right)^2 \frac{\cos^2(\theta/2)}{\sin^4(\theta/2)}. \tag{3}$$

We will use often σ_M for a charge of unity, called σ_p, which in the correct units can be written:

$$\sigma_p = \left(\frac{\alpha\hbar c}{2E}\right)^2 \frac{\cos^2(\theta/2)}{\sin^4(\theta/2)} \tag{4}$$

with E in MeV, $\hbar c = 197.32$ MeV fm, and σ_p in fm^2.

What happens now when one of the charges has a finite size? We use a non-relativistic plane-wave Born Approximation (PWBA) description to show the most important features. In PWBA the cross section for elastic scattering from a charge distribution described by the function $\rho(r)$ is given by

$$\frac{d\sigma}{d\Omega}^{PW} = \left(\frac{\mu}{2\pi\hbar^2}\right)^2 \frac{k_f}{k_i}\left|T^{PW}\right|^2, \tag{5}$$

where μ is the reduced mass of the projectile-target system and

$$T^{PW} = \int d\mathbf{r}\, e^{-i\mathbf{k}_f.\mathbf{r}} \int d\mathbf{r}'\, V(|\mathbf{r}-\mathbf{r}'|)\rho(\mathbf{r}')\, e^{i\mathbf{k}_i.\mathbf{r}}. \tag{6}$$

Here $\mathbf{r}$ is the coordinate of the projectile and $\mathbf{r}'$ a point of the charge distribution, $\mathbf{k}_i$ is the momentum vector of the initial, and $\mathbf{k}_f$ that of the final electron, while the interaction between projectile and charge distribution is described by $V(|\mathbf{r}-\mathbf{r}'|)$ (for symmetry reasons it can only depend on $|\mathbf{r}-\mathbf{r}'|$). This can be written (for convenience we will use here $q = |\mathbf{q}|$):

$$\begin{aligned} T^{PW} &= \int d\mathbf{r} \int d\mathbf{r}'\, e^{-i\mathbf{q}.(\mathbf{r}-\mathbf{r}')} e^{-i\mathbf{q}.\mathbf{r}'} V(|\mathbf{r}-\mathbf{r}'|)\rho(\mathbf{r}') \\ &= \int V(s) e^{i\mathbf{q}.\mathbf{s}} d\mathbf{s} \int \rho(\mathbf{r}') e^{-i\mathbf{q}.\mathbf{r}'} d\mathbf{r}' \equiv V(q)F_0(q). \end{aligned} \tag{7}$$

Here $V(q)$ and $F_0(q)$ are the Fourier transforms of the interaction and the charge distribution, respectively. (Notice that this formula is rather general, it applies also to e.g. proton scattering from a nucleon distribution). It is easy to derive that, if $V(r) = V_0 e^{-ar}/r$, $V(q) = 4\pi V_0/(q^2 + a^2)$. For the Coulomb force $V_0 = \alpha$ and $a = 0$, so

$$T^{PW} = 4\pi \frac{\alpha}{q^2} F_0(q). \tag{8}$$

Using this in (5) together with $\mu = m$, $k_i = k_f = k$ (scattering from a fixed charge) gives:

$$\frac{d\sigma}{d\Omega}^{PW} = \left(\frac{\mu}{2\pi\hbar^2}\right)^2 (4\pi)^2 \frac{\alpha^2}{q^4} F_0^2(q) = \frac{\alpha^2}{4m^2 v^4} \frac{1}{\sin^4(\theta/2)} F_0^2(q). \tag{9}$$

This is the Rutherford formula, modified by the so called form factor $F_0(q)$. The equivalent formula for electrons is

$$\frac{d\sigma}{d\Omega} = \sigma_p F_0^2(q). \tag{10}$$

For a point charge with charge Z one has $F_0(q) = Z$ (usually the value e of the elementary charge is not contained in ρ). For a charge with a finite size $F_0(q)$ will be smaller than Z, because different parts of $\rho(r)$ will give destructive contributions in the integral that constitues $F_0(q)$.

Often one finds the following formula for the form factor for elastic scattering:

$$F_0(q) = \frac{1}{Z} \frac{4\pi}{q} \int_0^\infty \rho(r) \sin(qr) r dr. \tag{11}$$

This can easily be derived from the definition of F_0 if one puts the factor Z in σ_M and not anymore in F_0, so $F_0(0) = 1$.

It is a good exercise to calculate $F_0(q)$ and $\frac{d\sigma}{d\Omega}$ for a uniformly charged sphere with radius R. One can compare the results with, e.g., the data from Phys. Lett. B 37 (1971) 282. It is also instructive to calculate the form factors for Yukawa, Gaussian, and exponential distributions.

2.3 Kinematics

Now we discuss the general kinematics of electron scattering in a relativistic covariant way. The momenta in the electromagnetic scattering process of an electron off a target are shown in Fig. 1. The energy ω ($= q_0$) and three-momentum $\mathbf{q}$ can be varied independently, so the exchanged photon in the

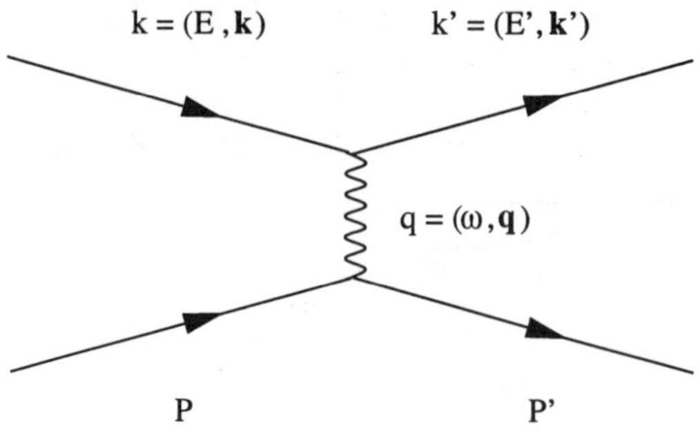

Figure 1: Kinematics of the inclusive electron scattering process. In the LAB system $P = (M, \mathbf{0})$.

process may be virtual, i.e., off mass-shell: $q^2 = q_\mu q^\mu = \omega^2 - \mathbf{q}^2 \neq 0$. Since $q^2 \leq 0$ in electron scattering, one usually defines $Q^2 \equiv -q^2$, which is positive. Since there are three independent four-vectors in the process, one can construct six Lorentz invariants. Two of them yield the masses of the electron and the target, two other ones are determined by the energy of the electron-target system and the scattering angle, so two are left that contain the physics of the probed target system. The commonly used ones in inclusive electron scattering experiments are:

$$Q^2 \equiv -q^2 \stackrel{\mathrm{LAB}}{=} 2EE' - 2|\mathbf{k}||\mathbf{k}'|\cos\theta \stackrel{\mathrm{m=0}}{=} 4EE'\sin^2(\theta/2) \geq 0 \qquad (12)$$

$$2M\nu \equiv 2P.q = Q^2 + W^2 - M^2 \stackrel{\mathrm{LAB}}{=} 2M\omega, \qquad (13)$$

where θ is the laboratory electron scattering angle, M the target mass, and $W = P'^2$ the invariant mass of the hadronic final state. The value of Q^2 can be obtained by measuring the electron scattering angle θ and the energy E' of the outgoing electron. The calculation of ν is especially simple in the LAB frame, where the target is at rest, since then obviously $\omega = \nu$. In collider experiments one can best use the basic definition of ν and express all variables, like the scattering angle and the parameter ϵ to be introduced later, in terms of the energies and masses of the colliding particles and the invariants Q^2 and ν.

For elastic scattering one has $W = M$ and thus

$$Q^2 = 2M\nu. \qquad (14)$$

This is the direct (relativistic) generalization of the non-relativistic result $\nu = |\mathbf{q}|^2/2M$. The relation $Q^2 = 2M\nu$ yields a straight line in the Q^2-ν plot given in Fig. 2. It is the boundary of the physical region in inelastic electron scattering,

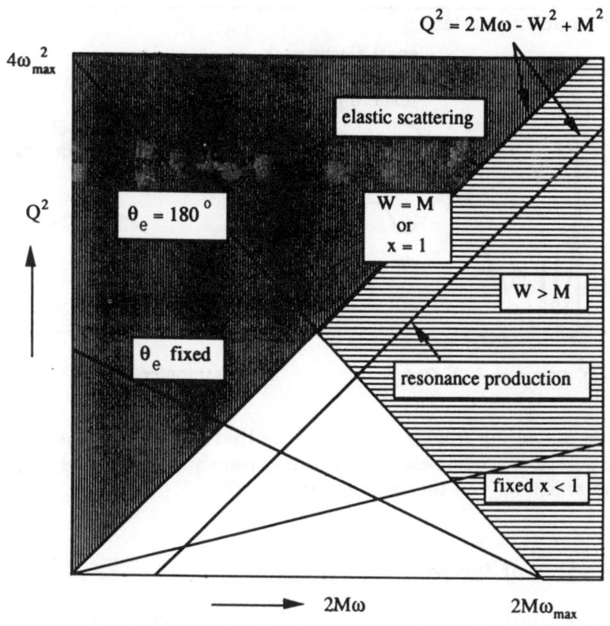

Figure 2: Kinematics of electron scattering in a Q^2 vs. ν plot. The dark shaded area is unphysical. In the LAB frame $\nu_{max} = E$.

since for any final state that can be produced $W \geq M$. The relation between Q^2 and ν for fixed electron scattering angle θ is $Q^2 = 4\nu_{max}(\nu_{max} - \nu)\sin^2(\theta/2)$. This yields also straight lines in Fig. 2. The line corresponding to $\theta = 180°$ gives the boundary of the region in which measurements with a given beam (or collision) energy are possible.

In deep-inelastic electron scattering the Bjorken scaling variable

$$x = Q^2/2M\nu \tag{15}$$

is often used. The range of this variable is $0 \leq x \leq 1$. Lines for constant x are also indicated in Fig. 2. The case $x = 1$ corresponds to elastic scattering. Note that there are several other variables that are commonly used in electron scattering, In quasi-elastic scattering one may encounter the dimensionful (GeV/c) parameter y, which is a function of Q^2 and ν, whereas in deep-inelastic scattering one uses y for the (dimensionless) variable ν/E, which ranges between 0 and 1.

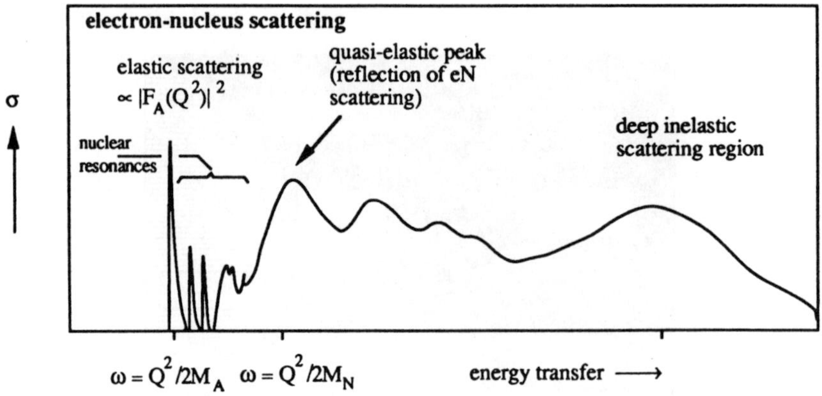

Figure 3: Electron scattering from a nucleus. Note that the plot is schematic: the contributions of elastic scattering, quasi-elastic scattering, resonance excitation and deep-inelastic scattering depend sensitively on the value of Q^2.

2.4 Global features of the response function

Let us now look globally how the cross section depends on Q^2 and ν (or ω). For a point-like particle the cross section is a δ-function in ν: $\sigma(\nu) = \delta(\nu - Q^2/2M)$. As a function of Q^2 it is, apart from the $1/Q^4$ dependence resulting from the square of the photon propagator (see Sect. 2.5), constant. For a composite particle there are two effects:
a. excited states give rise to discrete peaks or even a continuum at $\nu > Q^2/2M$,
b. the internal structure of the particles gives rise to form factors $F(Q^2)$ (see Sect. 2.2 and 2.5-6).
This is illustrated in Fig. 3. The picture depends strongly on the value of Q^2, i.e., on the wave length with which one probes the nucleus. At small Q^2 one will see elastic and inelastic scattering from the nucleus as a whole. At higher Q^2 these peaks disappear as the form factors get small, and the most prominent structure is the peak due to quasi-elastic scattering from the nucleons in the nucleus. The peak is shifted to $\nu > Q^2/2M$ due to binding-energy effects. Furthermore it is broadened due to Fermi motion of the nucleons. Also excited states of the nucleon like the Δ become visible. When Q^2 gets much larger than R_N^{-2}, the nucleon form factor kills the cross section and the structure seen then is thought to arise from the quarks in the nucleons (deep-inelastic scattering). Like the quasi-elastic peak it is broadened because the quarks are not at rest.

2.5 The general cross section formula

Scattering from point particles

Apart from phase space factors the cross section for scattering of two (unpolarised) pointlike spin $\frac{1}{2}$ fermions of charge e, with momenta $k(k')$ and $p(p')$ and spin projections $m(m')$ and $s(s')$ in the initial (final) state, is given in the one-photon exchange approximation by

$$|\mathcal{M}|^2 = \frac{1}{4} \sum_{s,m,s',m'} \left| \langle k', s' | \gamma_\mu | k, s \rangle \frac{e^2}{q^2} \langle p', m' | \gamma^\mu | p, m \rangle \right|^2 . \tag{16}$$

The two matrixelements, together with the factor e^2, reflect the interaction currents in the two vertices, while the factor $1/q^2$ results from the propagation of the virtual photon. The factor 1/4 comes from the spin averaging in the initial state. This formula can be written as

$$\left| \frac{\mathcal{M}}{4\pi} \right|^2 = \frac{\alpha^2}{Q^4} L_{\mu\nu}(k,k')\, L^{\mu\nu}(p,p'), \tag{17}$$

where the so-called lepton tensor is given by

$$L_{\mu\nu}(k,k') = \frac{1}{2} \sum_{s,s'} \langle k, s | \gamma_\mu | k', s' \rangle \langle k', s' | \gamma_\nu | k, s \rangle. \tag{18}$$

Using γ-matrix algebra this leads to the following formula for the cross section

$$\begin{aligned} \frac{d\sigma}{d\Omega\, dE'} &= \frac{4\,\alpha^2\, E'^2}{Q^4} \left(\cos^2(\theta/2) + \frac{Q^2}{2M^2} \sin^2(\theta/2) \right) \delta\left(\nu - \frac{Q^2}{2M} \right) \\ &= \sigma_p \left(1 + \frac{Q^2}{2M^2} \tan^2(\theta/2) \right) \delta\left(\nu - \frac{Q^2}{2M} \right), \end{aligned} \tag{19}$$

with σ_p the Mott cross section for unit charge (see Eqs. 3,4).

The differential cross section for the electromagnetic interaction between two charged pointlike fermions is purely elastic, which is reflected in the δ-function $\delta(\nu - Q^2/2M)$. It is therefore common also to integrate over the energy of outgoing particle 1, or equivalently over the energy transfer ν. This yields the Mott cross section times a recoil term and times a sum of two terms, of which the first one is due to the interaction of the charges and the second one ($\propto Q^2/2M^2$) to the interaction of the magnetic moments

$$\frac{d\sigma}{d\Omega} = \sigma_p \left(1 - \frac{Q^2}{2ME} \right) \left(1 + \frac{Q^2}{2M^2} \tan^2(\theta/2) \right). \tag{20}$$

Scattering from nucleons

For a particle with a charge distribution, usually indicating that it is a composite system, the charge operator involves form factors. In the example discussed in Sect. 2.2 we have seen the following charge operator for an extended charge distribution: $e^{i\mathbf{q}\cdot\mathbf{r}}$. In a relativistic covariant formulation one has to consider the full electromagnetic current J_μ. The most general form consistent with the requirements of Poincaré invariance, gauge invariance and invariance under parity and time reversal yields for a spin 0 system (e.g., a pion or a ^{12}C nucleus) the form

$$\langle p'|J_\mu|p\rangle = (p_\mu + p'_\mu)\, F(Q^2), \tag{21}$$

involving one invariant function, the form factor F, which depends on the (only) invariant in the elastic scattering process, Q^2. For spin 1/2 systems (such as a nucleon, ^{3}H or ^{3}He) the most general expression for the expectation value of the current can be written using free Dirac spinors $u_s(p)$, where s stands for a specific spin component.

$$\langle p', s'|J_\mu|p, s\rangle = \bar{u}_{s'}(p')\Gamma_\mu(p, p')u_s(p) \tag{22}$$

with

$$\Gamma_\mu(p, p') = \gamma_\mu F_1(Q^2) + \frac{i\sigma_{\mu\nu}q^\nu}{2M} F_2(Q^2), \tag{23}$$

with $\sigma_{\mu\nu} = \frac{i}{2}[\gamma_\mu, \gamma_\nu]$. The functions F_1 and F_2 are the Dirac and Pauli form factors. For an elementary fermion one has $F_1(Q^2) = 1$ and $F_2(Q^2) = 0$. For the nucleon one has for the proton $F_1^p(0) = e_p = 1$ and $F_2^p(0) = \kappa_p = 1.793$ and for the neutron $F_1^n(0) = e_n = 0$ and $F_2^n(0) = \kappa_n =$ -1.913 (where κ is the anomalous magnetic moment). The (Sachs) electric and magnetic form factors $G_E(Q^2)$ and $G_M(Q^2)$, which physically are most closely related to the electric charge and magnetic current distributions, are combinations of the Dirac and Pauli form factors,

$$G_E = F_1 - \frac{Q^2}{4M^2} F_2, \qquad G_M = F_1 + F_2. \tag{24}$$

Roughly one has for nucleons the dipole-like behavior

$$G_E^p(Q^2) \approx \frac{G_M^p}{\mu_p} \approx \frac{G_M^n}{\mu_n} \approx \frac{1}{(1 + Q^2/\Lambda^2)^2}, \tag{25}$$

with $\Lambda^2 \approx 0.71$ GeV^2 and $\mu_{p,n} = e_{p,n} + \kappa_{p,n}$ the nucleon's magnetic moment.

In order to obtain the expression for the differential cross section of electron-nucleon scattering, one must replace $L_{\mu\nu}$ for one of the pointlike fermions in (17) by

$$L_{\mu\nu}(k,k') = \frac{1}{2}\sum_{s,s'} \langle k,s|J_\mu|k',s'\rangle\langle k',s'|J_\nu|k,s\rangle. \tag{26}$$

This yields for the differential cross section for elastic electron-nucleon scattering

$$\frac{d\sigma}{d\Omega} = \sigma_p\left(1-\frac{Q^2}{2ME}\right)\left(\frac{G_E^2+Q^2G_M^2/4M^2}{1+Q^2/4M^2}+\frac{Q^2\,G_M^2}{2M^2}\tan^2(\theta/2)\right). \tag{27}$$

This result is the Rosenbluth formula. (Note that the term $1/(1+Q^2/4M^2)$ can be written as $Q^2/|\mathbf{q}|^2$, a factor that will appear more often, depending on the choice one makes for the longitudinal current in defining form factors and structure functions.) The essential feature is that the differential cross section divided by the Mott cross section and the recoil term is given by $A(Q^2)+B(Q^2)\tan^2(\theta/2)$. If one plots the cross section, measured at different combinations of incident energy and scattering angle such that Q^2 is constant, against $\tan^2(\theta/2)$, a linear dependence should be obtained. The only assumption underlying this result is that the interaction is mediated by the exchange of a single photon.

Scattering from composite particles

The generalization of the Rosenbluth formula is the following expression for inclusive (in)elastic electron scattering off some target. In the LAB-frame

$$\frac{d\sigma}{d\Omega\,dE'} = \frac{\alpha^2}{Q^4}\frac{E'}{E}\,L_{\mu\nu}W^{\mu\nu}, \tag{28}$$

where $L_{\mu\nu}$ is the leptonic tensor and $W_{\mu\nu}$ is the hadronic tensor, which contains the information on the hadronic part of the scattering process.

On the basis of Poincaré invariance, and invariance under parity and time reversal, the most general form allowed for the hadronic tensor in the case of unpolarized scattering is

$$W_{\mu\nu} = \left(\frac{q_\mu q_\nu}{q^2}-g_{\mu\nu}\right)W_1+\frac{\tilde{P}_\mu\tilde{P}_\nu}{M^2}W_2, \tag{29}$$

where $\tilde{P}_\mu = P_\mu - (P\cdot q/q^2)q_\mu$ and the structure functions W_1 and W_2 only depend on the invariants in the scattering process, ν and Q^2.

Working out the contraction between leptonic and hadronic tensor gives the cross section, which in the laboratory frame is

$$\frac{d\sigma}{d\Omega\, dE'} = \frac{4\alpha^2 E'^2}{Q^4} \left[\cos^2(\theta/2)\, W_2(\nu, Q^2) + 2\,\sin^2(\theta/2)\, W_1(\nu, Q^2)\right]. \qquad (30)$$

Often one encounters the expression in terms of longitudinal and transverse structure functions:

$$\frac{d\sigma}{d\Omega\, dE'} = \sigma_p \frac{Q^2}{|\mathbf{q}|^2} \left[W_L(\nu, Q^2) + \epsilon^{-1}\, W_T(\nu, Q^2)\right], \qquad (31)$$

where the polarization of the virtual photon is given by

$$\epsilon = \left[1 + 2\,(|\mathbf{q}|^2/Q^2)\,\tan^2(\theta/2)\right]^{-1}. \qquad (32)$$

One finds $W_L = (|\mathbf{q}|^2/Q^2)W_2 - W_1$ and $W_T = W_1$. Instead of W_L and W_T one sometimes encounters slightly different structure functions, for instance $R_L = (|\mathbf{q}|^2/Q^2)W_L$ and $R_T = 2W_T$. Similar as explained for the Rosenbluth formula for elastic electron-nucleon scattering, a longitudinal-transverse separation can be made to obtain W_L and W_T for any fixed value of ν and Q^2 by performing measurements with different beam energies and scattering angles.
One should note that for a virtual photon ($Q^2 > 0$) $\epsilon \to 0$ if $\theta \to 180°$. A real photon ($Q^2 = 0$) has $\epsilon = 0$, which corresponds to $\theta = 0°$, where $|\mathbf{q}| = E - E'$.

These general considerations clearly show that one can get in unpolarised (e,e$'$) scattering (at most) *two* (structure) functions, which depend on the variables Q^2 and ν, that give information about the target system. By using polarization one can obtain more information. For instance, elastic scattering of (longitudinally) polarized electrons from polarized nucleons is described by four structure functions [6], one of them depending on $G_E G_M$, which is used at present to determine the small electric form factor of the neutron, see e.g., Ref. [7].

2.6 *Cross section in terms of form factors*

It is useful to do a multipole decomposition of the current operator J_μ in the general case. For a transition to a discrete final state, like elastic scattering on ^{2}H, inelastic scattering to some excited state of a nucleus, or excitation of a nucleon to a Δ, this leads in the Plane Wave Born Approximation (PWBA), in which the incoming and outgoing electrons are described by plane waves, to the following formula[8,9]

$$\frac{d\sigma}{d\Omega} = \sigma_p \eta \left[\sum_{\lambda=0}^{\infty} \frac{Q^4}{q^4} |F_\lambda^C(q)|^2 + (\frac{Q^2}{2q^2} + \tan^2 \frac{\theta}{2}) \sum_{\lambda=1}^{\infty} \{ |F_\lambda^E(q)|^2 + |F_\lambda^M(q)|^2 \} \right]. \tag{33}$$

NB. For convenience we will use the letter q in this subsection to indicate the length of the three-vector: $q = |\mathbf{q}|$, and not the four-vector.
The recoil factor η, which is given by

$$\eta = \left(1 + \frac{2E \sin^2 \frac{\theta}{2}}{Mc^2} \right)^{-1} \tag{34}$$

with M being the mass of the target nucleus, takes into account that the final nucleus is moving with a certain recoil velocity. It is instructive to compare this formula to the Rosenbluth formula (27) and to Eq. 31. For nuclear targets the difference between Q^2 and q^2 is usually very small, so in that case often Q^2/q^2 is set to unity.

The F_λ's, which are functions of the momentum transfer q only, are the form factors [c] with multipolarity λ. F_λ^C is called the longitudinal or charge form factor. There are two transverse form factors: the (transverse) electric form factor F_λ^E and the magnetic form factor F_λ^M. The terms longitudinal and transverse have their origin in the fact that one can decompose the nuclear current $\mathbf{J}$ into a component parallel to $\mathbf{q}$ and two components perpendicular to $\mathbf{q}$. Because of the continuity equation the component parallel to $\mathbf{q}$ combines with the scattering from the charge of the nucleus and gives rise to F_λ^C , while linear combinations of the two components perpendicular to $\mathbf{q}$ yield F_λ^E and F_λ^M . In analogy to γ-decay one speaks of C_λ, E_λ and M_λ transitions. Whereas the multipolarity of the electric and magnetic transitions, like in γ-decay, starts at $\lambda = 1$, for a charge transition also $\lambda = 0$ is allowed. As mentioned already, the contribution from the longitudinal and from the transverse form factors can be determined separately by taking data at the same value of q, but different values of θ (and thus of E)(Rosenbluth separation).

One may wonder why the form factors defined here depend on $q = |\mathbf{q}|$, whereas those introduced in the previous subsection depend on the invariant Q^2. The reason is that those here are effectively defined in the Breit frame, i.e., the frame in which there is no energy transfer: $q_0 = \omega = 0$. Only in this

[c] Generally, but against the strict definition, the term form factor is used within the electron scattering community for $\frac{d\sigma}{d\Omega}/\sigma_p\eta$ or $\frac{d\sigma}{d\Omega}/\sigma_p\eta/(\frac{Q^2}{2q^2} + \tan^2\frac{\theta}{2})$. In PWBA (and only then) this corresponds to the 'real' form factor squared, provided one takes $Q^2/q^2 = 1$.)

frame one can interpret a form factor as a Fourier transform of, e.g., a charge distribution.

The form factors F^C, F^E and F^M contain the nuclear structure information. They are the (reduced) matrix elements of complicated operators between the initial and final nuclear wave functions, see Refs. [8,9]. The charge form factor F_C is relatively simple. It is given by (compare Eq. 11)

$$F_\lambda^C(q) = \int_0^\infty \rho_{if}^\lambda(r) j_\lambda(qr) r^2 dr, \tag{35}$$

where $\rho_{if}^\lambda(r)$ is a *transition* density and $j_\lambda(x)$ the spherical Bessel function of order λ. Here we are only concerned with the angular momentum and parity character of these operators. The selection rules for the various terms are such that for natural-parity or "electric" transitions, where $\pi_i \pi_f = (-1)^\lambda$, the magnetic form factor F_λ^M is zero. Similarly for unnatural-parity or "magnetic" transitions, where $\pi_i \pi_f = (-1)^{\lambda-1}$, F_λ^C and F_λ^E are zero. Furthermore, in the case of elastic scattering time-reversal invariance tells that F^E is zero for odd values of λ.

With these rules it is easy to find that the nucleon has two elastic form factors, a C0 (charge monopole) and an M1 (magnetic dipole), which can be determined from the two structure functions that one can measure. However, the deuteron has in addition a third form factor: C2 (charge quadrupole). The cross section for unpolarised elastic scattering on a deuteron is usually written as (compare Eqs. 27 and 30)

$$\frac{d\sigma}{d\Omega} = \sigma_p \left[A(Q^2) + B(Q^2) \tan^2(\theta/2) \right], \tag{36}$$

where

$$A = G_C^2 + \frac{8}{9}\eta^2 G_Q^2 + \frac{2}{3}\eta G_M^2 \tag{37}$$

$$B = \frac{4}{3}\eta(1+\eta) G_M^2 \qquad (\eta = Q^2/4M_d^2). \tag{38}$$

Obviously G_M can be separated from the charge (monopole) form factor G_C and the charge quadrupole form factor G_Q by a Rosenbluth separation, but separation of the latter two requires the use of polarization, either by using a tensor polarized deuterium target, or measuring the tensor polarization of the recoiling deuterons (T_{20} measurements [10]).

In a similar way one needs polarization to determine the three (M1, C2 and E2) (transition) form factors for the $N \to \Delta$ transition, or one can use the (e,e$'\pi^0$) reaction (see Sect. 7).

3 (e,e′x) reactions

Here we will describe the general aspects of A(e,e′x)B reactions, where x can be a proton, deuteron or α-particle, but also a pion or even a real photon. Usually B is the undetected residual nucleus or nucleon, but one may reverse the situation experimentally by detecting the latter, B being, for instance, a π^0, a neutron, or a γ, like in Virtual Compton Scattering. Thus one avoids the difficult detection of neutral particles.

3.1 Kinematical variables

The kinematical variables for an (e,e′x) reaction are shown in Fig. 4. The angle ϕ (whenever there is no danger of misunderstanding, for convenience we will drop the subscript x) represents the angle between the electron-scattering plane and the plane defined by the momentum transfer $\mathbf{q}$ and the three-momentum $\mathbf{p}$ of particle x. The missing energy E_m and missing momentum $\mathbf{p}_m$ are defined [d] as:

$$E_m = E_e - E_{e'} - E_x \tag{39}$$

$$\mathbf{p}_m = \mathbf{q} - \mathbf{p}\,. \tag{40}$$

Using these one can calculate the missing mass

$$M_m = \sqrt{E_m^2 - \mathbf{p}_m^2}, \tag{41}$$

which is equal to M_B. In many cases this has discrete values, e.g., the mass of a nucleus in one of its discrete excited states, or the mass of a π^0 or γ, but it may also be continuous as in a ^{1}H(e,e′p)$\pi\pi$ reaction or an (e,e′p) reaction to the continuum above the break-up threshold of final nucleus B. One should not confuse this missing mass with the previously defined invariant mass W, which is a property of the combined B+x system (see Sect. 7 for an illustration).

Compared to the inclusive case there is now a new independent four-vector, viz. the four-momentum p of particle x. Its square just gives the mass of x. By contracting p with the three independent four-vectors one has in (e,e′), one can construct three new Lorentz invariants. One of these is related to the angle ϕ. Another one gives the (invariant) mass M_B of the residual state. For the last one different variables (not always Lorentz invariants) are used, depending on the sort of reaction, like the missing momentum $p_m = |\mathbf{p}_m|$ or the angle θ, or the Mandelstam variable t (see Sect. 3.3).

[d] In knock-out reactions like (e,e′p/d/α) the missing energy is usually defined differently, see Sect. 4.

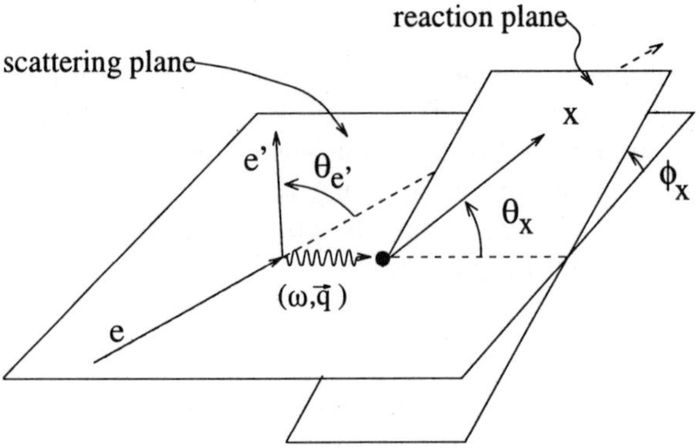

Figure 4: Kinematics of the (e,e′x) reaction.

3.2 General cross section formula

In the case of the (unpolarized) (e,e′x) reaction the most general gauge invariant form for $W_{\mu\nu}$ can be written as

$$W_{\mu\nu}(P, p_x, q) = \left(-g_{\mu\nu} + \frac{q_\mu q_\nu}{q^2}\right) W_1 + \frac{\tilde{P}_\mu \tilde{P}_\nu}{M^2} W_2 + \frac{\tilde{p}_{x\mu} \tilde{P}_\nu + \tilde{P}_\mu \tilde{p}_{x\nu}}{M m_x} W_3 + \frac{\tilde{p}_{x\mu} \tilde{p}_{x\nu}}{m_x^2} W_4, \tag{42}$$

with $\tilde{p}_{x\mu} = p_{x\mu} - (p_x \cdot q/q^2)\, q_\mu$ in analogy to the definition of $\tilde{P}_\mu$. Thus, while in (e,e′) there were two functions W describing the (hadronic) part of the most general cross section, in (e,e′x) there are four such functions. These structure functions depend on the four invariants discussed in the previous section (e.g., W, Q^2, M_B and t).

One can find explicit expressions for the factors in front of the W's, but as in the inclusive case it is convenient to use the virtual photon polarization. The cross section can then generally be written in the form:

$$\frac{d^6\sigma}{dE'_e\, d\Omega'_e\, dE_x\, d\Omega_x} = |\mathbf{p}_x| E_x\, \sigma_p \frac{Q^2}{|\mathbf{q}|^2} \frac{1}{\epsilon} \left[W_T + \epsilon\, W_L + \epsilon\, W_{TT} \cos 2\phi + \sqrt{\frac{\epsilon(\epsilon+1)}{2}}\, W_{LT} \cos\phi \right]. \tag{43}$$

Note that two terms in this expression involve the azimuthal angle ϕ. If one integrates over all ϕ, those terms do not contribute, and one is left with only the longitudinal and the transverse structure functions known from the inclusive case (but now for the "inclusive" (e,e′x) reaction).

It should be realized that this formula is derived in the one-photon exchange approximation. This means that the incoming and outgoing electrons are described as plane waves. The formula can thus not be used for, e.g., (e,e′p) reactions on heavier nuclei, where Coulomb distortions of the electron waves as described for instance by solving the Dirac equation with a Coulomb potential, are not negligible. Apart from this caveat the formula is completely general, and thus encompasses all kinds of reaction mechanisms: - reactions driven by one-body currents/operators, or by two-body currents like Meson Exchange Currents or Isobar Currents, where the virtual photon interacts with an exchanged meson or produces a Δ that by interaction with another nucleon decays back to a nucleon, - quasi-free scattering with plane waves, distorted waves, or even coupled-channels wave functions for the final hadronic state (but as stressed above always with plane waves for the electrons), in which many partial waves of the latter are important, and - resonance reactions, where only a few partial waves play a role.

Another form that is used often for discrete final states (note that it is a five-fold cross section then) is the following:

$$\frac{d^5\sigma}{dE_{e'}\, d\Omega_{e'}\, d\Omega_x} = \Gamma_v \; \frac{d^2\sigma}{d\Omega_x}$$

$$\frac{d^2\sigma}{d\Omega_x} = \epsilon\,\sigma_L + \sigma_T + \sqrt{2\epsilon(\epsilon+1)}\;\sigma_{LT}\;\cos\phi + \epsilon\,\sigma_{TT}\;\cos 2\phi, \tag{44}$$

where

$$\Gamma_v = \frac{\alpha}{2\pi^2}\frac{E_{e'}}{E_e}\frac{q_L}{Q^2}\frac{1}{1-\epsilon} \tag{45}$$

is the flux of virtual photons. The factor $q_L = (W^2 - M_A^2)/(2M_A)$ is the equivalent real-photon energy, the laboratory energy a real photon needs to have to produce a system with invariant mass W. The cross section σ_X, which is a shorthand for $d^2\sigma_X/d\Omega_x$, depends on Q^2, the invariant mass W and the Mandelstam variable t. The solid angle $d\Omega_x$ can be in the lab frame or in the centre-of-mass (CM) frame. The virtue of writing the cross section in this form is that there is a direct relation with the cross section for real photons, which is just given by σ_T (clearly, for $Q^2 = 0$). Furthermore, σ_{TT} is related to the cross section for linearly polarized photons.

The LT and TT terms in (43) and (44) are called the longitudinal-transverse and transverse-transverse interference structure functions (cross sections), respectively. These names originate from the fact that while the L and T terms result from squares of the corresponding components of the current J_μ, the LT and TT terms result from interference between different components. The interference terms have a specific dependence on the angle θ: W_{LT} and W_{TT} are zero for $\theta = 0$ (parallel kinematics), since to leading order they depend on $p_\perp$ and $p_\perp^2$, i.e., $\sin\theta$ and $\sin^2\theta$, respectively, where $p_\perp$ is the component of $\mathbf{p}_x$ perpendicular to $\mathbf{q}$ (they have to be zero since for $\theta = 0$ the angle ϕ is undetermined).

It is good to realize what one can measure and change in different kinematics in (e,e′x). As mentioned, for the transition to a discrete final state, the cross sections and structure functions depend on the three variables W, Q^2 and t (or $|\mathbf{q}|$, ω and p_m). In parallel kinematics one has only two structure functions, which can be separated via a Rosenbluth separation. However, when one wants to change the value of p_m in parallel kinematics, either the value of $|\mathbf{q}|$ or of ω (and hence of p_x), or in terms of W and Q^2, either one or both of these, has to change, since one has already chosen the last, third variable ($\theta = 0$). If one chooses so called (q, ω) constant kinematics, one varies p_m by changing θ, but then also the LT and TT structure functions will start to contribute. One has also to be aware that the L and T structure functions in (q, ω) kinematics, even at the same p_m, are not the same as those in parallel kinematics, since one can never match both $\mathbf{q}$, ω and p_m in the two different kinematics.

3.3 The Mandelstam variables

A rather useful set of Lorentz invariants are the Mandelstam variables. These variables are extensively used in high-energy hadronic reactions (see e.g. Ref.[11]) to describe and study the structure of the scattering amplitude, like in Regge theory. They are defined for a 2-body reaction $1 + 2 \to 3 + 4$, in our case: virtual photon plus target A leading to final state B plus particle x. Then one can make the following invariants (see Fig. 5):

$$\begin{aligned} s &= (q + p_A)^2 = (p + p_B)^2 \\ t &= (p - q)^2 = (p_B - p_A)^2 \\ u &= (p - p_A)^2 = (p_B - q)^2. \end{aligned} \tag{46}$$

Only two of those are independent; in most cases s and t are used. The variable s should be familiar already: it is just W^2. The quantity t is the four-momentum transfer (squared) to the nuclear system (going from A to B). It depends on the "scattering angle", the angle between the direction of the virtual photon and that of the produced particle x (the angle θ in Fig. 4). It thus describes the angular distribution of x. In (e,e′x) reactions t plays a similar

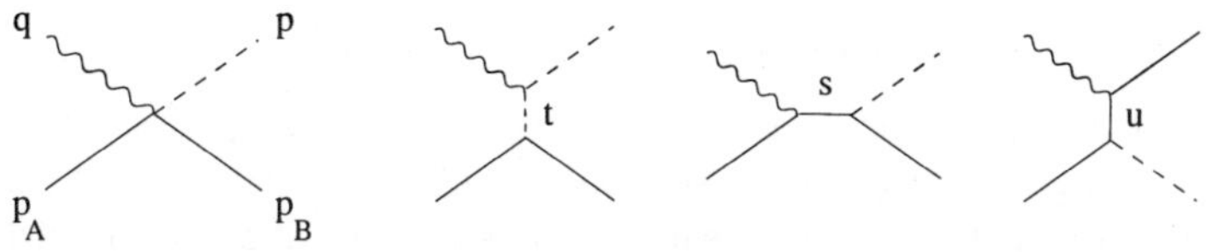

Figure 5: Kinematics diagram for the Mandelstam variables. The diagrams to the right *illustrate* them. Note that these Feynman diagrams represent specific reaction mechanisms, whereas the Mandelstam variables themselves are universal.

role as Q^2 in inclusive (e,e′) scattering, yielding for instance (transition) form factors that depend on t. Thus in such reactions one should clearly distinguish between Q^2, the mass of the (virtual) photon, and the "momentum transfer" t. Consequently also in photoproduction reactions one will have nuclear form factors, depending on t (which unfortunately sometimes is written then as Q^2.)

3.4 Separation of structure functions

As presented in the previous section, the cross section contains four structure functions. Those can be separated by performing measurements at different values of ϵ and/or ϕ, while keeping $|\mathbf{q}|$ and ω (i.e., Q^2 and W) constant. In parallel kinematics one can determine W_L and W_T by taking data at two beam energies (the normal Rosenbluth separation). For the determination of W_{LT} two measurements at the same value of θ are required: one at $\theta < \theta_\mathbf{q}$ corresponding to $\phi = 0°$ and the other one at $\theta > \theta_\mathbf{q}$, corresponding to $\phi = 180°$, where $\theta_\mathbf{q}$ is the angle between the electron beam and $\mathbf{q}$. For a separation of W_T and $W_L + W_{TT}$ an additional measurement at a different beam energy and scattering angle is needed at $\phi = 0°$ or $180°$. After subtracting the W_{LT} contribution from the cross section $W_L + W_{TT}$ and W_T can be determined. The contribution of W_{TT} can only be determined by "out-of-plane" measurements ($\phi \neq 0°$ or $180°$).

As mentioned in Sect. 3.2, with the inclusion of electron distortions Eqs. 43 and 44 are not valid anymore. When the electron distortions are not too large, one can still define effective structure functions (which will depend on the electron kinematics, however) by assuming that one can write the cross section in the form of Eq. 43. E.g., W_{LT} is then *defined* as the difference between cross sections left and right of $\mathbf{q}$ divided by the appropriate factor. Those should be compared then to the effective structure functions given by some model calculation, evaluated with the same procedure.

4 Formalism for quasi-elastic (e,e′p) reactions

In an (e,e′p) reaction the transfer of a virtual photon with energy ω and momentum $\mathbf{q}$ (in this and the next section we will again use q for the length of the three vector $\mathbf{q}$) to a nucleus leads to the knockout of a proton with kinetic energy T_p and momentum $\mathbf{p}$. In (e,e′p) reactions the missing energy E_m and missing momentum $\mathbf{p}_m$ are usually defined as:

$$E_m = \omega - T_p - T_{A-1} \tag{47}$$

$$\mathbf{p}_m = \mathbf{q} - \mathbf{p}\,. \tag{48}$$

The definition of the missing energy differs from the one given before in Eq. 39 by including T_{A-1} and leaving out M_p. The one here had better be called "missing mass" (though with an offset), since it is equal to $M_p + M_{A-1} - M_A$. Thus, unless the hit proton (de)excites the residual nucleus to another (excited) state before it is detected, which is a second order process and normally can be neglected, E_m is the separation energy of the hit proton with respect to the A-1 nucleus. For instance, knock-out from different orbitals will give different values for E_m.

The missing momentum is the momentum of the residual A-1 nucleus. The kinetic energy T_{A-1} can be calculated from it. If the A-1 nucleus is a spectator, the missing momentum is minus the momentum of the proton before it was knocked out of the nucleus [e]. This is never exactly correct, however, since the proton always feels the interaction with the residual nucleus (see the discussion on PWIA/DWIA below).

If one describes the process as quasi-elastic scattering, which means that the scattering is described as if it were on a free (moving) proton (so two-body currents like MEC and IC are neglected) and uses plane waves for the electrons and the outgoing proton (PWIA), the cross section can be written as

$$\frac{d^6\sigma}{dE_{e'}\,d\Omega_{e'}\,dE_p\,d\Omega_p} = K\sigma_{ep}(q)S(\mathbf{p}_m, E_m), \tag{49}$$

where $K = |\mathbf{p}|E_p$ is a kinematical factor, σ_{ep} is the relativistic off-shell elastic electron-proton scattering cross section [f] and $S(\mathbf{p}_m, E_m)$ is the spectral function, which is the probability of finding a proton in the parent nucleus

[e] Several groups define the missing momentum as $\mathbf{p} - \mathbf{q}$, so that $\mathbf{p}_m$ is the momentum of the struck proton.

[f] Several prescriptions exist for this off-shell cross section, see e.g. Ref. [12]. One should also be aware that when one uses DWIA calculations, such a cross section is not used at all, since a DWIA program uses a current operator. Furthermore there are current conservation problems. See for a discussion of some of these aspects Ref. [13].

with binding energy E_m and momentum $\mathbf{p}_m$ (note that this probability does not depend on the direction of $\mathbf{p}_m$). For a transition with quantum numbers $\alpha = (n)lj$ to a discrete final state at energy E_α the spectral function can be written as

$$S(\mathbf{p}_m, E_m) = \rho_\alpha(\mathbf{p}_m)\delta(E_m - E_\alpha)\,. \tag{50}$$

The five-fold cross section for such a discrete transition then reads

$$\frac{d^5\sigma}{dE_{e'}\, d\Omega_{e'}\, d\Omega_p} = K R\, \sigma_{ep}(q)\rho_\alpha(\mathbf{p}_m), \tag{51}$$

where the recoil factor

$$R = \frac{\partial E_p}{\partial E_m} = \left(1 - \frac{E_p}{E_B}\frac{\mathbf{p}_m \cdot \mathbf{p}_p}{|\mathbf{p}_p|^2}\right)^{-1} \tag{52}$$

results form the difference between integrating over E_p and over E_m.

The momentum distribution $\rho_\alpha(\mathbf{p}_m)$ contains the spectroscopic factor [14] (the number of protons in orbital α participating in this transition) and the shape of the $\langle A-1|A\rangle$ overlap function (bound-state wave function) $\phi_\alpha(\mathbf{r})$. In the independent-particle shell model the latter is the wave function of the single-particle orbital in which the proton moved. For a completely filled shell the corresponding spectroscopic factor is $S_\alpha = 2j+1$.

At this stage it is instructive to make the relation with the general formalism described in the previous section. Clearly in PWIA all structure functions W_X contain the same spectral function, which can be factorized out, so the L, T etc. character is wholly contained in σ_{ep} (since one scatters from a moving proton the electron-proton cross section contains now also LT and TT parts!).

For scattering from the charge of the proton one can easily see that in PWIA the cross section contains the momentum distribution. In PWIA the total wave function of the final state is written as

$$\Psi_{A'} = \mathrm{e}^{i\mathbf{k}.\mathbf{r}_p}\Psi_{A-1}. \tag{53}$$

The (e,e′p) equivalent of the form factor F_0 in (7) is then

$$F(\mathbf{p}_m) = \int \phi_\alpha(\mathbf{r})\mathrm{e}^{-i\mathbf{p}_m.\mathbf{r}}d\mathbf{r}, \tag{54}$$

which is the Fourier transform of the overlap function $\phi_\alpha(\mathbf{r})$, the square of which gives the momentum distribution. (This formula was "derived" for a charge operator, but one finds the same (in PWIA!) for scattering from currents.) Taking into account the angular momentum character of ϕ_α (one has to square and sum over m-states to get a cross section) one finds that the cross section contains

$$\rho(\mathbf{p}_m) = |\int_0^\infty R_{nl}(r) j_l(p_m r) r^2 dr|^2, \tag{55}$$

where $R_{nl}(r)$ is the radial part of ϕ_α. Thus while one measures in (e,e′) a charge distribution in momentum space, one probes with (e,e′p) the wave function of a proton in some orbital in a nucleus in momentum space. (This statement needs some comments. The wave function ϕ_α is an overlap function $\langle A-1|A\rangle$. Only in the IPSM this is the wave function of orbital α. In general the single-particle strength is fragmented and the overlap function is influenced by the mechanism causing this fragmentation.)

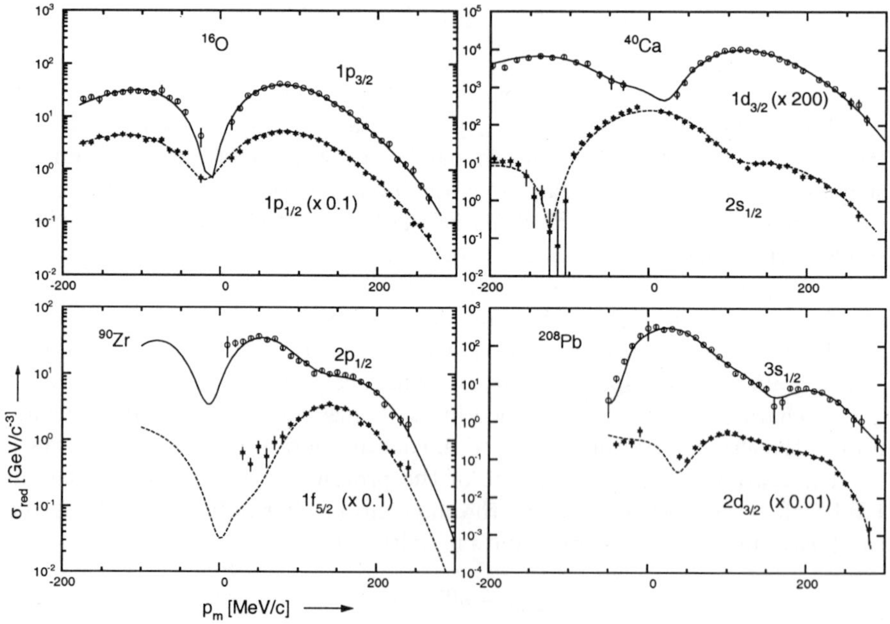

Figure 6: Reduced cross sections in parallel kinematics for (e,e′p) on various nuclei.

Just like shapes in r-space, the shape of F and ρ depends on the quantum numbers n and l. At $p_m = 0$ $F = 0$ except for s-states, and the first maximum moves to larger values of p_m with increasing l. Furthermore, the value of n determines the number of nodes in F. See Fig. 6 for some examples.

PWIA is in principle not a good approximation, since the outgoing proton always feels the interaction with the residual nucleus. Moreover, the initial and final states are not orthogonal in the PW approximation (initial and final states are not for the same Hamiltonian, since the initial state of the proton is a bound state in some potential well). In general PWIA works better for low values of the missing momentum and the higher the momentum transfer q is.

A much better and usually sufficiently good description is provided by the Distorted-Wave Impulse Approximation (DWIA). There one uses

$$\Psi_{A'} = \chi(\mathbf{r}_p)\Psi_{A-1}, \tag{56}$$

where the distorted wave χ is usually an optical model wave function describing p-(A-1) elastic scattering. (At high proton energies ($T_p > 1$ GeV) one can use the Glauber model to describe the proton distortions. These are then included on the level of the spectral function.) The optical model potential has a real part, describing the (usually attractive) interaction between the proton and the nucleus, and an imaginary (absorptive) part accounting in a global way for loss of protons due to reactions. As a result of this final-state interaction (FSI) between the outgoing proton and the residual A-1 nucleus the missing momentum p_m is no longer equal to the momentum of the proton before the reaction.

In the DWIA formalism the coincidence cross section does not factorize anymore into an electron-proton part and a nuclear structure part, like in Eq. 49. However, one defines in analogy to the PWIA formula a reduced cross section (or "distorted spectral function") according to:

$$\sigma_{red} = \frac{d^6\sigma}{dE_{e'}\, d\Omega_{e'}\, dE_p\, d\Omega_p} / K\sigma_{ep} = S^D(\mathbf{p}_m, E_m, \mathbf{p}) \tag{57}$$

and for a transition to a discrete final state the corresponding "distorted momentum distribution" ρ^D_α. Both depend now also on $\mathbf{p}$, i.e., on the kinematics. Or in other words: the "distorted spectral functions" are different for the L, T, LT and TT parts of the cross section (compare PWIA). (One should avoid using the words spectral function and momentum distribution in relation to experimental data, and use the term reduced cross section, as only in the limit of no distortions the latter can be identified with a spectral function or momentum distribution.)

The main effects of distortions are a reduction of the cross section (typically 20-40%) as a result of absorption of the protons, and changes in $\mathbf{p}$ and thus $\mathbf{p}_m$ as a result of the potential between the proton and the residual nucleus. The global structure of the momentum distribution is usually conserved, though [g].

[g] Especially for heavier (larger Z) nuclei one has also to investigate the influence of Coulomb distortions on the electron (plane) waves. In first order one can take them into account by using effective momenta [15,16].

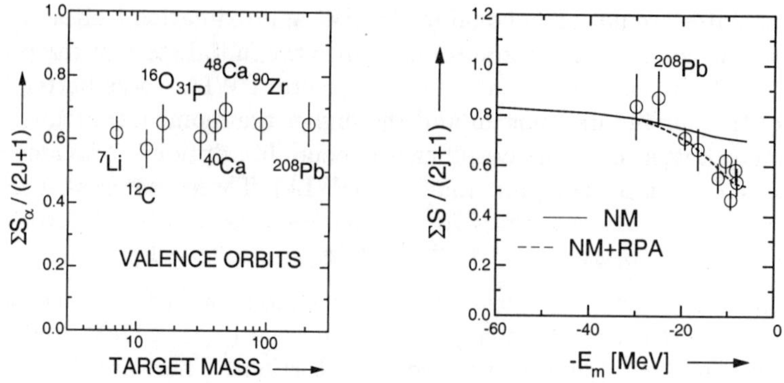

Figure 7: Occupation of shell model orbitals as determined from (e,e′p) studies [17].

Some examples are shown in Fig. 6. Note the different behaviour for knockout of protons from different orbitals and the filling of the minima as a result of FSI.

One of the main results that came out of (e,e′p) studies on a range of nuclei is that the occupation of shell model orbitals is only some 60-70% for valence orbitals, and even for deeply bound orbitals only 85-90%, see Fig. 7. This is attributed to the effects of long and short range correlations between nucleons. The latter will be discussed in the next section.
Further details about the formalism can be found in Refs. [18,19,20]. See Ref. [17] for a discussion of (e,e′p) results.

5 Study of NN correlations with (e,e′d) and (e,e′pp) reactions

5.1 Introduction

The motivation for the study of (e,e′d) and (e,e′pp) reactions is to investigate correlations between nucleons in a nucleus. It is well known that such correlations are at the origin of alpha decay and play, for instance, a large role in pion absorption, but also the reduction of strength in the quasi-elastic peak observed in inclusive (e,e′) studies is thought to be largely due to correlations. Finally, the results of the (e,e′p) studies mentioned in the previous section have been interpreted as reflecting correlations.

We will discuss first the (e,e’d) reaction as most results exist for this reaction. Deuteron type correlations can and have also been studied by other reactions like (d,α), (α,^{6}Li) or (p,pd). The (e,e′d) reaction has several advantages, however. The interaction driving the reaction is known and weak, so two-step processes induced by the projectile are negligible. Furthermore, in the usual description only a two-body hadronic final state with one optical-model potential (for a deuteron) is present instead of two such potentials for composite nuclei (in the case of the pickup reactions) or a three-body hadronic final state (for the knockout reaction). Inconsistencies in the results from hadron induced reactions are undoubtedly related to the complications arising from the latter points.

Compared to the (e,e′pn) reaction, which in principle is most suited to study (pn) correlations, the (e,e′d) reaction has some advantages. First of all it is experimentally much simpler and it has a better resolution (typically 250 keV instead of at least a few MeV). The latter is of crucial importance, for instance, when one wants to study the correlations in the important system ^{4}He, as one wants to separate the final ^{2}H state, which has $T = 0$, from the (pn) breakup continuum, which just above threshold is mainly of $T = 1$ character. Furthermore, the description of the (e,e′pn) reaction is complicated by having a three-body hadronic final state. And perhaps the most useful feature of the (e,e′d) reaction is that it provides an isospin filter, as one knows that the final (pn) pair is in a $T = 0$ state.

Recently the first (e,e′pp) experiments have been performed. Because of their kinematic flexibility those are in principle most suited to study correlations, provided the reaction mechanism is sufficiently clear. Some of these issues will be discussed briefly in Sect. 5.5.

5.2 Formalism for (e,e′d)

Analogous to the (e,e′p) reaction one defines the missing energy and missing momentum as:

$$E_m = \omega - T_{A-2} - T_d \tag{58}$$

$$\mathbf{p}_m = \mathbf{q} - \mathbf{p}_d\,. \tag{59}$$

The missing energy is now the separation energy of the deuteron with respect to the A-2 nucleus.

The simplest (not necessarily correct) way to describe the (e,e′d) reaction is as quasi-elastic knock-out of a pre-existing deuteron. Like in the (e,e′p) case, including the final-state interaction (FSI), e.g., by describing the outgoing deuteron with an optical model, the DWIA cross section can be written as

$$\frac{d^6\sigma}{dE_{e'}\,d\Omega_{e'}\,dE_d\,d\Omega_d} = K\sigma_{ed}(q)S^D(\mathbf{p}_m, E_m, \mathbf{p}_d), \tag{60}$$

where $\sigma_{ed}(q)$ is the elastic electron deuteron cross section. One might wonder why the interaction of an electron with a nucleus, which is mainly a (sum of) one-body operator(s), can lead to the knockout of two nucleons (the question also applies to the (e,e′NN) case). One should realize that in a correlated system there is a finite chance that if one hits one nucleon, another nucleon moves in such a way that in the final state it is not bound to A-2 anymore (i.e., there is a non-zero overlap between A-1 plus a knocked out nucleon and A-2 plus two nucleons in the continuum). The simplest example of this is (e,e′d) on a deuteron, i.e., elastic scattering. The higher the value of q the smaller the chance that the remaining neutron has such a momentum that it has overlap with the hit proton within the deuteron (momentum space) wave function. The relevant amplitude is given by

$$\int \phi_d^*(\mathbf{q}/2+\mathbf{p}', \mathbf{q}/2-\mathbf{p}')\phi_d(\mathbf{p}+\mathbf{q}, -\mathbf{p})d\mathbf{p}d\mathbf{p}', \tag{61}$$

where the variables of ϕ_d are the momenta of the proton and neutron inside the deuteron. This gives rise to the deuteron form factor $F_d(q)$, which falls off with q much faster than the proton form factor. In a similar way one can get an (e,e′d) reaction on a nucleus, the cross section of which will depend on the (internal) form factor of the (pn) pair inside the nucleus and the probability to have a (pn) pair with a certain centre-of-mass (c.o.m.) momentum (the spectral function or momentum distribution).

This can be seen from a microscopic description of the (e,e′d) reaction. We assume only one-body operators and take for simplicity only charge interaction. The T-matrix then reads:

$$T = \langle \phi_d(\mathbf{r})\chi_d(\mathbf{R})\phi_{A-2}|F_p(q)e^{i\mathbf{q}\cdot\mathbf{r}_p}|\phi_A\rangle, \tag{62}$$

where the ϕ's are (internal) wave functions of the indicated particles, χ_d is the d-(A-2) relative wave function and $F_p(q)$ is the proton form factor. By making an A→ A-2 expansion one can write

$$\langle A-2|A\rangle = \Phi(\mathbf{r}_1, s_1, \mathbf{r}_2, s_2) = \sum_{S,T,\lambda,\Lambda} \left[\phi_{pn}^{\lambda,S,T}(\mathbf{r})\psi_\Lambda(\mathbf{R})\right], \tag{63}$$

where we have made a transformation to relative and c.o.m. coordinates of the (pn) pair, λ and Λ being the corresponding angular momenta. Since $\mathbf{r}_p = \mathbf{R}+\mathbf{r}/2$, one finds that T contains the factors [h]

$$F(q) = \langle \phi_d(\mathbf{r})|F_p(q)e^{i\mathbf{q}\cdot\mathbf{r}/2}|\phi_{pn}(\mathbf{r})\rangle \tag{64}$$

[h] Because of the sum in Eq. 63 T will generally be a coherent sum.

and

$$\langle \chi_d(\mathbf{R}) | e^{i\mathbf{q}\cdot\mathbf{R}} | \psi(\mathbf{R}) \rangle. \tag{65}$$

The latter leads to the spectral function (in PWIA it is the momentum distribution for the c.o.m motion of the (pn)-pair), whereas the first one is a sort of transition form factor: it contains both the wave function of the (pn) pair inside the nucleus and that of a free deuteron. Provided the reaction mechanism is known well enough one thus can study both the c.o.m. motion (by varying p_m; be aware that (65) is essentially a function of p_m) and the relative motion (by varying q) of a (pn) pair in a nucleus.

It is now easily seen that one gets the quasi-elastic formula only in the case that $\phi_{pn}(r) = \phi_d(r)$, i.e., the (pn) pair in the initial nucleus is already a (preformed) deuteron. In that case Eq. 64 gives the deuteron form factor $F_d(q)$. If this is not the case, e.g., if the radial wave function is different, or the quantum numbers of the (pn) pair are different from those of a deuteron, the cross section will not follow $K\sigma_{ed}$.

5.3 ^{6}Li(e,e′d)^{4}He and ^{12}C(e,e′d)^{10}B

The first case studied was the ^{6}Li(e,e′d)^{4}He$_{gs}$ reaction. As a result of the large binding energy of ^{4}He (28.296 MeV) and the relatively loose binding of the proton and neutron to this ^{4}He core in ^{6}Li (the binding energy of ^{6}Li is only 31.994 MeV), ^{6}Li can be well described as a ^{4}He + d cluster (the ^{4}He - d binding energy is only 1.474 MeV). Experiments have been performed at rather low deuteron energies at Saclay [21] and at higher energies at NIKHEF [22] and Saclay [23] (see these references for the spectroscopic results).

In order to investigate whether the reaction can be described as quasi-elastic knockout, data have been taken at different values of the momentum transfer q, while keeping the missing momentum p_m constant. According to Eq. 60 (after correction for differences in FSI since $\mathbf{p}_d$ changes) the cross section should follow $K\sigma_{ed}(q)$. The results obtained at NIKHEF (the Saclay results are similar) are shown in Fig. 8. The change in cross section by a factor of nearly 40 is reproduced to within the uncertainty (about 10%) of the data points. It is hard to imagine that a different reaction process would yield the same observed q-dependence.

Next the (e,e′d) reaction was studied [24] on a 'real' nucleus, viz. ^{12}C. The first ^{12}C(e,e′d) spectrum gave a surprising result: the $0^+, T = 1$ state at 1.74 MeV in ^{10}B was excited about as strongly as the 3^+ ground state and the 1^+ state at 0.72 MeV, which are both $T = 0$ states. However, the transition to the $T = 1$ state is isospin forbidden for quasi-elastic knockout, since the deuteron has isospin zero.

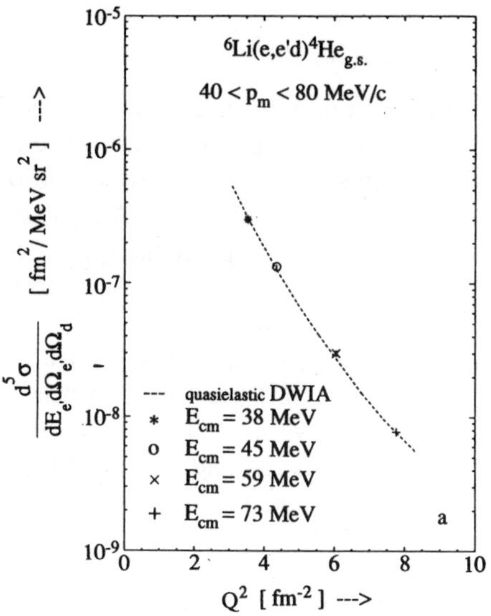

Figure 8: Measured cross sections for the ^{6}Li(e,e$'$d)^{4}He reaction compared to the behaviour (corrected for distortion effects and normalised at the highest q) of $K\sigma_{ed}(q)$.

The following mechanism may explain the excitation: at the values of q used, the cross section for electrodesintegration of the deuteron, which just above threshold leads to an $S = 0, T = 1$ state, is comparable to elastic scattering. In the nucleus the inverse process may occur: the electron deexcites an $S = 0, T = 1$ (pn) pair into a deuteron with $S = 1, T = 0$. This mechanism can be investigated in the following way. Generally the cross section for scattering of an electron from a deuteron has a longitudinal (charge) and a transverse (magnetic) part (see Sect. 2):

$$\sigma_{ed}(q) = \sigma_p \frac{Q^2}{q^2} \left[R_L(q) + \epsilon^{-1} R_T(q)\right] . \tag{66}$$

For elastic scattering both R_L and R_T are nonzero, but the transition to the $S = 0, T = 1$ state is purely transverse. A longitudinal/transverse separation has been performed for the ^{12}C(e,e$'$d)^{10}B reaction[24] by taking data at the same value of q, but different values of $\theta_{e'}$ (Rosenbluth separation) . The results are shown in Fig. 9. The data clearly indicate that whereas the transitions to the $T = 0$ states are dominantly longitudinal, the transition to the $T = 1$ state is within error bars purely transverse. Also the q-dependence of the cross section has been measured. It agrees with the one for electrodesintegration of a

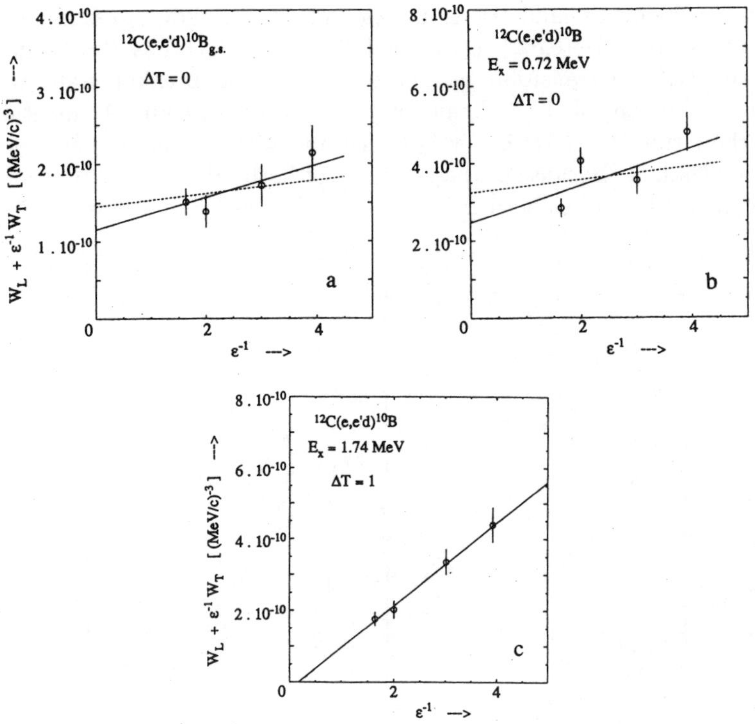

Figure 9: Results of the Rosenbluth separation for the ^{12}C(e,e′d) reaction.

deuteron just above threshold. One can make use of this "deuteron integration" process to study singlet (pn) correlations in a nucleus, an important example being the ^{4}He(e,e'd) reaction to the final (pn) state just in the continuum, see the next subsection.

5.4 ^{3}He(e,e′d)p and ^{4}He(e,e′d)

It is of interest to study the (e,e'd) reaction on ^{3}He and ^{4}He. In principle exact Faddeev calculations are available for ^{3}He, both for the ground state and the p+d final state. Thus FSI is treated exactly, and also two-body currents can be incorporated. By switching those ingredients off and on in the calculations, one can get an estimate how important they are, and thus calibrate the (e,e′d) reaction mechanism.

For ^{4}He exact calculations for the continuum state do not yet exist, unfortunately. On the other hand ^{4}He is interesting, since it allows to study a pure $\Delta T = 0$ transition: ^{4}He(e,e$'$d) to the final ^{2}H state, which has $T = 0$, and an almost pure $\Delta T = 1$ transition, leading to the the (pn) breakup state, which near the breakup threshold is dominantly of $T = 1$ character. As mentioned already, the latter transition is purely transverse, whereas the former is dominantly longitudinal. Because of the isospin $\frac{1}{2}$ of both ^{3}He and the proton, the ^{3}He(e,e$'$d)p transition has both $\Delta T = 0$ and $\Delta T = 1$ contributions.

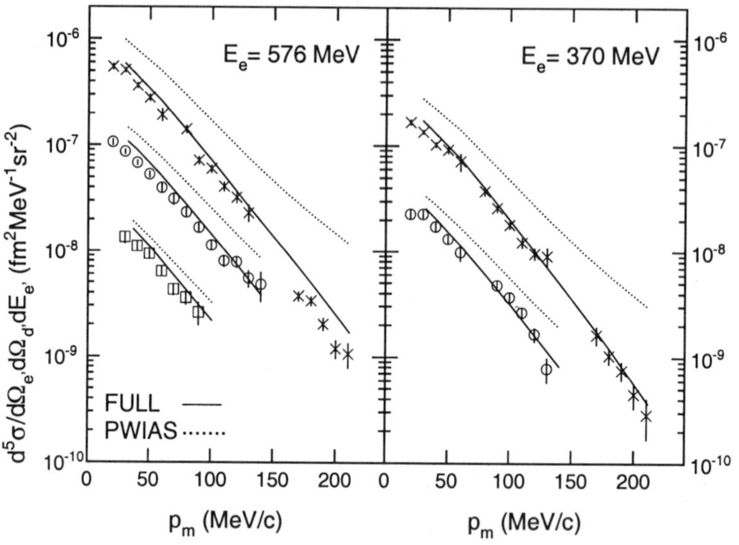

Figure 10: Cross sections for the ^{3}He(e,e$'$d)p reaction in parallel kinematics.

The reacions were studied [25] for missing momenta up to 200 MeV/c, and values of the transferred momentum q of 412, 504 and 604 MeV/c, both in parallel and (q, ω) kinematics. Since the data comprise both the q and p_m dependence of the reaction, different parts of the pn motion can be studied. The cross sections were measured at two beam energies, so that a structure function separation could be performed.

The cross sections measured in parallel kinematics are shown in Fig. 10 in comparison with the results of plane-wave and full Faddeev calculations performed by the Bochum group using the Bonn-B potential. Two-body currents were not yet included. Without going into details one can notice the following. FSI reduces the cross sections by about a factor of two, but the global behaviour is preserved. The dependence on p_m, which is well described by the calculations, is similar to what is found in ^{3}He(e,e$'$p)d, and reflects the mo-

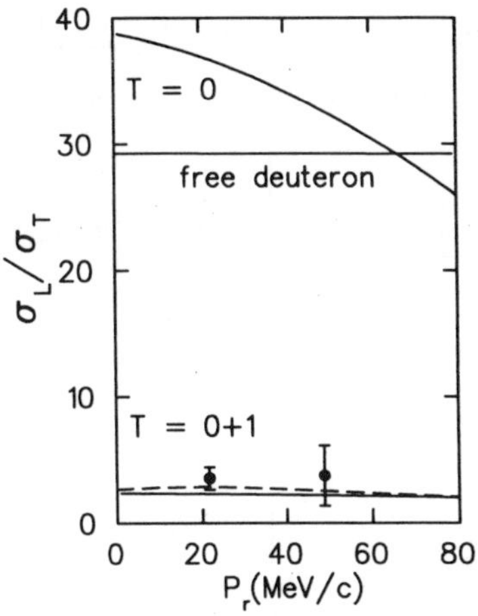

Figure 11: Measured σ_L/σ_T ratio for the ^{3}He(e,e′d)p reaction in parallel kinematics at $q = 420$ MeV/c compared to the results of two types of calculations (see further Ref. [26]).

mentum distribution. The strong dependence on q reflects the relative wave function of the (pn) pair. By comparing with calculations with and without it, it is found that at the highest value of q the contribution of the D-state of the (pn) pair and the deuteron is non-negligible.

The separated W_L and W_T structure functions turned out to be of comparable magnitude, indicating the presence of sizeable contributions from the $\Delta T = 1$ transition. This had already been observed for $q = 420$ MeV/c at low values of p_m in an experiment performed at MIT/Bates [26], see Fig. 11. For a pure $\Delta T = 0$ transition the L/T ratio would have been much larger, see the figure.

In Fig. 12 we show the separated structure functions W_L and W_T for the ^{4}He(e,e′d)pn reaction in parallel kinematics [27]. The lowest two p_m points for $q = 406$ MeV/c have rather large error bars. This is due to the fact that the integrated charge for every kinematics was chosen on account of the ^{3}He cross sections, that were measured simultaneously, and which at these kinematics are much larger than those for ^{4}He(e,e′d)pn. Anyway the data for W_L are consistent with zero for all kinematics, whereas those for W_T are clearly non-zero for the kinematics with better statistical accuracy. These observations are fully consistent with the discussed mechanism of an isovector transition, for which obviously W_L is identical to zero. The curve represents the results of calculations by Nagorny in a diagrammatic approach. See Ref. [27] for further details.

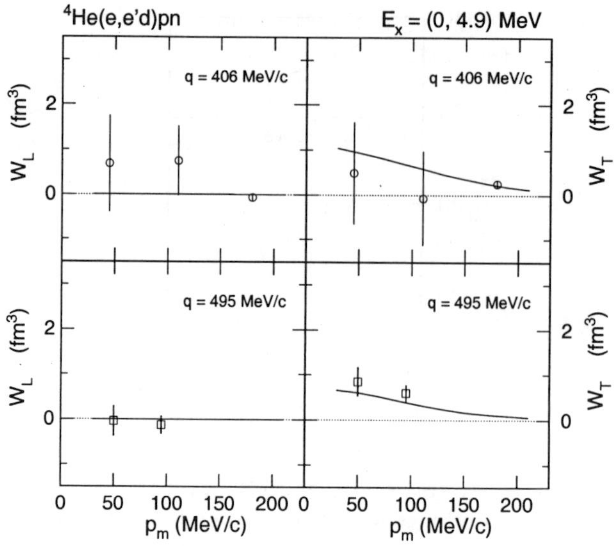

Figure 12: Separated structure functions W_L and W_T for the ^{4}He(e,e′d)pn reaction in parallel kinematics.

It is interesting to look into (e,e′p) and (e,e′d) reactions with polarized electrons on polarized ^{3}He. Let's assume that the polarization is completely carried by the neutron, and that the protons are coupled to $S = 0$, and that FSI does nothing to polarizations. In ^{3}He(e,e′p)d the proton hit by the virtual photon then has its spin opposite to that of the neutron, i.e., of ^{3}He, since the proton that forms the deuteron has to have its spin parallel to the neutron to make a spin 1 state. On the other hand the hit proton in (e,e′p) to the (pn) breakup state with $S = 0$ has to have its spin in the same direction as the neutron. One thus expects opposite asymmetries when comparing the two reaction channels (and very small asymmetries if one does not separate them!). Finally, in the ^{3}He(e,e′d)p reaction the hit proton should have its spin also parallel to that of the neutron. An experiment studying the asymmetries in these reactions has been performed with the internal beam of the storage ring AmPS at NIKHEF [28].

5.5 The (e,e′pp) reaction

As mentioned already in Sect. 4 the missing strength in (e,e′p) to discrete states was an indirect indication for strong nucleon-nucleon correlations. A more direct one comes from (e,e′p) reactions leading to the continuum. There a peak in the E_m spectrum is observed approximately at a value $p_m^2/2M$ with M the nucleon mass. The interpretation is that this peak results from absorption of the virtual photon on the proton of a correlated pN pair (N=p or n), where the proton and the other nucleon have opposite momenta, that nucleon leaving the nucleus, too. In that case the missing momentum is carried by that nucleon, and not by the whole residual nucleus. One thus expects a peak at an apparent E_m value of $p_m^2/2M$ minus the assumed $T_{A-1} = p_m^2/2(A-1)M$. The peak will be broadened by Fermi motion of the c.o.m. of the pair. This is just what has been observed [29,30]. Theoretical calculations of the spectral function [31,32] had already shown that high nucleon momenta in a nucleus are found at high E_m values, supporting the picture that (especially short-range) correlations between nucleons are responsible for those.

Detailed studies of such correlations can be done with the (e,e′NN) reaction. This can can be seen from formulae analogous to (62-65) for this process. Taking plane waves for the outgoing nucleons one can easily find that the T-matrix for (e,e′NN) is (a coherent sum of)

$$T_{nlNL} = F_{nl}(p_{rel})F_{NL}(P_m), \tag{67}$$

where $F_{NL}(P_m)$ is the Fourier transform of the c.o.m. motion of the pair, having quantum numbers N and L, and $F_{nl}(p_{rel})$ similar for the relative motion. P_m is minus the c.o.m. momentum of the nucleon pair in the initial state, and p_{rel} half the relative momentum between them. They are given by

$$\mathbf{P}_m = \mathbf{q} - \mathbf{p}_1' - \mathbf{p}_2' \tag{68}$$

$$\mathbf{p}_{rel} = (\mathbf{p}_1 - \mathbf{p}_2)/2 = (\mathbf{p}_1' - \mathbf{q} - \mathbf{p}_2')/2. \tag{69}$$

Here (un)primed quantities are for the (initial) final state, and we have assumed that the virtual photon coupled to nucleon 1 and that there is no FSI. The c.o.m. motion with corresponding momentum distribution, e.g., the dependence of $F(P_m)$ on the values of N and L, is similar to that discussed in Sect. 4 for a proton, and has typical momenta of 0 - 200 MeV/c. The relative motion of the two nucleons, as described by $F(p_{rel})$, especially the high-momentum parts of it ($p_{rel} \approx 200 - 400$ MeV/c), depends strongly on the character of the short-range correlations. In this PW picture the cross section has the structure

$$\sigma \propto \sigma_{ep}(q)S(\mathbf{p}_{rel}, \mathbf{P}_m, E_m), \tag{70}$$

or, taking the correlations into σ_{ep} (Ref. [33]):

$$\sigma \propto \sigma_{epp}(\mathbf{q}, \mathbf{p}_{rel})S(\mathbf{P}_m, E_m). \tag{71}$$

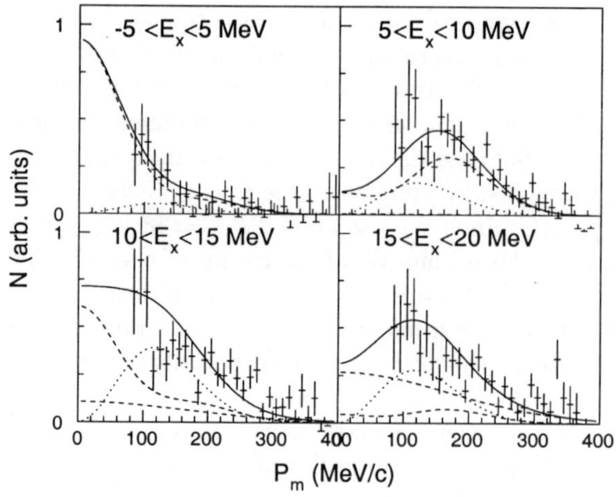

Figure 13: Missing momentum distributions from Ref. [34] for the $^{16}O(e,e'pp)^{14}C$ reaction. The dashed (dotted) curves give the contributions of ^1S (^3P) correlated (pp) pairs. The dot-dashed line is the assumed contribution from 3N final states.

Unfortunately the mechanism where the virtual photon interacts with one nucleon of a correlated pair is not the only one that can lead to two-nucleon emission. MECs may contribute, too, since when the virtual photon hits, for instance, a pion that is being exchanged by two nucleons, both of them may be knocked out. Also excitation of a nucleon to a Δ, followed by decay of the Δ with another nucleon into a 2N final state, will contribute. Taking this into account, the (e,e′pp) reaction is better suited (also on experimental grounds) than the (e,e′pn) reaction (although $T = 0$ (pn) correlations are expected to be stronger than (pp) correlations) to study two-nucleon correlations. In leading order MECs are zero for a (pp) pair, since the exchanged meson has no charge, and for the dominant $^1S_0$ (i.e., $l = 0, S = 0$) NN correlations the $\gamma_v + NN \to \Delta + N \to N + N$ process is forbidden for the dominant M1 excitation of the Δ (the Δ+N, and hence the final N+N state, has $J^\pi = 1^+, T = 1$, while the latter state should have $l+S+T=$ odd, and l is even because of parity). MECs and isobar currents can further be suppressed by separating the longitudinal structure function, since they are transverse. The (e,e′pp) reaction is in general described by 6 structure functions, but in superparallel kinematics, where both $\mathbf{p}'_1$ and $\mathbf{p}'_2$ are (anti)parallel to $\mathbf{q}$, only W_L and W_T are nonzero.

Because of the coherent sum in Eq. 67 the description of two-nucleon knockout is pretty complicated, also from the nuclear structure side. Most nuclear wave functions are given in a single-particle basis. Those have to be converted to relative and c.o.m. variables of the involved pair. The correlations then enter the relative wave function. The most important correlations are for relative s-waves ($l = 0$, ${}^1S_0$ for $T = 1$ and ${}^3S_1$ for $T = 0$), but $l = 1$ correlations are non-negligible (and for both one has to take into account more values of n). An interesting consequence of the latter is that even for a $0^+ \rightarrow 0^+$ transition like ^{16}O(e,e$'$pp)^{14}C(gs) the c.o.m. motion has not only an $L = 0$ (when $l = 0$), but also an $L = 1$ (when $l = 1$) contribution, which influences the P_m distribution, see Fig. 13. See further Ref. [35].

6 The ^{1}H(e,e$'\pi^0$) reaction near threshold and Chiral Perturbation Theory

6.1 Introduction

The pion plays a central role in strong-interaction physics. First of all, it is the particle that mediates (part of) the strong interaction between nucleons. Secondly, many nucleon resonances, for instance the delta, have a large width for decay into a pion plus a nucleon. Finally, because of the low mass of the pion the production of a pion is the lowest-energy reaction that produces an extra hadron. For these reasons the study of pion production, both on nucleons and nuclei, is of fundamental importance.

If one wants to study the production process itself, one likes to start with investigating the production on a nucleon and use a well known interaction like the electromagnetic interaction, i.e., one should study pion production with real and virtual photons on the proton and neutron and thus determine the elementary amplitudes. The study of these processes near the threshold for the reaction has the special feature that effective theories, like Chiral Perturbation Theory (ChPT), that represent the low-energy limit of quantum chromodynamics, can be used to describe the process. Especially π^0 production provides a very sensitive test of these theories, since the first order term in the production amplitude is zero because of the zero charge of the neutral pion. After introducing the amplitudes used for describing pion production we will discuss in this section π^0 electroproduction on the proton near threshold. In Sect. 7 and 8 we will present examples of other studies making use of pion production.

6.2 The pion production amplitudes

In Sect. 2.6 we have encountered already the Cλ, Eλ and Mλ transitions and form factors for electron scattering. Pion production on a nucleon is described by amplitudes denoted by $X_{l\pm}$, in which X indicates the type of (virtual) photon initiating the reaction: L (instead of C) for longitudinal photons [i], E for electric ones and M for magnetic ones, l stands for the pion angular momentum with respect to the nucleon in the final state, and $\pm$ denotes the total angular momentum $j = l \pm \frac{1}{2}$ in the final state.

Take as example pion photoproduction near threshold. The pions have then $l = 0$ (s-waves) and obviously the total spin of the pion-nucleon system is $\frac{1}{2}$. Because of the intrinsic negative parity of the pion, the parity of that system is negative. The transition thus is from a $\frac{1}{2}^+$ to a $\frac{1}{2}^-$ state, which needs an E1 photon (in electroproduction also a C1 virtual photon is possible). Thus the pion ampltude is E_{0+}. See Sect. 7 for the amplitudes playing a role in the $N \to \Delta$ transition.

6.3 π^0 electroproduction on the proton near threshold

At threshold the ^{1}H(e,e$'\pi^0$)p reaction is described by the pion s-wave multipoles E_{0+} and L_{0+} only. Away from threshold the p-wave multipoles E_{1+}, M_{1+}, M_{1-}, L_{1+} and L_{1-} start contributing. The L multipoles contribute only to electroproduction. The interest in this reaction stems from the fact that there exist rather fundamental predictions for some multipoles from Chiral Perturbation Theory (ChPT) (see Ref. [36] for a discussion of Low Energy Theorems and ChPT, and the history of photoproduction neat threshold). Here we focus on what one can learn by using virtual photons.

There are several reasons why one would like to investigate also π^0 electroproduction on the proton near threshold. Firstly, one can study the L_{0+} multipole, which is absent in photoproduction. Secondly, one can investigate the Q^2 dependence of, e.g., the E_{0+} multipole. Finally, four structure functions (in the case of no polarisation) contribute to electroproduction instead of only one in photoproduction. By separating (some of) these structure functions one can enhance the sensitivity to the different contributing multipoles. Hence the study of electroproduction enables a much more detailed comparison with the results from various theoretical models.

The first π^0 electroproduction experiment near threshold was performed at NIKHEF with the 1% duty factor beam available at that time[37]. A much more detailed experiment[38] was performed at NIKHEF in 1994 with the high duty-factor beam from the Pulse Stretcher ring AmPS. The E_{0+} and L_{0+} multipoles were studied separately, including their energy dependence, by measuring the angular distribution for π^0 production for values of the invariant energy W

---

[i] Some authors use S. The difference bewteen L and S is the so-called longitudinal polarization parameter ϵ_L, which is equal to $\epsilon Q^2/\omega^{*2}$, where ω^* is the energy of the virtual photon in the center of mass frame.

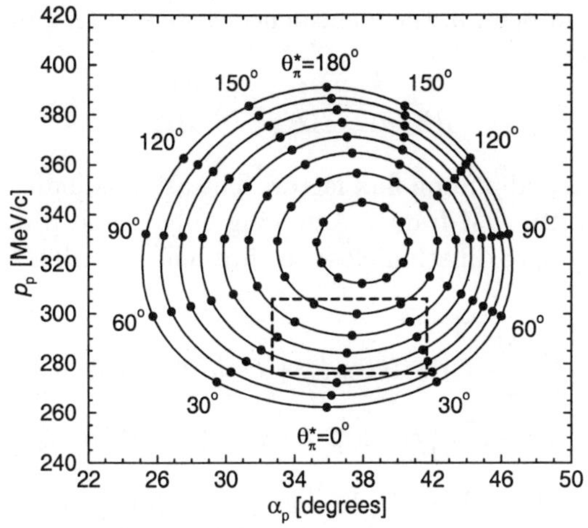

Figure 14: Proton momentum p_p versus angle with the electron beam, α_p, in the lab frame, for protons emitted in the scattering plane. The curves are for values of the invariant energy W of 1, 3, 5, 7, 9, 11, 13, and 15 MeV above threshold. The innermost curve corresponds to the lowest value of W. The numbers indicate the polar angle of the pion in the CM frame. The right-hand part of the curves corresponds to $\phi^*_\pi = 0°$ and the left-hand part to $\phi^*_\pi = 180°$. These curves were calculated with a constant angle of the scattered electron, which explains why the centre of the curves moves to the left with increasing values of W. The dashed rectangle indicates the momentum and angular acceptance of the QDQ spectrometer in the $\theta^*_\pi = 0°$ kinematic (see text for details).

from 1 to 14 MeV above the threshold at $Q^2 = 0.10$ (GeV/c)^2. The electron energy was 525 MeV.

The scattered electrons, which determine the values of Q^2 and W, were detected in the QDD spectrometer. The detection of the π^0 was avoided by measuring the recoil proton instead. The protons are emitted in a cone around $\mathbf{q}$ and were detected in the QDQ spectrometer. For every value of W (and a given Q^2) there is a relation between the direction of the proton (which is directly related to the direction $(\theta^*_\pi, \phi^*_\pi)$ of the emitted pion) and its momentum, see Fig. 14. Data were taken in four kinematics with central values at $W = 6$ MeV above threshold for θ^*_π of 0°, 90° (for both $\phi^*_\pi = 0°$ and 180°) and 180°. Since the momentum bite of the QDQ spectrometer is $\pm$ 5% and its opening angle $\pm$ 4°, globally the region $W = 2 - 12$ MeV above threshold and $\theta^*_\pi = 0° - 180°$ was covered, with data around $\theta^*_\pi = 90°$ for both $\phi^*_\pi = 0°$ and 180°.

As discussed in Sect. 3, the coincidence cross section can be written as

$$\frac{d^5\sigma}{dE_{\mathrm{e}'}d\Omega_{\mathrm{e}'}d\Omega_\pi^*} = \Gamma_{\mathrm{v}}\frac{d\sigma}{d\Omega_\pi^*}, \tag{72}$$

where Γ_v is the virtual photon flux factor. The asterisk indicates that the pion cross section is in the CM frame. Assuming only s- and p-waves for the pion (which is a good approximation close to threshold) $d\sigma/d\Omega_\pi^*$ can be written [39] as:

$$\begin{aligned}\frac{d\sigma}{d\Omega_\pi^*} = \frac{p_\pi^*}{q_{\mathrm{L}}}\frac{W}{m_{\mathrm{p}}}\{&A + B\cos\theta_\pi^* + C\sin\theta_\pi^*\cos\phi_\pi^* + D\cos^2\theta_\pi^* \\ &+E\sin^2\theta_\pi^*\cos 2\phi_\pi^* + F\sin\theta_\pi^*\cos\theta_\pi^*\cos\phi_\pi^*\},\end{aligned} \tag{73}$$

where the coefficients $A, B, C, D, E,$ and F, which depend on Q^2 and W, are combinations of s- and p-wave multipoles. The coefficients can be written as

$$\begin{aligned}A &= \epsilon_{\mathrm{L}}|L_{0+}|^2 + |E_{0+}|^2 + \tfrac{1}{2}|3E_{1+} - M_a|^2 + \tfrac{1}{2}|M_b|^2 \\ B &= 2\mathrm{Re}\{E_{0+}^*(3E_{1+} + M_a)\} \\ C &= -\sqrt{2\epsilon_{\mathrm{L}}(\epsilon+1)}\mathrm{Re}\{L_{0+}^*(3E_{1+} - M_a)\} \\ D &= |3E_{1+} + M_a|^2 - \tfrac{1}{2}|M_b|^2 - \tfrac{1}{2}|3E_{1+} - M_a|^2 \\ E &= 3\epsilon(\tfrac{3}{2}|E_{1+}|^2 - \tfrac{1}{2}|M_{1+}|^2 - \mathrm{Re}\{E_{1+}^*M_a + M_{1+}^*M_{1-}\}) \\ F &= 0,\end{aligned} \tag{74}$$

with M_a and M_b combinations of M_{1+} and M_{1-}:

$$\begin{aligned}M_a &= M_{1+} - M_{1-} \\ M_b &= 2M_{1+} + M_{1-}.\end{aligned} \tag{75}$$

In these formula's terms containing the L_{1-} and L_{1+} multipoles have been neglected, since in all models they are calculated to be small.

Eq. 73 shows that the value of the coefficient B can be determined by comparing the cross sections at $\theta_\pi^* = 0°$ and $180°$. The value of C can be determined by comparing the cross sections measured at $\phi_\pi^* = 0°$ and $180°$ provided $\sin\theta_\pi^* \neq 0$. Since the p-wave multipoles are almost purely real close to threshold, the real part of the E_{0+} and L_{0+} multipoles can be extracted from the values of B and C with the aid of Eq. 74, provided that reliable predictions for the relevant combination of p-wave multipoles are available. For this the values calculated within ChPT by Bernard, Kaiser and Meißner [41,42] were used. The thus extracted E_{0+} and L_{0+} multipoles as a function of ΔW are shown

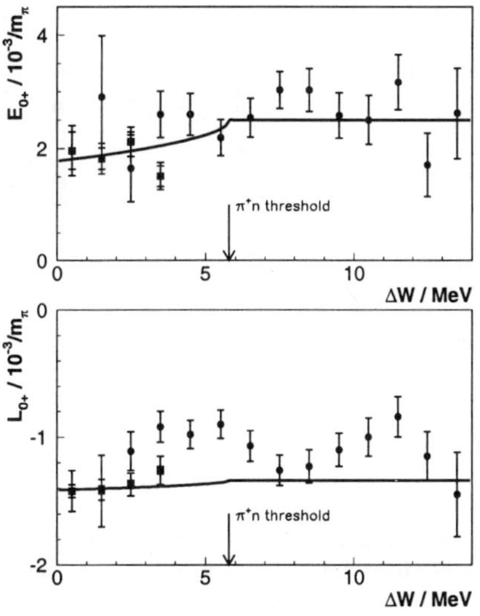

Figure 15: Values of the real parts of E_{0+} and L_{0+} at $Q^2 = 0.10$ (GeV/c)^2 from Ref. [38] (dots) and Ref. [43].

in Fig. 15, together with points determined from an analysis of longitudinal-transverse data measured very close to threshold at different beam energies in Mainz [43].

It is interesting to compare the experimental and calculated results for the multipoles and the combination

$$a_0 = |E_{0+}|^2 + \epsilon_{\mathrm{L}}|L_{0+}|^2, \tag{76}$$

which determines the cross section at threshold, as a function of Q^2. The results from Ref. [38] for a_0 (black circle), as extracted from the value of A, together with the values at threshold of $|E_{0+}|^2$ (up-triangle) and $\epsilon_{\mathrm{L}}|L_{0+}|^2$ (down-triangle), are presented in Fig. 16. The values from Ref. [37] (scaled for the different value of ϵ) are indicated by the open squares. The prediction for a_0 from ChPT [42,44] is given by the solid curve. Also shown are the separate contributions of $|E_{0+}|^2$ (dashed) and $\epsilon_{\mathrm{L}}|L_{0+}|^2$ (dotted) as calculated in ChPT. The value of E_{0+} is negative at $Q^2 = 0$, as determined in the photoproduction experiments, goes through zero near $Q^2 = 0.04$ (GeV/c)^2 and is positive at our value of Q^2. The value of L_{0+} is negative everywhere and goes from a value of -2.6 ± 0.5 (Ref. [37]) at $Q^2 = 0$ to -1.1 ± 0.1 at $Q^2 = 0.1$. It is interesting to note the decrease of $\epsilon_{\mathrm{L}}|L_{0+}|^2$, which means an even larger one of $|L_{0+}|^2$,

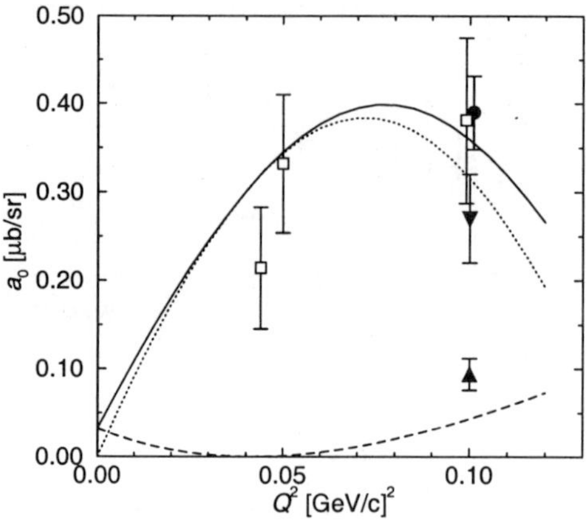

Figure 16: Values of a_0 as a function of Q^2. The value of $|E_{0+}|^2$ is indicated by the triangle, and that of $\epsilon_L|L_{0+}|^2$ by the diamond. The dashed, dotted and solid curves are the ChPT results for $|E_{0+}|^2$, $\epsilon_L|L_{0+}|^2$, and a_0, respectively.

in the ChPT calculation. However, the one-loop corrections in ChPT become quite large at $Q^2 = 0.10$ (GeV/c)^2, and therefore it is not clear if the one-loop expansion suffices.

7 Study of the $\Delta(1232)$ by π^0 production

The method to detect the recoil proton can also be used to study nucleon resonances like the $\Delta(1232)$ or the $S_{11}(1535)$ via single meson production channels ^{1}H(e,e$'$p)x, where x is a π^0 or η. The produced meson can be identified via the missing mass $m_x = \sqrt{E_m^2 - \mathbf{p}_m^2}$, see Eq. 41, while the resonance to be studied is selected by choosing the invariant mass of the final state W. An example is shown in Fig. 17, where one can see clearly π^0 particles coming from the Δ and η particles coming from the S_{11} resonance.

We will now discuss in some detail what can be learned about the Δ from the ^{1}H(e,e$'$p)π^0 reaction. In Sect. 2.6 we saw that the $N \to \Delta$ transition can proceed via C2, E2 and M1 transitions. The corresponding pion production amplitudes are L_{1+} (or S_{1+}), E_{1+} and M_{1+}. The large interest in those stems from the following. In a simple quark model, in which the three quarks are in in a relative s-state, the $N \to \Delta$ transition is just a spin-flip of one of the quarks, and therefore an M1 transition. Due to the colour hyperfine interaction the Δ (and also the proton) may have a D-state ("deformation of the Δ"), making (presumably small) C2 and E2 transitions possible. However, at very high

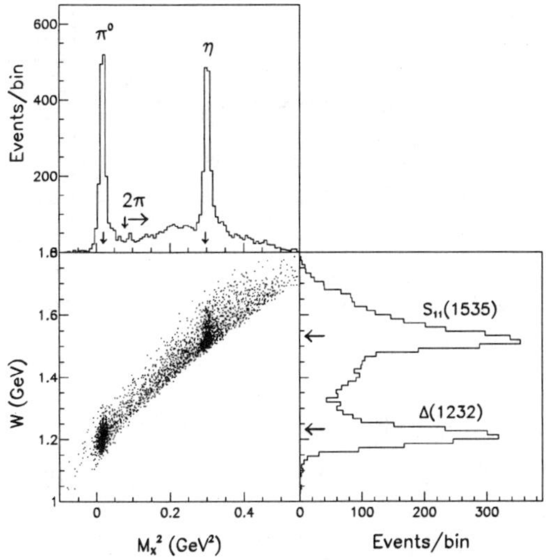

Figure 17: Invariant mass vs. missing mass squared as extracted from ^{1}H(e,e$'$p)x data[45].

values of Q^2 quark-helicity conservation[1,46] tells that the helicity-flip amplitude $A_{3/2}$ as well as the longitudinal amplitude $C_{1/2}$ goes to zero, only the non spin-flip amplitude $A_{1/2}$ surviving. In terms of these helicity amplitudes the pion multipole amplitudes are given by

$$\begin{aligned} E_{1+} &= a\,(A_{3/2} - \tfrac{1}{3}\sqrt{3}A_{1/2}) \\ M_{1+} &= a\,(A_{3/2} + \sqrt{3}A_{1/2}) \\ S_{1+} &= a\,C_{1/2}\,, \end{aligned} \tag{77}$$

where a is a constant. Hence E_{1+} should grow (relatively) with Q^2 to be comparable to M_{1+}, whereas S_{1+} should go to (or stay near) zero. The interesting question is at which value of Q^2 this perturbative behaviour sets in.

The values of S_{1+}, E_{1+} and M_{1+} can be determined from the π^0 angular distribution, similar to the threshold case. The cross section can be written as:

$$\frac{d\sigma}{d\Omega^*_\pi} = A + \epsilon B + \epsilon C \sin^2\theta^*_\pi \cos 2\phi^*_\pi + \sqrt{2\epsilon(\epsilon+1)}D sin\theta^*_\pi \cos\phi^*_\pi \tag{78}$$

with A, B, C and D polynomials in $\cos\theta^*_\pi$, the coefficients of which depend on

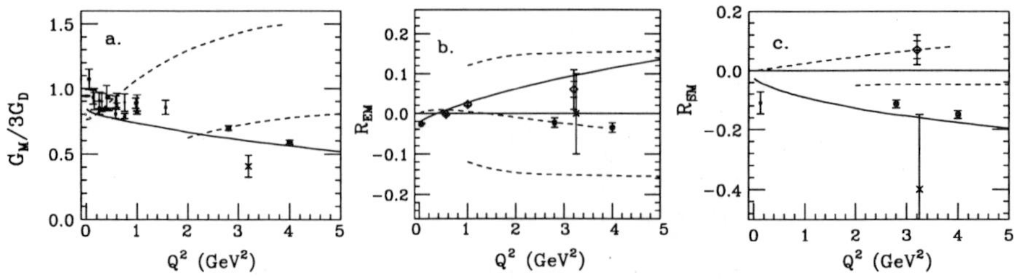

Figure 18: Results for the Δ form factor and R_{EM} and R_{SM}. See Ref. [48] for detailed explanations.

W and Q^2. Limiting oneself to pion s and p waves one gets:

$$\begin{aligned} A' &= A + \epsilon B = A_0 + A_1 \cos\theta_\pi^* + A_2 \cos^2\theta_\pi^* \\ C &= C_0 \\ D &= D_0 + D_1 \cos\theta_\pi^*, \end{aligned} \tag{79}$$

where the A and ϵB terms are taken together, since when having data at one value of ϵ only, they can not be separated. It is instructive to compare this formula with the general one Eq. 44. The coefficients A_0, A_2 and C_0 contain the interference term $\mathrm{Re}(E_{1+}^* M_{1+})$, while D_1 is dominated by the $\mathrm{Re}(S_{1+}^* M_{1+})$ interference term [47]. Thus by fitting the θ_π^* and ϕ_π^* dependence of the cross section one can determine the ratio's R_{EM}=$\mathrm{Re}(E_{1+}^* M_{1+}/|M_{1+}|^2$ and R_{SM}=$\mathrm{Re}(S_{1+}^* M_{1+}/|M_{1+}|^2$. However, one has to be careful in drawing conclusions about the properties of the Δ directly from R_{EM} and R_{SM}. The experiment measures the total amplitudes, which include the contribution from non-resonant ("background") terms. Furthermore, the angular distributions are also influenced by other amplitudes, of which especially the M_{1-} is pretty large. One can try to determine the background amplitudes away from the Δ peak, or scan over it, since they are not resonant, but the determination of the Δ amplitudes will always be model dependent.

Experiments have been performed with real photons at Mainz and LEGS, and with virtual photons at Bonn and Jlab [48]. The latter experiment obtained data at values of Q^2 of 2.8 and 4.0 (GeV/c)^2 using incoming electron energies of 3.245 and 4.045 GeV. The scattered electrons were detected in the S(hort) O(rbit) S(pectrometer) and the protons, which had momenta of appr. 1.8 and 2.5 GeV/c, were detected in the H(igh) M(omentum) S(pectrometer). Fig. 18 shows the results (black dots), together with those from other experiments, and with the results of several calculations. One can see that even at $Q^2 = 4$ (GeV/c)^2 R_{EM} is rather small, as is R_{SM}.

8 Study of the pion form factor with the ^{1}H(e,e$'\pi^+$)n-reaction

One of the key questions in hadron physics is how the strong interaction between quarks, as described by Quantum Chromo-Dynamics (QCD), gives rise to bound systems, the hadrons. This requires from the theoretical side the development of models and techniques to solve QCD in the non-perturbative regime. On the experimental side precise experimental data are needed to gauge the development of such non-perturbative QCD calculations. Precise data on the structure of the pion are of particular importance, since the pion, composed of a valence quark/anti-quark pair, is one of the simplest hadrons that can be studied, and thus its structure presumably can be calculated more reliably at present than that of e.g. the proton. For these reasons virtually all calculations of the structure of hadrons use the pion as its prime example, and calculate e.g. its charge form factor. Compared to the form factor of the nucleon, the pion form factor has the additional advantage that the value of the charge form factor for $Q^2 \to \infty$ is determined by the decay constant for the weak decay $\pi^+ \to \mu^+ + \nu_\mu$ [49].

Since there are no free-pion targets, one has to resort to special methods to measure the pion form factor. At low values of Q^2 one can use the inverse reaction: scattering of high-energy pions from atomic electrons. In this way the pion form factor has been determined [50] for values of $Q^2 < 0.3$ (GeV/c)^2. For higher values of Q^2 the only way to measure the pion form factor is to scatter high-energy electrons from a virtual π^+ particle in a proton, whereby the π^+ is knocked out of the proton and a neutron is left.

As for other (e,e$'$x) reactions the general cross section for the ^{1}H(e,e$'\pi^+$)n reaction contains four terms

$$\frac{d^5\sigma}{d\Omega_{e'}\, dE_{e'}\, d\Omega_\pi} = \Gamma_v \left[\epsilon\, \sigma_L + \sigma_T + \sqrt{2\epsilon(\epsilon+1)}\, \sigma_{LT} \cos\phi + \epsilon\, \sigma_{TT} \cos 2\phi \right]. \quad (80)$$

The cross sections σ_X depend on Q^2, the invariant mass W and the Mandelstam variable t. The longitudinal cross section σ_L is mainly due to the Coulomb interaction between the electron and the pion (t-pole diagram) and thus contains the (square of the) pion form factor. The other terms contain this t-pole term, but get also sizeable contributions from other mechanisms.

For small values of $|t|$, σ_L is dominated by the t-pole contribution, which leads to

$$\sigma_L \sim \frac{-t}{(t-m_\pi)^2}\, Q^2\, F_\pi^2(Q^2)\, g_{\pi NN}^2(t), \quad (81)$$

where m_π is the pion mass, $F_\pi(Q^2)$ the pion form factor and $g_{\pi NN}(t)$ the function that represents the probability amplitude to have a (virtual) π^+ inside

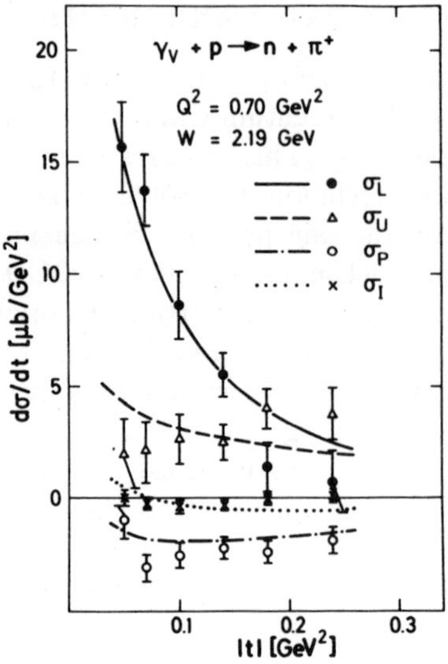

Figure 19: Separated cross sections for the (e,e$'\pi^+$)n reaction from Ref. [52].

the proton for a given t. It is generally taken as $g_{\pi NN}(0)/(1 - t/\Lambda^2)$ with $\Lambda \simeq 0.85$ (GeV/c)^2, but there is some discussion about the correct value of Λ. Other processes that may contribute to σ_L, similar to the ones that give rise to σ_T, are not forbidden, but are relatively unimportant for small values of $|t|$ and do not possess a pole at $t = m_\pi^2$. Here one should note that t is always negative, so that one can never reach the pole. One easily can find that when neglecting the proton-neutron mass difference the value of t is given by $-2M_nT_n \approx -p_n^2$, T_n being the kinetic energy of the residual neutron, and that the neutron can never be produced at rest. The lowest value of $|t|$ is obtained in parallel kinematics, and this t_{min} increases with Q^2 and decreases with W, so in order to enhance the pole part of the cross section, one should choose the highest possible W (so use high electron energies), but then the direction of $\mathbf{q}$ moves close to the beam line, which limits the maximum achievable value of W.

In practice the best way to determine the pion form factor probably is to use some theoretical model for the ^{1}H(e,e$'\pi^+$)n reaction that describes well all four structure functions and contains the pion form factor as a parameter, like the generalized Born model that has been used for determining F_π in the past (see below), or a recently developed Regge type model [51].

The four cross sections σ_X can be separated for a range in $|t|$ by measuring the pion angular distribution near parallel kinematics at different values of ϕ for two values of ϵ, hence at two values of the electron energy E. keeping W and Q^2 the same. Only one experiment has been performed [52] in which this was actually done. The results are shown in Fig. 19. A clear rise in σ_L for small $|t|$ is observed. The pion form factor was extracted by using a generalized Born model. The value is indicated by the black dot in Fig. 20, which gives the world's data. The points below $Q^2 = 0.3$ (GeV/c)^2 are due to the mentioned π-e$_{at}$ scattering experiment. The points at the other values of Q^2 result from measurements [53] in which even when sometimes data were obtained at two electron energies, only the total cross section was used for determining F_π. The contributions from σ_{TT} and σ_{LT} were removed by averaging over ϕ, while certain assumptions were made for σ_T, based on total photoproduction cross sections, which was then subtracted. Hence the results suffer from sizeable model uncertainties, in addition to the relatively large statistical uncertainties shown in Fig. 20. The total uncertainty in the existing data for $Q^2 > 1$ (GeV/c)^2 is so large, that a meaningful comparison with theoretical calculations is not possible.

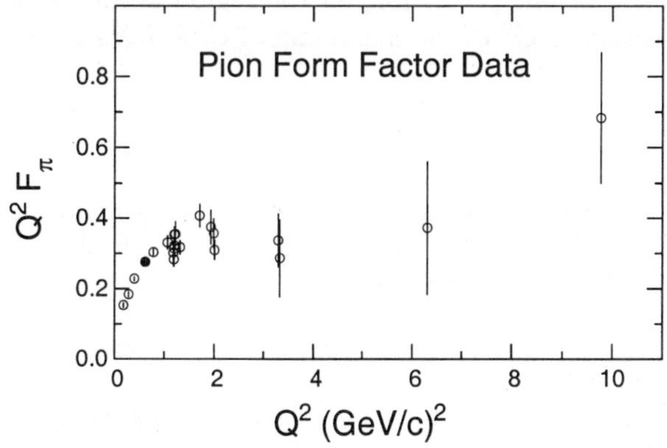

Figure 20: Existing data for the pion form factor. Error bars shown are statistical only.

With the high intensity and high resolution electron beam at Jlab it is now possible to measure the pion form factor with good statistical accuracy, and without the large systematic uncertainties inherent in the existing data [54]. Data for the ^{1}H(e,e$'\pi^+$)n reaction at values of Q^2 from 0.6 to 1.6 (GeV/c)^2 have already been taken in Hall C with a 100 μA beam of 2.5 to 4.0 GeV electrons incident on a liquid H_2 target with a length of 4 cm operated at 19 K. The scattered electrons were detected in the Short Orbit Spectrometer (SOS) and the knocked-out pions in the High Momentum Spectrometer (HMS). The very good resolution of the spectrometers, combined with the excellent beam conditions, enabled measurements with a precision that hitherto could not be reached. At the same time data were obtained for (quasi-free) π^+ and π^- production on ^{2}H. Those data are interesting in themselves, but will also help to constrain models for the ^{1}H(e,e$'\pi^+$)n reaction that will be used to determine F_π from the data.

The analysis of all these data is underway, but will take some time before final results will be available. The value of F_π will be extracted from the data with the help of a recently developed Regge-type model for the N(e,e$'\pi$) reaction[51]. An accuracy in $F_\pi(Q^2)$ of 7-12%, depending on the value of Q^2, will be reached, to be compared with the present uncertainties shown in Fig. 20. At the lowest Q^2 value, where F_π is accurately known, also $g_{\pi NN}(t)$ will be determined.

There exist quite a few calculations now for the pion form factor, see Ref.[55] and the contribution by A. Radyushkin in these proceedings. All these calculations globally agree for values of Q^2 below 1 (GeV/c)^2), since they are heavily constrained by the well known value for the pion radius. However, they give rather different predictions in the region above $Q^2 = 1$ (GeV/c)^2, e.g., for the position in Q^2 of the maximum of $Q^2 F_\pi(Q^2)$ or the height of the maximum.

Acknowledgements

I like to thank P.J. Mulders for providing part of the description of the formalism, R. Ent and L. Lapikás for useful comments about the manuscript, and D.L. Groep for help with the figures.

References

1. F. Halzen and A.D. Martin, *Quarks and Leptons* (Wiley)
2. P.J. Mulders, Physics Reports **185**, 83 (1990)
3. S. Boffi, C. Giusti, F.D. Pacati and M. Radici, *Electromagnetic Response of Atomic Nuclei* (Oxford, 1996)
4. E.A.J.M. Offerman *et al.*, Phys. Rev. Lett. **57**, 1546 (1986); N. Kalantar *et al.*, Phys. Rev. Lett. **63**, 2032 (1989)
5. J. Templon *et al.*, preprint nucl-ex/9909008; N.C.R. Makins *et al.*, submitted to Phys. Rev. C
6. T.W. Donnelly and A.S. Raskin, Annals of Physics **169**, 247 (1986)
7. I. Passchier *et al.*, Phys. Rev. Lett. **82**, 4988 (1999)
8. T. de Forest and J.D. Walecka, Advances in Physics **15**, 57 (1966)

9. H.P. Blok and J.H. Heisenberg in *Modern Topics in Electron Scattering*, eds. B. Frois and I. Sick (World Scientific, 1991)
10. E. Beise *et al.*, Few-Body Systems Suppl. **10**, 315 (1999); M. Bouwhuis *et al.*, Phys. Rev. Lett. **82**, 3755 (1999)
11. P.D.B. Collins and A.D. Martin, *Hadron Interactions* (Adam Hilger Ltd, 1984)
12. S. Pollock, H.W.L. Naus and J.H. Koch, Phys. Rev. C **53**, 2304 (1996)
13. H.P. Blok in *Proc. 6th Workshop on Perspectives in Nuclear Physics at Intermediate Energies*, eds. S. Boffi, C. Ciofi degli Atti, M. Giannini, Trieste, 1993 (World Scientific)
14. A. Bohr and B.R. Mottelson. *Nuclear Structure, Vol. I* (Benjamin, 1969) p. 423
15. Y. Jin, H.P. Blok and L. Lapikás, Phys. Rev. C **48**, R964 (1993)
16. K.S. Kim and L.E. Wright, Phys. Rev. C **56**, 302 (1997)
17. L. Lapikás, Nucl. Phys. A **553**, 297c (1993)
18. S. Frullani and J. Mougey, Adv. Nucl. Phys. **14**, 1 (1984)
19. S. Boffi, C. Giusti and F.D. Pacati, Physics Reports **226**, 1 (1993)
20. J. Kelly, Annals of Physics **23**, 75 (1996)
21. J.P. Genin *et al.*, Phys. Lett. B **52**, 46 (1974)
22. R. Ent *et al.*, Phys. Rev. Lett. **57**, 2367 (1986)
23. M. Jodice *et al.*, Phys. Lett. B **282**, 31 (1992)
24. R. Ent *et al.*, Phys. Rev. Lett. **62**, 24 (1989)
25. C.M. Spaltro *et al.*, Phys. Rev. Lett. **81**, 2870 (1998)
26. C. Tripp *et al.*, Phys. Rev. Lett. **76**, 885 (1996)
27. C.M. Spaltro *et al.*, Few-Body Systems **26**, 271 (1999)
28. J.F.J. van den Brand *et al.*, Nucl. Instr. Methods A **402**, 268 (1998)
29. C. Marchand *et al.*, Phys. Rev. Lett. **60**, 1703 (1988)
30. J.J. van Leeuwe *et al.*, Nucl. Phys. A **631**, 594c (1998)
31. C. Ciofi degli Atti *et al.*, Phys. Rev. C **44**, R7 (1991)
32. H. Müther, A. Polls and W.H. Dickhoff, Phys. Rev. C **51**, 3040 (1995)
33. J. Ryckebusch, Phys. Lett. B **383**, 1 (1995)
34. C.J.G. Onderwater *et al.*, Phys. Rev. Lett. **78**, 4893 (1997)
35. W.H.A. Hesselink *et al.*, in *Proceedings of the 10th Mini-Conference*, eds. H.P. Blok, E.Jans, G. van der Steenhoven (NIKHEF, Amsterdam, 1999) p. 80
36. U.-G. Meißner, in *Proceedings BARYONS '95*, Santa Fe, 1995
37. T.P. Welch *et al.*, Phys. Rev. Lett. **69**, 2761 (1992)
38. H.B. van den Brink *et al.*, Phys. Rev. Lett. **74**, 3561 (1995); H.B. van den Brink *et al.*, Nucl. Phys. A **612**, 391 (1997)
39. D. Drechsel and L. Tiator, J. Phys. G **18**, 449 (1992)

LATTICE QCD FOR NOVICES

G. PETER LEPAGE
Newman Laboratory of Nuclear Studies
Cornell University, Ithaca, NY 14853
E-mail: gpl@mail.lns.cornell.edu

These lectures are for novices to lattice QCD. They introduce a set of simple ideas and numerical techniques that can be implemented in a short period of time and that are capabable of generating nontrivial, nonperturbative results from lattice QCD. The simplest of these calculations can be completed on a standard workstation or high-end personal computer.

1 Introduction

These lectures are for novices who are interested in learning how to do lattice QCD simulations. My intent is to describe in detail everything that one needs to know in order to create and run a simple lattice QCD simulation. My focus here is not on lengthy derivations or detailed comparisons of algorithms. Rather I want to introduce a set of simple ideas and techniques that one can implement in a relatively short time and that are capable of generating nontrivial results from lattice QCD.

We begin in Section 2 with simple one-dimensional quantum mechanics. Most of the simulation techniques can be applied to ordinary quantum mechanics, and the simulations require only seconds or minutes of computer time, rather than the hours or days needed for QCD simulations. Consequently such applications are ideal for learning the simulation technology. We broaden the discussion to include quantum field theories in Section 3. Here we discuss the theoretical techniques needed to obtain accurate results on the relatively coarse lattices well suited to smaller computers. Finally, in Section 4, we adapt our techniques for simulations of gluon dynamics in QCD.

2 Numerical Path Integrals

2.1 Discretizing the Path Integral

We begin with numerical techniques for evaluating *path integrals.* Recall what path integrals tell us. In one-dimensional quantum mechanics, for example, the evolution of a position eigenstate $|x_\mathrm{i}\rangle$ from time t_i to time t_f can be computed using a path integral[1]:

$$\langle x_\mathrm{f}|\,\mathrm{e}^{-\tilde{H}(t_\mathrm{f}-t_\mathrm{i})}\,|x_\mathrm{i}\rangle = \int \mathcal{D}x(t)\,\mathrm{e}^{-S[x]}. \tag{1}$$

Here the $\int \mathcal{D}x(t)$ designates a sum over all possible particle paths

$$\{x(t) \quad \text{for} \quad t = t_\mathrm{i} \to t_\mathrm{f}\} \tag{2}$$

with

$$x(t_\mathrm{i}) = x_\mathrm{i} \qquad\qquad x(t_\mathrm{f}) = x_\mathrm{f}. \tag{3}$$

The hamiltonian is $\tilde{H}$, and $S[x]$ is the classical action,

$$S[x] \equiv \int_{t_\mathrm{i}}^{t_\mathrm{f}} dt\, L(x,\dot{x}) \equiv \int_{t_\mathrm{i}}^{t_\mathrm{f}} dt \left[\frac{m\,\dot{x}(t)^2}{2} + V(x(t))\right], \tag{4}$$

evaluated for each path $x(t)$. There are no i's in these formulas because we are using "euclidean" path integrals. These are derived in the same way as standard path integrals but with $it \to t$. Euclidean path integrals are much better for numerical work since the integrands do not oscillate wildly in sign.

Knowledge of the propagator, Eq. (1), as a function of $x_\mathrm{i}, t_\mathrm{i}, x_\mathrm{f}, t_\mathrm{f}$ gives us complete information about the quantum theory. For example, we can easily determine the groundstate energy and wavefunction. Setting

$$x_\mathrm{i} = x_\mathrm{f} \equiv x \qquad\qquad t_\mathrm{f} - t_\mathrm{i} \equiv T, \tag{5}$$

the propagator can be rewritten

$$\langle x|\, \mathrm{e}^{-\tilde{H}T}\, |x\rangle = \sum_n \langle x|\, E_n\rangle\, \mathrm{e}^{-E_nT}\, \langle E_n|\, x\rangle \tag{6}$$

where $|E_n\rangle$ is the energy eigenstate with eigenvalue E_n. The sum is dominated by the lowest-energy states when T is large, because of the exponentials, and in the limit of very large T only the groundstate, $|E_0\rangle$, contributes:

$$\langle x|\, \mathrm{e}^{-\tilde{H}T}\, |x\rangle \overset{T\to\infty}{\longrightarrow} \mathrm{e}^{-E_0T}\, |\langle x|\, E_0\rangle|^2 . \tag{7}$$

We extract the groundstate energy E_0 by integrating over x,

$$\int dx\, \langle x|\, \mathrm{e}^{-\tilde{H}T}\, |x\rangle \overset{T\to\infty}{\longrightarrow} \mathrm{e}^{-E_0T}, \tag{8}$$

and then, going back to the previous equation, we determine the groundstate wavefunction $\psi_{E_0}(x) \equiv \langle x|\, E_0\rangle$.

Our goal therefore is to develop a numerical procedure for evaluating the propagator using a path integral. There are two issues we must address. First we must find a way to represent an arbitrary particle path $\{x(t),\, t_\mathrm{i} \le t \le t_\mathrm{f}\}$

in the computer. A path is specified by a function $x(t)$ which, in principle, can be infinitely complex (and therefore too much for any computer). We approximate this function by specifying $x(t)$ only at the nodes or sites on a discretized t axis:

$$t_j = t_{\rm i} + j\,a \qquad \text{for } j = 0, 1 \ldots N \tag{9}$$

where a is the grid spacing,

$$a \equiv \frac{t_{\rm f} - t_{\rm i}}{N}. \tag{10}$$

Then a path is described by a vector of numbers,

$$x = \{x(t_0), x(t_1) \ldots x(t_N)\}. \tag{11}$$

It is common practice to refer to such a path as a "configuration". The integral over all paths in this approximation becomes an ordinary integral over all possible values for each of the $x(t_j)$'s: that is,

$$\int \mathcal{D}x(t) \to A \int_{-\infty}^{\infty} dx_1\, dx_2 \ldots dx_{N-1} \tag{12}$$

where we have adopted the notation $x_j \equiv x(t_j)$. We don't integrate over the endpoints since they are held fixed; for example, for boundary conditions (5),

$$x_0 = x_N = x. \tag{13}$$

We won't need the normalization factor A for most of our work, but for our one-dimensional problem it is[1]

$$A \equiv \left(\frac{m}{2\pi a}\right)^{N/2} \tag{14}$$

The second issue we must address concerns the evaluation of the action given only a discretized path $\{x_j\}$. Focusing just on the contribution from $t_j \le t \le t_{j+1}$, the obvious approximation is

$$\int_{t_j}^{t_{j+1}} dt\, L \;\approx\; a \left[\frac{m}{2} \left(\frac{x_{j+1} - x_j}{a} \right)^2 + \frac{1}{2} \left(V(x_{j+1}) + V(x_j) \right) \right] \tag{15}$$

With this approximation, our numerical representation of the path integral is complete, and we have an approximate expression for the quantum mechanical propagator: for example,

$$\langle x |\, \mathrm{e}^{-\tilde{H}T}\, | x \rangle \approx A \int_{-\infty}^{\infty} dx_1 \ldots dx_{N-1}\, \mathrm{e}^{-S_{\rm lat}[x]} \tag{16}$$

where

$$S_{\text{lat}}[x] \equiv \sum_{j=0}^{N-1} \left[\frac{m}{2a}(x_{j+1} - x_j)^2 + aV(x_j) \right], \tag{17}$$

$x_0 = x_N = x$, and $a = T/N$. We have reduced quantum mechanics to a problem in numerical integration.

One might worry about approximating $\dot{x}$ with $(x_{j+1} - x_j)/a$ in our formula for the lattice action $S_{\text{lat}}[x]$. It is not obvious that this is a good approximation given that $x_{j+1} - x_j$ can be arbitrarily large in our path integral; that is, paths can be arbitrarily rough. While not so important for our one-dimensional problem, this becomes a crucial issue for four-dimensional field theories. It is dealt with using renormalization theory, which we discuss in later sections.

Exercise: Set $a = 1/2$ and $N = 8$ in approximate formula (16) for the propagator, and integrate the right-hand side numerically. The seven dimensional integral that results is easily evaluated using standard routines, such as **vegas**[2]. Do this first for the one-dimensional harmonic-oscillator potential

$$V(x) = \frac{x^2}{2} \qquad \text{with} \qquad m = 1. \tag{18}$$

Evaluate the propagator for several values of $x_0 = x_N = x$, and compare your results with those of standard quantum mechanics:

$$\langle x | \, \mathrm{e}^{-\tilde{H}T} \, | x \rangle \approx |\langle x | \, E_0 \rangle|^2 \, \mathrm{e}^{-E_0 T} \tag{19}$$

where $E_0 = 1/2$ and

$$\langle x | \, E_0 \rangle = \frac{\mathrm{e}^{-x^2/2}}{\pi^{1/4}}. \tag{20}$$

Extract the energy and wavefunction from your numerical result. Repeat this exercise for $V(x) = x^4/2$. (If you wish, you may restrict x integrations to the region $-5 \to 5$ rather than $-\infty \to \infty$; this has negligible effect on the results of this exercise).

My results for the harmonic oscillator case are shown in Fig. 1.

2.2 Monte Carlo Evaluation of Path Integrals

Our analysis in the previous section focused on the groundstate. In quantum field theory, where the groundstate is the vacuum, we are generally interested in excited states. To analyze excited states using path integrals, we interrupt the propagation of the groundstate by introducing new operators at intermediate times. Consider, for example, the quantity

$$\langle\langle x(t_2)x(t_1) \rangle\rangle \equiv \frac{\int \mathcal{D}x(t) \, x(t_2)x(t_1) \, \mathrm{e}^{-S[x]}}{\int \mathcal{D}x(t) \, \mathrm{e}^{-S[x]}} \tag{21}$$

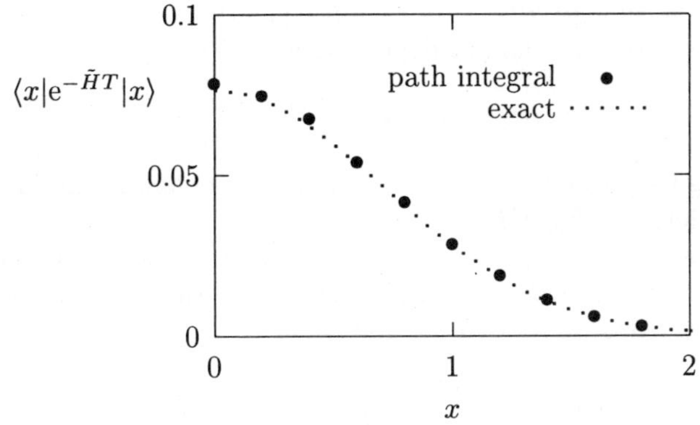

Figure 1: Euclidean harmonic-oscillator propagator at large time ($T = 4$) computed exactly (dotted line), and computed using numerical integration to evaluate the discretized path integral (points). The path integral was approximated by an 8 dimensional integral which was evaluated numerically, using **vegas**, at the points indicated. The exact result is approximated by the square of the ground state wavefunction multiplied by $\exp(-E_0T)$.

where now we integrate over all $x_i = x_f = x$ as well as the intermediate $x(t)$'s. This quantity is a weighted average of $x(t_2)x(t_1)$ over all paths, with weight $\exp(-S[x])$. The numerator on the right-hand side equals, in quantum mechanics,

$$\int dx\ \langle x|\, e^{-\tilde{H}(t_f-t_2)}\, \tilde{x}\, e^{-\tilde{H}(t_2-t_1)}\, \tilde{x}\, e^{-\tilde{H}(t_1-t_i)}\, |x\rangle\,. \tag{22}$$

Setting $T = t_f - t_i$ and $t = t_2 - t_1$ we can rewrite the full expression as

$$\langle\langle x(t_2)x(t_1)\rangle\rangle = \frac{\sum e^{-E_nT}\,\langle E_n|\,\tilde{x}\, e^{-(\tilde{H}-E_n)t}\,\tilde{x}\,|E_n\rangle}{\sum e^{-E_nT}}. \tag{23}$$

If $T \gg t$ and large, then the groundstate $|E_0\rangle$ dominates the sums and

$$G(t) \equiv \langle\langle x(t_2)x(t_1)\rangle\rangle \to \langle E_0|\,\tilde{x}\, e^{-(\tilde{H}-E_0)t}\,\tilde{x}\,|E_0\rangle\,. \tag{24}$$

In our harmonic oscillator example, the state propagating between the two $\tilde{x}$'s cannot be $|E_0\rangle$ since $\tilde{x}$ switches the parity of the state. Thus if we now make t large (but still $\ll T$)

$$G(t) \stackrel{t\ \text{large}}{\longrightarrow} |\langle E_0|\,\tilde{x}\,|E_1\rangle|^2\ e^{-(E_1-E_0)t} \tag{25}$$

where $|E_1\rangle$ is the first excited state. Consequently we can extract the first excitation energy from the large-t dependence of $G(t)$,

$$\log(G(t)/G(t+a)) \to (E_1 - E_0)a, \tag{26}$$

and then, going back to $G(t)$, we can determine the quantum mechanical transition matrix element $\langle E_0|\,\tilde{x}\,|E_1\rangle$.

In principle, path integral averages $\langle\langle \Gamma[x]\rangle\rangle$ of arbitrary functionals $\Gamma[x]$ can be used to compute any physical property of the excited states in the quantum theory. Also we note in passing that

$$\langle\langle \Gamma[x]\rangle\rangle = \frac{\sum \mathrm{e}^{-E_n T}\,\langle E_n|\,\Gamma[\tilde{x}]\,|E_n\rangle}{\sum \mathrm{e}^{-E_n T}} \tag{27}$$

becomes a (quantum mechanical) thermal average if we replace

$$T \to \beta \equiv 1/k_B\,T_{\mathrm{temp}}. \tag{28}$$

Thus any computer code designed to compute path integral averages can be used for thermal physics as well. Here we focus on the zero-temperature limit of large T.

We could evaluate the path integrals in $\langle\langle \Gamma[x]\rangle\rangle$ using a standard multidimensional integration code like `vegas`, at least for one-dimensional systems. Here, instead, we employ a more generally useful Monte Carlo procedure. Noting that

$$\langle\langle \Gamma[x]\rangle\rangle = \frac{\int \mathcal{D}x(t)\,\Gamma[x]\,\mathrm{e}^{-S[x]}}{\int \mathcal{D}x(t)\,\mathrm{e}^{-S[x]}}, \tag{29}$$

is a weighted average over paths with weight $\exp(-S[x])$, we generate a large number, N_{cf}, of random paths or configurations,

$$x^{(\alpha)} \equiv \{x_0^{(\alpha)} x_1^{(\alpha)} \dots x_{N-1}^{(\alpha)}\} \qquad \alpha = 1, 2 \dots N_{\mathrm{cf}}, \tag{30}$$

on our grid in such a way that the probability $P[x^{(\alpha)}]$ for obtaining any particular path $x^{(\alpha)}$ is

$$P[x^{(\alpha)}] \propto \mathrm{e}^{-S[x^{(\alpha)}]}. \tag{31}$$

Then an unweighted average of $\Gamma[x]$ over this set of paths approximates the weighted average over uniformly distributed paths:

$$\langle\langle \Gamma[x]\rangle\rangle \approx \overline{\Gamma} \equiv \frac{1}{N_{\mathrm{cf}}} \sum_{\alpha=1}^{N_{\mathrm{cf}}} \Gamma[x^{(\alpha)}]. \tag{32}$$

$\overline{\Gamma}$ is our "Monte Carlo estimator" for $\langle\langle\Gamma[x]\rangle\rangle$ on our lattice. Of course the estimate will never be exact since the number of paths $N_{\rm cf}$ will never be infinite. The Monte Carlo uncertainty $\sigma_{\overline{\Gamma}}$ in our estimate is a potential source of error; it is estimated in the usual fashion[3]:

$$\sigma^2_{\overline{\Gamma}} \approx \frac{1}{N_{\rm cf}} \left\{ \frac{1}{N_{\rm cf}} \sum_{\alpha=1}^{N_{\rm cf}} \Gamma^2[x^{(\alpha)}] - \overline{\Gamma}^2 \right\}. \tag{33}$$

This becomes

$$\sigma^2_{\overline{\Gamma}} = \frac{\langle\langle\Gamma^2\rangle\rangle - \langle\langle\Gamma\rangle\rangle^2}{N_{\rm cf}} \tag{34}$$

for large $N_{\rm cf}$. Since the numerator in this expression is independent of $N_{\rm cf}$ (in principle, it can be determined directly from quantum mechanics), the statistical uncertainties vanish as $1/\sqrt{N_{\rm cf}}$ when $N_{\rm cf}$ increases.

We need some sort of specialized random-vector generator to create our set of random paths $x^{(\alpha)}$ with probability (31). Possibly the simplest procedure, though not always the best, is the Metropolis Algorithm[4]. In this procedure, we start with an arbitrary path $x^{(0)}$ and modify it by visiting each of the sites on the lattice, and randomizing the x_j's at those sites, one at a time, in a particular fashion that is described below. In this way we generate a new random path from the old one: $x^{(0)} \to x^{(1)}$. This is called "updating" the path. Applying the algorithm to $x^{(1)}$ we generate path $x^{(2)}$, and so on until we have $N_{\rm cf}$ random paths. This set of random paths has the correct distribution if $N_{\rm cf}$ is sufficiently large.

The algorithm for randomizing x_j at the $j^{\rm th}$ site is:

- generate a random number ζ, with probability uniformly distributed between $-\epsilon$ and ϵ for some constant ϵ;

- replace $x_j \to x_j + \zeta$ and compute the change ΔS in the action caused by this replacement (generally only a few terms in the lattice action involve x_j, since lagrangians are local; only these need be examined);

- if $\Delta S < 0$ (the action is reduced) retain the new value for x_j, and proceed to the next site;

- if $\Delta S > 0$ generate a random number η unformly distributed between 0 and 1; retain the new value for x_j if $\exp(-\Delta S) > \eta$, otherwise restore the old value; proceed to the next site.

An implementation of this algorithm, in the Python computer language[5], is shown in Fig. 2. The code examples and Python are discussed in the Appendix.

```
def update(x):
    for j in range(0,N):
        old_x = x[j]                          # save original value
        old_Sj = S(j,x)
        x[j] = x[j] + uniform(-eps,eps) # update x[j]
        dS = S(j,x) - old_Sj                  # change in action
        if dS>0 and exp(-dS)<uniform(0,1):
            x[j] = old_x                      # restore old value

def S(j,x):                                   # harm. osc. S
    jp = (j+1)%N                              # next site
    jm = (j-1)%N                              # previous site
    return a*x[j]**2/2 + x[j]*(x[j]-x[jp]-x[jm])/a
```

Figure 2: Python code for one Metropolis update of path $\{x_j, j = 0 \ldots N-1\}$. The path is stored in array `x[j]`. Function `S(j,x)` returns the value of the part of the action that depends on x_j. Function `uniform(a,b)` returns a random number between a and b. A sample `S(j,x)` is shown, for a harmonic oscillator with $x_N = x_0$.

There are two important details concerning the tuning and use of this algorithm. First, in general some or many of the x_j's will be the same in two successive random paths. The amount of such overlap is determined by the parameter ϵ: when ϵ is very large, changes in the x_j's are usually large and most will be rejected; when ϵ is very small, changes are small and most are accepted, but the new x_j's will be almost equal to the old ones. Neither extreme is desirable since each leads to very small changes in x, thereby slowing down the numerical exploration of the space of all important paths. Typically ϵ should be tuned so that 40%–60% of the x_j's are changed on each pass (or "sweep") through the lattice. Then ϵ is of order the typical quantum fluctuations expected in the theory. Whatever the ϵ, however, successive paths are going to be quite similar (that is "highly correlated") and so contain rather similar information about the theory. Thus when we accumulate random paths $x^{(\alpha)}$ for our Monte Carlo estimates we should keep only every $N_{\rm cor}$-th path; the intervening sweeps erase correlations, giving us configurations that are statistically independent. The optimal value for $N_{\rm cor}$ depends upon the theory, and can be found by experimentation. It also depends on the lattice spacing a, going roughly as

$$N_{\rm cor} \propto \frac{1}{a^2}. \tag{35}$$

Other algorithms exist for which $N_{\rm cor}$ grows only as $1/a$ when a is reduced, but since our interest is in large a's we will not discuss these further.

The second detail concerns the procedure for starting the algorithm. The very first configuration used to seed the whole process is usually fairly atypical. Consequently we should discard some number of configurations at the beginning, before starting to collect $x^{(\alpha)}$'s. Discarding $5N_{\rm cor}$ to $10N_{\rm cor}$ configurations is usually adequate. This is called "thermalizing the lattice."

To summarize, a computer code for a complete Monte Carlo calculation of $\langle\langle\Gamma[x]\rangle\rangle$ for some function $\Gamma[x]$ of a path x consists of the following steps:

- initialize the path, for example, by setting all x_j's to zero;
- update the path $5N_{\rm cor}$–$10N_{\rm cor}$ times to thermalize it;
- update the path $N_{\rm cor}$ times, then compute $\Gamma[x]$ and save it; repeat $N_{\rm cf}$ times.
- average the $N_{\rm cf}$ values of $\Gamma[x]$ saved in the previous step to obtain a Monte Carlo estimate $\overline{\Gamma}$ for $\langle\langle\Gamma[x]\rangle\rangle$.

A Python implementation of this procedure is shown in Fig. 3.

Exercise: Write a computer program to implement the Metropolis Monte Carlo algorithm for a one dimensional harmonic oscillator (Eq. (18)), and calculate

$$G(t) = \frac{1}{N}\sum_j \langle\langle x(t_j + t)x(t_j)\rangle\rangle \tag{36}$$

for all $t = 0, a, 2a \ldots (N-1)a$; that is calculate

$$G_n = \frac{1}{N}\sum_j \langle\langle x_{(j+n){\rm mod}N}\, x_j\rangle\rangle \tag{37}$$

for $n = 0 \ldots N-1$. The $(j+n){\rm mod}N$ in this last expression reflects the periodic boundary conditions. Try $N = 20$ lattice sites with lattice spacing $a = 1/2$, and set $\epsilon = 1.4$ and $N_{\rm cor} = 20$. Try $N_{\rm cf}$'s of 25, 100, 1000 and 10000. Use the results to compute the excitation energy from

$$\Delta E_n \equiv \log(G_n/G_{n+1}) \stackrel{n\ \text{large}}{\longrightarrow} (E_1 - E_0)a \tag{38}$$

Try this for the harmonic oscillator potential and also for anharmonic potentials. Vary the various parameters.

My results for the harmonic oscillator potential with $N = 1000$ configurations are shown in Fig. 4. These results required less than a minute of personal computer time.

```
def compute_G(x,n):
    g = 0
    for j in range(0,N):
        g = g + x[j]*x[(j+n)%N]
    return g/N

def MCaverage(x,G):
    for j in range(0,N):                    # initialize x
        x[j] = 0
    for j in range(0,5*N_cor):              # thermalize x
        update(x)
    for alpha in range(0,N_cf):             # loop on random paths
        for j in range(0,N_cor):
            update(x)
        for n in range(0,N):
            G[alpha][n] = compute_G(x,n)
    for n in range(0,N):                    # compute MC averages
        avg_G = 0
        for alpha in range(0,N_cf):
            avg_G = avg_G + G[alpha][n]
        avg_G = avg_G/N_cf
        print "G(%d) = %g" % (n,avg_G)
```

Figure 3: Sample Python code for a Monte Carlo evaluation of of $G(t)$ (Eq. (36)). Function `computeG(x,t)` computes $G(t)$ for a given path `x`. Function `MCaverage(x,G)` computes the Monte Carlo average over random paths `x`. The results for path $x^{(\alpha)}$ are stored in the array `G[alpha][t]`, and the averages are computed and printed. Function `update(x)` does one Metropolis sweep through the lattice (see Fig. 2).

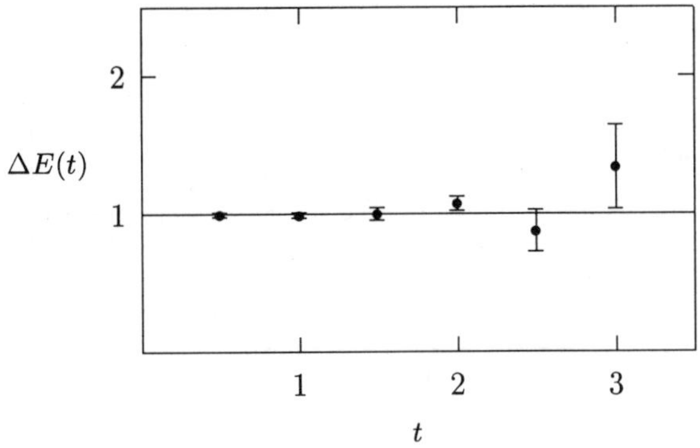

Figure 4: Monte Carlo values $\Delta E(t) \equiv \log(G(t)/G(t+a))/a$ plotted versus t for an harmonic oscillator. The exact asymptotic result, $\Delta E(\infty) = 1$, is indicated by a line. Results are for a one dimensional lattice with $N = 20$ sites, lattice spacing $a = 1/2$, and $N_{\rm cf} = 1000$ configurations, keeping configurations only every $N_{\rm cor} = 20$ sweeps. The Metropolis step size ϵ was 1.4, resulting in a Metropolis acceptance ratio of 0.5.

Exercise: Redo the previous exercise but propagator

$$G(t) = \frac{1}{N} \sum_j \langle\langle x^3(t_j + t) x^3(t_j) \rangle\rangle. \tag{39}$$

Here we use x^3 rather than x to create and destroy the excited stated; that is, we use x^3 rather than x as the "source" and the "sink". Note that $\Delta E(t)$ converges to the same result, but only at much larger t's than before. Different sources and sinks often lead to different asymptotic behavior. Choices that result in fast convergence as t increases are usually preferable because statistical errors are smaller at smaller t's. Compare your best estimate of the asymptotic value obtained using x^3 with that obtained using x.

Note also that $\Delta E(t)$ approaches its its asymptotic value from above. Prove that this must be true in general, provided source and sink are the same operator, using Eq. (23). This result is useful because it implies that each $\Delta E(t)$ gives a rigorous upper bound on the asymptotic value, even at small t's before convergence.

My results with an x^3 source and sink are shown in Fig. 5.

2.3 Statistical Errors

A Monte Carlo estimate $\overline{\Gamma}$ of some expectation value $\langle\langle \overline{\Gamma} \rangle\rangle$ is never exact; there are always statistical errors that vanish only in the limit where infinitely many

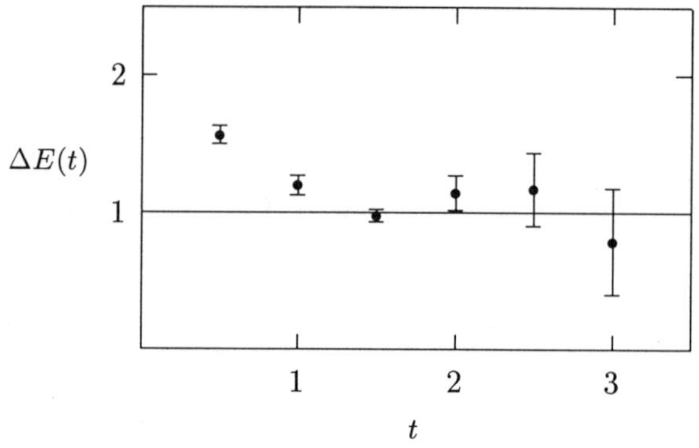

Figure 5: Monte Carlo values $\Delta E(t) \equiv \log(G(t)/G(t+a))/a$ plotted versus t for an harmonic oscillator using x^3 as the source and sink. Parameters are the same as used to generate Fig. 4. The energies take longer to reach their asymptotic value.

configurations are employed ($N_{\rm cf} \to \infty$). An important part of any Monte Carlo analysis is the estimation of these statistical errors. There is a simple but very powerful method, called the "statistical bootstrap," for making such estimates.

In the previous exercises, for example, we assemble an "ensemble" of measurements of the propagator $G^{(\alpha)}$, one for each configuration $x^{(\alpha)}$. These are averaged to obtain $\overline{G}$, and, from it, an estimate for ΔE_n (Eq. (38)). An obvious way to check the statistical errors on this estimate for ΔE_n is to redo the whole calculation, say, 100 times, each time with different random numbers to generate different random paths. With 100 copies of the entire calculation, we could analyze the distribution of the 100 random ΔE_n's obtained, and deduce the statistical uncertainty in our original estimate. This, however, is exceedingly expensive in computer time. The bootstrap procedure provides new, almost *zero-cost* random ensembles of measurements by synthesizing them from the original ensemble of $N_{\rm cf}$ measurements.

Given an ensemble $\{G^{(\alpha)}, \alpha = 1 \ldots N_{\rm cf}\}$ of Monte Carlo measurements, we assemble a "bootstrap copy" of that ensemble by selecting $G^{(\alpha)}$'s at random from the original ensemble, taking $N_{\rm cf}$ in all while allowing duplications and omissions. The resulting ensemble of G's might have two or three copies of some $G^{(\alpha)}$'s, and no copies of others. This new ensemble can be averaged and a new estimate obtained for ΔE_n. This procedure can be repeated to generated as many bootstrap copies of the original ensemble as we wish, and from each

```
def bootstrap(G):
    N_cf = len(G)
    G_bootstrap = []                      # new ensemble
    for i in range(0,N_cf):
        alpha = int(uniform(0,N_cf))      # choose random config
        G_bootstrap.append(G[alpha])      # keep G[alpha]
    return G_bootstrap
```

Figure 6: Sample Python code for producing a bootstrap copy of an ensemble of measurements `G`. The original ensemble consists of individual measurements `G[alpha]`, one for each configuration. The function `bootstrap(G)` returns a single bootstrap copy of ensemble `G`, consisting of `N_cf` measurements. Function `uniform(a,b)` returns a random number between `a` and `b`.

we can generate a new estimate for ΔE_n. The distribution of these ΔE_n's approximates the distribution of ΔE_n's that would have been obtained from the original Monte Carlo, and so can be used to estimate the statistical error in our original estimate. A Python implementation of this procedure is shown in Fig. 6.

Another useful procedure related to statistical errors is "binning." At the end of a large simulation we might have 100's or even 100,000's of configurations $x^{(\alpha)}$, and for each a set of measurements like $G^{(\alpha)}$, our propagator. The measurements will inevitably be averaged, but we want to save the separate $G^{(\alpha)}$'s for making bootstrap error estimates and the like. We can save a lot of disk space, RAM, and CPU time by partially averaging or binning the measurements: For example, instead of storing each of

$$G^{(1)} \quad G^{(2)} \quad G^{(3)} \quad G^{(4)} \quad G^{(5)} \ldots \tag{40}$$

we might instead store

$$\begin{aligned} \overline{G}^{(1)} &\equiv \frac{G^{(1)} + G^{(2)} + G^{(3)} + G^{(4)}}{4} \\ \overline{G}^{(2)} &\equiv \frac{G^{(5)} + G^{(6)} + G^{(7)} + G^{(8)}}{4} \\ &\ldots \end{aligned} \tag{41}$$

The $\overline{G}^{(\beta)}$'s are far less numerous but have the same average, standard deviation, and other statistical properties as the original set. Typically the bin size is adjusted so that there are only 50–100 $\overline{G}^{(\beta)}$'s. A Python implementation of this procedure is shown in Fig. 7.

```
def bin(G,binsize):
    G_binned = []                          # binned ensemble
    for i in range(0,len(G),binsize):      # loop on bins
        G_avg = 0
        for j in range(0,binsize):         # loop on bin elements
            G_avg = G_avg + G[i+j]
        G_binned.append(G_avg/binsize)     # keep bin avg
    return G_binned
```

Figure 7: Sample Python code for producing a binned copy of an ensemble of measurements `G`. The original ensemble consists of individual measurements `G[alpha]`, one for each configuration. The function `bin(G,binsize)` bins the ensemble into bins of size `binsize`, averages the `G`'s within each bin, and returns an ensemble consisting of the averages.

Binning has an important side effect: it reduces or can even remove the effects of correlations between different configurations. If, when generating configurations, $N_{\rm cor}$ is too small, successive Monte Carlo estimates are statistically correlated. This leads to error estimates, using Eq. (33), that can be much smaller than the true errors — a very bad situation. If, however, the Monte Carlo estimates are binned with sufficiently large bins, the majority of estimates in one bin will be uncorrelated from the majority in adjacent bins. Consequently the bin averages will be uncorrelated, and standard statistical formulas, like Eq. (33), are reliable. To determine the bin size required to remove correlations, first bin the measurements and compute the errors. Then rebin with double the bin size and recompute the errors. There are no correlations if the statistical errors are roughly independent of bin size; if, on the other hand, the statistical errors grow substantially with bin size (for example, proportional to the square root of the bin size), then there are strong correlations between bins. Continuing doubling the bin size until the statistical errors stop growing. Note that measurements of different physical quantities decorrelate at different rates; different things may require different bin sizes.

Exercise: Rerun your Metropolis simulation of the harmonic oscillator with $N_{\rm cor} = 1$. Do several different runs and compare your results. Do they agree within statistical errors? Try binning the results from each of the runs in bins of 20 and recompute the statistical errors. Verify that different runs now agree within the errors computed from the binned results.

In Fig. 8 I show results from $N = 1000$ configurations using $N_{\rm cor} = 1$. Compare this with results in Fig. 4, which come from the same number of configurations but with $N_{\rm cor} = 20$. The errorbars in the $N_{\rm cor} = 1$ plot are obviously unreliable.

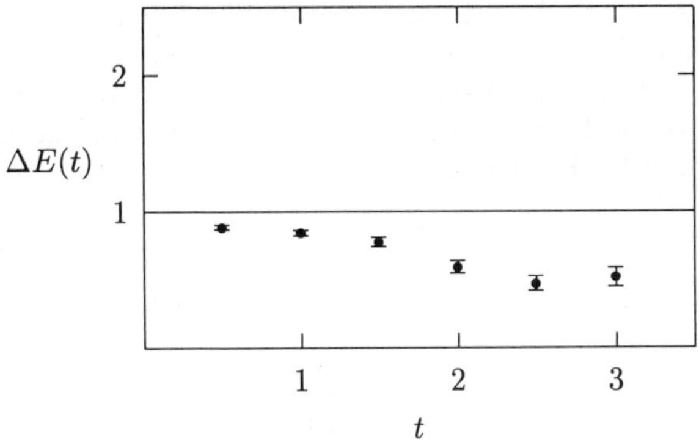

Figure 8: Monte Carlo values $\Delta E(t) \equiv \log(G(t)/G(t+a))/a$ plotted versus t for an harmonic oscillator, as in Fig. 4 but with $N_{\text{cor}} = 1$. The errorbars are unreliable.

3 Field Theory on a Lattice

3.1 From Quantum Mechanics to Field Theory

Field theories of the sort we are interested in have lagrangian formulations and so can be quantized immediately using path integrals. The procedure is precisely analogous to what we do in the previous section when quantizing the harmonic oscillator. The analogues of the coordinates $x(t)$ in quantum mechanics are just the fields $\phi(x)$ or $A_\mu(x)$ where $x = (t, \vec{x})$ is a space-time point. Indeed our quantum mechanical examples can be thought of as field theory examples in 0 spatial and 1 temporal dimension: $x(t) \to \phi(t) \to \phi(x)$. The analogue of the ground state in quantum field theory is the vacuum state, $|0\rangle$, while the analogues of the excited states, created when $\phi(x)$ or ϕ^3 or ... acts on $|0\rangle$, correspond to states with one or more particles create in the vacuum.

In the lattice approximation both space and time are discrete:

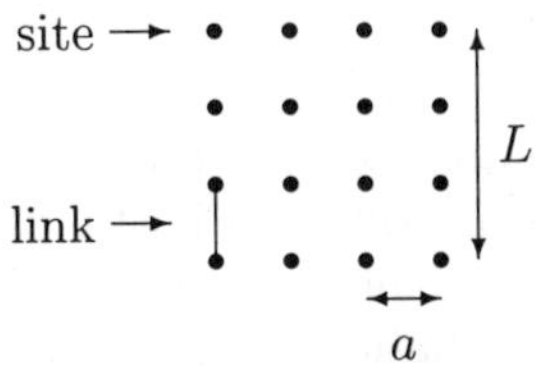

The nodes or "sites" of the grid are separated by lattice spacing a, and the length of a side of the grid is L; the lines joining adjacent sites are called "links." The quantum field is specified by its values at the grid sites: a configuration is describe by the set of numbers $\{\phi(x_j),\ \forall\, x_j \,\epsilon\, \text{grid}\}$. The path integral generalizes in the obvious fashion:

$$\langle\langle\Gamma[\phi]\rangle\rangle \equiv \frac{1}{\mathcal{Z}} \int \mathrm{e}^{-S[\phi]}\, \Gamma[\phi] \prod_{x_j \,\epsilon\, \text{grid}} d\phi(x_j) \tag{42}$$

where

$$\mathcal{Z} \equiv \int \mathrm{e}^{-S[\phi]} \prod_{x_j} d\phi(x_j). \tag{43}$$

Here the action $S[\phi]$ is the continuum action with spatial and temporal derivatives replaced by differences between field values at the grid sites. We study excitations of the field theory using operators like

$$\Gamma(t) \equiv \frac{1}{\sqrt{N}} \sum_{\vec{x}_j} \phi(\vec{x}_j, t) \tag{44}$$

where the sum over the N spatial $\vec{x}_j$'s enforces zero three-momentum. Since the excitations correspond to particle creation, their energies are the energies of particles: for example,

$$\langle\langle\Gamma(t)\Gamma(0)\rangle\rangle \overset{t\,\text{large}}{\longrightarrow} |\langle 0|\,\Gamma(0)\,|\phi : \vec{p} = 0\rangle|^2\, \mathrm{e}^{-m_\phi t}, \tag{45}$$

where $|\phi : \vec{p} = 0\rangle$ is a one ϕ-particle state with zero three momentum, and m_ϕ is the mass of the ϕ particle.

Exercise: Show that

$$\langle 0|\,\Gamma(0)\,|\phi : \vec{p} = 0\rangle = \frac{Z_2}{2m_\phi} \tag{46}$$

where Z_2 is the wavefunction renormalization parameter for the ϕ field.

3.2 Coarse Lattices

We have, through the lattice approximation to the path integral, turned the problem of solving a nonperturbative relativistic quantum field theory, once again, into a problem of numerical integration. This is a major development for theories, like QCD, where perturbation theory doesn't suffice (at low energies).

Early enthusiasm for such an approach to QCD, back when QCD was first invented, quickly gave way to the grim realization that very large computers would be needed to numerically integrate the path integral. In recent years, however, two developments have made QCD simulations far more accessible. One is that small computers have become much faster; the other is that QCD simulations have become much faster — 10^3 to 10^6 times faster. These developments imply that the simplest QCD simulations can be done using no more than a single personal computer or even a laptop.

What has changed to make QCD simulations faster? The cost of a QCD calculation is given roughly by the formula

$$\text{cost} \approx \left(\frac{L}{a}\right)^4 \frac{1}{a} \frac{1}{m_\pi^2 a}, \tag{47}$$

where the first factor is the number of lattice sites, while the second and third factors are due to "critical-slowing-down" of the algorithms used for the simulation. From this formula, the single most important determinant of the cost is the lattice spacing: the cost is proportional to $1/a^6$. This means that one wants to keep a as large as possible. Until recently it was thought that $a < 0.05$–$0.1\,\text{fm}$ would be essential for accurate QCD simulations. As we shall see $a \approx 0.3$–$0.4\,\text{fm}$ works quite well. Given that the cost varies as $1/a^6$, the coarser lattices should be 10^3 to 10^6 times cheaper to simulate.

The size of the lattice spacing is limited by discretization errors. The challenge is to make the lattice spacing as large as possible while keeping the discretization errors of order, say, a few percent or less. These errors have two sources: first, the lattice forces us to use approximate derivatives, and, second, it imposes an ultraviolet cutoff. We consider each in turn.

In the lattice approximation, we only know the fields at the lattice sites. Thus all derivatives in field equations, the action, and the like must be converted to differences. For example, the second derivative of field ϕ at some point x_j on the lattice is given approximately by

$$\frac{\partial^2 \phi(x_j)}{\partial x^2} = \Delta_x^{(2)}\, \phi(x_j) + \mathcal{O}(a^2) \tag{48}$$

where

$$\Delta_x^{(2)}\, \phi(x) \equiv \frac{\phi(x+a) - 2\phi(x) + \phi(x-a)}{a^2}. \tag{49}$$

We generally want more accurate approximations for work on coarse lattices: for example, the approximation

$$\frac{\partial^2 \phi(x_j)}{\partial x^2} = \Delta_x^{(2)}\, \phi(x_j) - \frac{a^2}{12}\,(\Delta_x^{(2)})^2\, \phi(x_j) + \mathcal{O}(a^4) \tag{50}$$

is accurate to within a few percent even when acting on structures in $\phi(x)$ that are only four or five lattice spacings across. With such precision one might expect that lattice spacings as large as a quarter the diameter of hadron, or about 0.4 fm, would still be quite useful. Our theories are quantum theories, however, and therefore there is a second important consideration.

The shortest wavelength oscillation that can be modeled on a lattice is one with wavelength $\lambda_{\min} = 2a$; for example, the function $\phi(x) = +1, -1, +1 \ldots$ for $x = 0, a, 2a \ldots$ oscillates with this wavelength. Thus gluons and quarks with momenta $p = 2\pi/\lambda$ larger than π/a are excluded from the lattice theory by the lattice; that is, the lattice functions as an ultraviolet cutoff. In simple classical field theories this is often irrelevant: short-wavelength ultraviolet modes are either unexcited or decouple from the long-wavelength infrared modes of interest. However, in a noisy nonlinear theory, like an interacting quantum field theory, ultraviolet modes strongly affect infrared modes. Thus we cannot simply discard all particles with momenta larger than π/a; we must somehow mimic their effects on infrared states. This is done by changing or "renormalizing" the parameters in our discretized theory and by adding new local interactions.

The new interactions complicate the improved discretizations discussed above. For example, an interacting scalar theory on the lattice would have a discretized kinetic lagrangian

$$\sum_\mu \tfrac{1}{2}\phi^\dagger \partial_\mu^2 \phi \to \sum_\mu \tfrac{1}{2}\left(\phi^\dagger \Delta_\mu^{(2)}\phi + a^2 c\, \phi^\dagger (\Delta_\mu^{(2)})^2 \phi\right) \tag{51}$$

where parameter c has two parts: $-1/12$ from numerical analysis (Eq. (50)), and an additional renormalization due to the cutoff. Typically the renormalization is completely context dependent — for example, it is different for QED and QCD, or for particles of different spin, and so on. It cannot be looked up in a numerical analysis book; rather, it must be computed using quantum field theory. In QCD these renormalizations can be computed using (weak-coupling) perturbation theory, since the renormalizations are due to QCD physics at large momenta, $p > \pi/a$, where the theory is perturbative:

$$c = -\frac{1}{12} + c_1\alpha_s(\pi/a) + c_2\alpha_s^2(\pi/a) + \cdots. \tag{52}$$

This is true, that is, provided the lattice spacing is small enough that momentum π/a is perturbative. Work in continuum QCD suggests that lattice spacings of 0.5 fm or smaller should suffice, but, until recently, lattice simulations seemed to suggest that perturbation theory only started to work for lattice spacings smaller than 0.05–0.1 fm.

Exercise: Our action for one-dimensional quantum mechanics, Eq. (4), can be rewritten

$$S[x] \equiv \int_{t_i}^{t_f} dt \left[-\tfrac{1}{2}m\, x(t)\ddot{x}(t) + V(x(t))\right], \tag{53}$$

by integrating by parts (assuming $x(t_i) = x(t_f) = x$). To discretize we replace

$$\ddot{x}(t_j) \rightarrow \Delta^{(2)} x_j \equiv \frac{x_{j+1} - 2x_j + x_{j-1}}{a^2}, \tag{54}$$

where the x_j's are periodic ($x_0 = x_N$, $x_{-1} = x_{N-1}$, and so on); this gives the same lattice action we used earlier. We can improve the discretization, however, by using the corrected approximation, Eq. (50), for the second derivative:

$$S_{\mathrm{imp}}[x] \equiv \sum_{j=0}^{N-1} a\left[-\tfrac{1}{2}m\, x_j\left(\Delta^{(2)} - a^2(\Delta^{(2)})^2/12\right)x_j + aV(x_j)\right] \tag{55}$$

Modify your Monte Carlo code for the harmonic oscillator to include the correction term. Run high-statistics comparisons ($N_{\mathrm{cf}} = 10^4$ or 10^5) with and without the correction term.

Exercise: Discretized, euclidean classical equations of motion can be derived from the actions in the previous exercise by setting

$$\frac{\partial S[x]}{\partial x_j} = 0. \tag{56}$$

Using the improved action, for example, we obtain

$$m\left(\Delta^{(2)} - a^2(\Delta^{(2)})^2/12\right)x_j = \frac{dV(x_j)}{dx_j}. \tag{57}$$

Setting

$$V(x) = \tfrac{1}{2}m\,\omega_0^2 x^2, \tag{58}$$

find solutions of the form $x_j = \exp(-\omega t_j)$, the euclidean-time version of an oscillatory solution with frequency ω. Show that the frequency is given by

$$\omega^2 = \omega_0^2\left[1 - \frac{(a\omega_0)^2}{12} + \mathcal{O}((a\omega)^4)\right] \tag{59}$$

for the unimproved action. The $(a\omega_0)^2$ correction is the error caused by the finite lattice spacing.

Repeat the exercise for the improved action and show that it has two solutions. One,

$$\omega^2 = \omega_0^2\left[1 + \frac{(a\omega_0)^4}{90} + \mathcal{O}((a\omega)^6)\right], \tag{60}$$

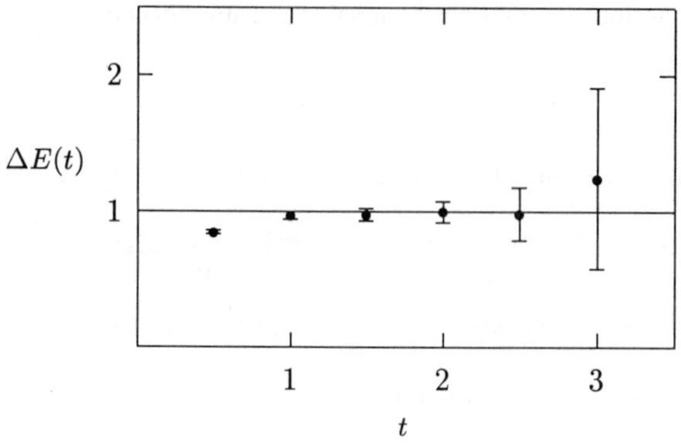

Figure 9: Monte Carlo values $\Delta E(t) \equiv \log(G(t)/G(t+a))/a$ plotted versus t for an harmonic oscillator, as in Fig. 4 but with an improved action. The energies approach their asymptotic value from below.

is an improved version of the previous result; its errors are fourth order in $a\omega_0$ rather than second order. The other solution, however, is

$$\omega^2 \approx \left(\frac{2.6}{a}\right)^2. \tag{61}$$

It corresponds to a new oscillation mode that does not appear in the continuum; it is an artifact of the improved lattice theory. This new mode is sometimes called a "numerical ghost." In a quantum field theory it would be a new, very massive particle ($m \propto 1/a$).

Our lattice theory was designed to be accurate for low-energies, and so we should not be surprised when unphysical modes appear at high energies. These ghost modes, being high-energy, typically decouple from low-energy physics and so can usually be ignored. However, they can have one unfortunate effect on the numerical analysis. Returning to the previous exercise, note that the ΔE_n's for the improved action are below the asymptotic result when n is small (see Fig. 9), in contradiction of the general result discussed in Section 2. The general result ignored the possibility that spurious states might be induced by the numerical analysis. Here these states have negative norm (impossible for real quantum states), which is why the energies rise from below. This is incovenient because it means that the ΔE_n's cannot be used to rigorously bound the true answer — they may be either above or below it — unlike the case for the unimproved action, where they must always be above.

Ghost modes always arise when improved discretizations are used for temporal derivatives. There is a trick, however, for correcting temporal difference operators that avoids extra states. This is to change integration variables in the path integral: for the harmonic oscillator we replace $x_j \to \tilde{x}_j$ where

$$x_j = \tilde{x}_j + \delta\tilde{x}_j \tag{62}$$

and

$$\delta\tilde{x}_j \equiv \xi_1\, a^2 \Delta^{(2)} \tilde{x}_j + \xi_2\, a^2 \omega_0^2 \tilde{x}_j. \tag{63}$$

Substituting this into the action we obtain

$$\begin{aligned} S[x] &= S[\tilde{x} + \delta\tilde{x}] \\ &= S[\tilde{x}] + \sum_j \delta\tilde{x}_j \,\frac{\partial S[\tilde{x}]}{\partial \tilde{x}_j} + \mathcal{O}(a^4) \\ &\equiv \tilde{S}[\tilde{x}] \end{aligned} \tag{64}$$

Find values for the ξ_i such that the improved harmonic-oscillator action, in terms of $\tilde{x}$, is

$$\tilde{S}_{\rm imp}[\tilde{x}] = \sum_{j=0}^{N-1} a \left[-\tfrac{1}{2} m\, \tilde{x}_j \Delta^{(2)} \tilde{x}_j + \tilde{V}_{\rm imp}(x_j) \right] \tag{65}$$

where

$$\tilde{V}_{\rm imp}(\tilde{x}) \equiv \tfrac{1}{2} m \omega_0^2 \tilde{x}_j^2 \left(1 + \frac{(a\omega_0)^2}{12} \right) \tag{66}$$

and all $\mathcal{O}(a^4)$ terms are ignored. This action has no a^2 errors, but also has no ghosts. The value of the path integral is not changed by a simple change of variables (provided that the jacobian is included — show that in this case it has no effect on expectation values). Rerun your numerical tests from the last exercise using this action.

Exercise: The field transformation trick in the previous exercise is particularly simple for the harmonic oscillator. Generalize the trick for the case of an anharmonic oscillator with, for example,

$$V(x) = \tfrac{1}{2} m\, \omega_0^2 x^2 (1 + c m \omega_0 x^2). \tag{67}$$

where c is a dimensionless parameter. Try a variable change with

$$\delta\tilde{x}_j \equiv \xi_1\, a^2 \Delta^{(2)} \tilde{x}_j + \xi_2\, a^2 \omega_0^2 \tilde{x}_j + \xi_3\, a^2 m \omega_0^3 \tilde{x}_j^3. \tag{68}$$

The resulting action is as above but with a new $\tilde{V}_{\rm imp}$:

$$\begin{aligned} \tilde{V}_{\rm imp}(\tilde{x}) &= \frac{m\omega_0^2}{2}\, \tilde{x}^2 (1 + c m \omega_0 \tilde{x}^2) + \frac{a^2 m \omega_0^4}{24} \left(\tilde{x} + 2 c m \omega_0 \tilde{x}^3 \right)^2 \\ &- a\delta v(\tilde{x}) + \frac{a^3}{2}\, \delta v(\tilde{x})^2 \end{aligned} \tag{69}$$

where the terms involving

$$\delta v(\tilde{x}) \equiv c m \omega_0^3 \tilde{x}^2/4 \tag{70}$$

are due to the jacobian (from the change of integration variables; the jacobian matters here because the change is nonlinear). Test this improved lattice action against the original unimproved action using $m = \omega_0 = 1$ and $c = 2$. (The asymptotic value for ΔE is 1.933 with these parameters.)
One might expect errors of order a^4 with the improved action. However renormalization effects arise when interactions are anharmonic. In particular the coefficients of the x^2 and x^4 interactions are renormalized away from their naive values. Such effects enter at the same order as the corrections from the jacobian. In our one-dimensional theory, unlike in QCD, these corrections vanish like powers of a when $a \to 0$. The leading such correction is an $\mathcal{O}(a)$ shift in the coefficient of the x^2 potential. Run high-precision simulations at $a = 1/2$ and $a = 1/4$ to compute the $\mathcal{O}(a)$ error due to renormalization. Try adjusting the coefficient of x^2 to remove this error. (Alternatively one could try using perturbation theory to compute the shift needed to eliminate the error.)

3.3 Perturbation Theory and Tadpole Improvement[a]

Improved discretizations and large lattice spacings are old ideas, pioneered by Wilson, Symanzik and others[7]. However, perturbation theory is essential; the lattice spacing a must be small enough so that $p \approx \pi/a$ QCD is perturbative. This was the requirement that drove lattice QCD towards very costly simulations with tiny lattice spacings. Traditional perturbation theory for lattice QCD begins to fail at distances of order 1/20 to 1/10 fm, and therefore lattice spacings must be at least this small before improved actions are useful. This seems very odd since phenomenological applications of continuum perturbative QCD suggest that perturbation theory works well down to energies of order 1 GeV, which corresponds to a lattice spacing of 0.6 fm. The breakthrough, in the early 1990's, was the discovery of a trivial modification of lattice QCD, called "tadpole improvement," that allows perturbation theory to work even at distances as large as 1/2 fm[6,8,9].

One can readily derive Feynman diagram rules for lattice QCD using the same techniques as in the continuum, but applied to the lattice lagrangian[11]. The particle propagators and interaction vertices are usually complicated functions of the momenta that become identical to their continuum analogues in the low-momentum limit. All loop momenta are cut off at $p_\mu = \pm\pi/a$.

Testing perturbation theory is also straightforward. One designs short-distance quantities that can be computed easily in a simulation (i.e., in a

[a]This section is based upon work with Paul Mackenzie that is described in [6].

Monte Carlo evaluation of the lattice path integral). The Monte Carlo gives the exact value which can then be compared with the perturbative expansion for the same quantity. An example of such a quantity is the expectation value of the Wilson loop operator,

$$W(\mathcal{C}) \equiv \langle 0|\tfrac{1}{3}\,\mathrm{Re}\,\mathrm{Tr}\,\mathcal{P}\mathrm{e}^{-ig\oint_{\mathcal{C}} A\cdot dx}|0\rangle, \tag{71}$$

where A is the QCD vector potential, $\mathcal{P}$ denotes path ordering, and $\mathcal{C}$ is any small, closed path or loop on the lattice. $W(\mathcal{C})$ is perturbative for sufficiently small loops $\mathcal{C}$. We can test the utility of perturbation theory over any range of distances by varying the loop size while comparing numerical Monte Carlo results for $W(\mathcal{C})$ with perturbation theory.

Fig. 10 illustrates the highly unsatisfactory state of traditional lattice-QCD perturbation theory. It shows the "Creutz ratio" of $2a \times 2a$, $2a \times a$ and $a \times a$ Wilson loops,

$$\chi_{2,2} \equiv -\ln\left(\frac{W(2a\times 2a)\,W(a\times a)}{W^2(2a\times a)}\right), \tag{72}$$

plotted versus the size $2a$ of the largest loop. Traditional perturbation theory (dotted lines) underestimates the exact result by factors of three or four for loops of order $1/2\,\mathrm{fm}$; only when the loops are smaller than $1/20\,\mathrm{fm}$ does perturbation theory begin to give accurate results.

The problem with traditional lattice-QCD perturbation theory is that the coupling it uses is much too small. The standard practice was to express perturbative expansions of short-distance lattice quantities in terms of the bare coupling α_{lat} used in the lattice lagrangian. This practice followed from the notion that the bare coupling in a cutoff theory is approximately equal to the running coupling evaluated at the cutoff scale, here $\alpha_s(\pi/a)$, and therefore that it is the appropriate coupling for quantities dominated by momenta near the cutoff. In fact the bare coupling in traditional lattice QCD is much smaller than true effective coupling at large lattice spacings: for example,

$$\alpha_{\mathrm{lat}} = \alpha_V(\pi/a) - 4.7\,\alpha_V^2 + \cdots \tag{73}$$

$$\leq \tfrac{1}{2}\alpha_V(\pi/a) \qquad \text{for } a > .1\,\mathrm{fm} \tag{74}$$

where $\alpha_V(q)$ is a continuum coupling defined by the static-quark potential,

$$V_{\mathrm{Q\overline{Q}}}(q) \equiv -4\,\pi\,C_{\mathrm{F}}\,\frac{\alpha_V(q)}{q^2}. \tag{75}$$

Consequently α_{lat} expansions, though formally correct, badly underestimate perturbative effects, and converge poorly.

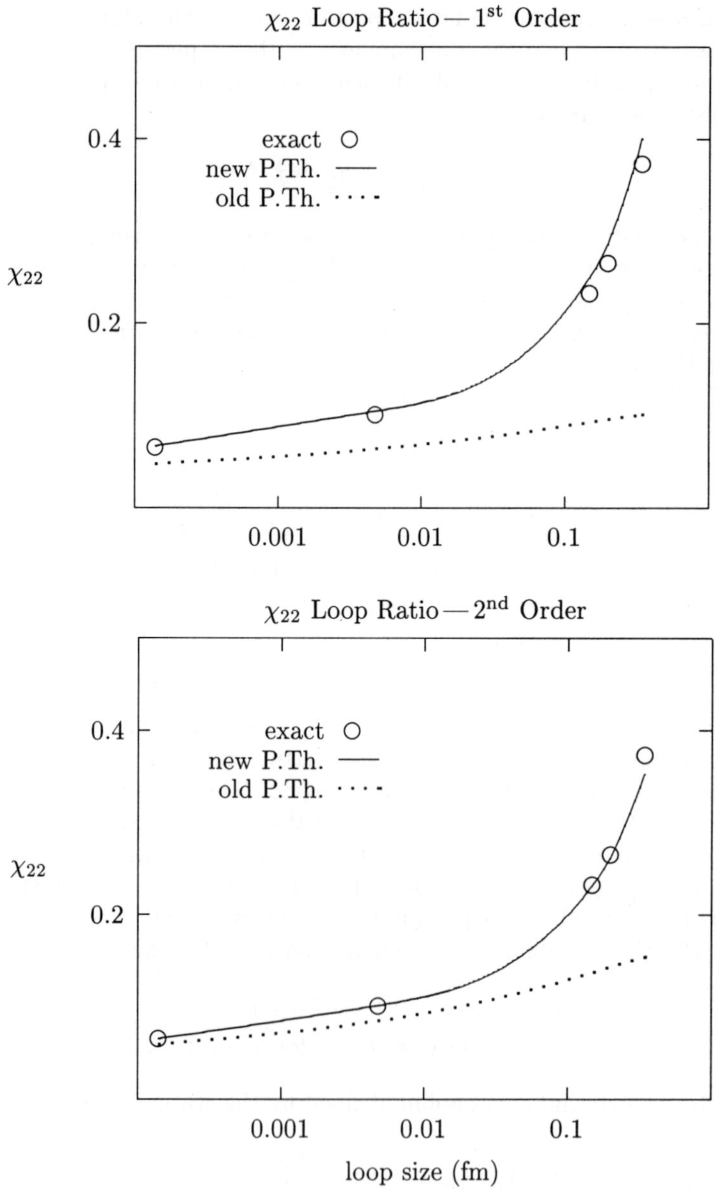

Figure 10: The χ_{22} Creutz ratio of Wilson loops versus loop size. Results from Monte Carlo simulations (exact), and from tadpole-improved (new) and traditional (old) lattice perturbation theory are shown.

The anomalously small bare coupling in the traditional lattice theory is a symptom of the "tadpole problem". As we discuss later, all gluonic operators in lattice QCD are built from the link operator

$$U_\mu(x) \equiv \mathcal{P}\mathrm{e}^{-i\int_x^{x+a\hat{\mu}} gA\cdot dx} \approx \mathrm{e}^{-iagA_\mu} \tag{76}$$

rather than from the vector potential A_μ. Thus, for example, the leading term in the lagrangian that couples quarks and gluons is $\overline{\psi}U_\mu\gamma_\mu\psi/a$. Such a term contains the usual $\overline{\psi}gA\cdot\gamma\psi$ vertex, but, in addition, it contains vertices with any number of additional powers of agA_μ. These extra vertices are irrelevant for classical fields since they are suppressed by powers of the lattice spacing. For quantum fields, however, the situation is quite different since pairs of A_μ's, if contracted with each other, generate ultraviolet divergent factors of $1/a^2$ that precisely cancel the extra a's. Consequently the contributions generated by the extra vertices are suppressed by powers of g^2 (not a), and turn out to be uncomfortably large. These are the tadpole contributions.

The tadpoles result in large renormalizations — often as large as a factor of two or three — that spoil naive perturbation theory, and with it our intuition about the connection between lattice operators and the continuum. However tadpole contributions are generically process independent and so it is possible to measure their contribution in one quantity and then correct for them in all other quantities.

The simplest way to do this is to cancel them out. The mean value u_0 of $\frac{1}{3}\,\mathrm{Re}\,\mathrm{Tr}\,U_\mu$ consists of only tadpoles and so we can largely cancel the tadpole contributions by dividing every link operator by u_0. That is, in every lattice operator we replace

$$U_\mu(x) \to \frac{U_\mu(x)}{u_0} \tag{77}$$

where u_0 is computed numerically in a simulation.

The u_0's cancel tadpole contributions, making lattice operators and perturbation theory far more continuum-like in their behavior. Thus, for example, the only change in the standard gluon action when it is tadpole-improved is that the new bare coupling α_{TI} is enhanced by a factor of $1/u_0^4$ relative to the coupling α_{lat} in the unimproved theory:

$$\alpha_{\mathrm{TI}} = \frac{\alpha_{\mathrm{lat}}}{u_0^4}. \tag{78}$$

Since $u_0^4 < .6$ when $a > .1\,\mathrm{fm}$, the tadpole-improved coupling is typically more than twice as large for coarse lattices. Expressing α_{TI} in terms of the continuum coupling α_V, we find that now our intuition is satisfied:

$$\alpha_{\mathrm{TI}} = \alpha_V(\pi/a) - .5\,\alpha_V^2 + \cdots \tag{79}$$
$$\approx \alpha_V(\pi/a). \tag{80}$$

Perturbation theory for the Creutz ratio Eq. (72) converges rapidly to the correct answer when it is reexpressed in terms of α_{TI}. An even better result is obtained if the expansion is reexpressed as a series in a coupling constant defined in terms of a physical quantity, like the static-quark potential, where that coupling constant is measured in a simulation. By measuring the coupling we automatically include any large renormalizations of the coupling due to tadpoles. It is important that the scale q^* at which the running coupling constant is evaluated be chosen appropriately for the quantity being studied[6,8]. When these refinements are added, perturbation theory is dramatically improved, and, as illustrated in Fig. 10, is still quite accurate for loops as large as $1/2$ fm.

This same conclusion follows from Fig. 11 which shows the value of the bare quark mass needed to obtain zero-mass pions using Wilson's lattice action for quarks. This quantity diverges linearly as the lattice spacing vanishes, and so should be quite perturbative. Here we see dramatic improvements as the tadpoles are removed first from the gluon action, through use of an improved coupling, and then also from the quark action.

The Creutz ratio and the critical quark mass are both very similar to the couplings we need to compute for improved lagrangians. Tadpole improvement has been very successful in a wide range of applications.

Asymptotic freedom implies that short-distance QCD is simple (perturbative) while long-distance QCD is difficult (nonperturbative). The lattice separates short from long distances, allowing us to exploit this dichotomy to create highly efficient algorithms for solving the entire theory: $p > \pi/a$ QCD is included via corrections $\delta\mathcal{L}$ to the lattice lagrangian that are computed using perturbation theory; $p < \pi/a$ QCD is handled nonperturbatively using Monte Carlo integration. Thus, while we wish to make the lattice spacing a as large as possible, we are constrained by two requirements. First a must be sufficiently small that our finite-difference approximations for derivatives in the lagrangian and field equations are sufficiently accurate. Second a must be sufficiently small that π/a is a perturbative momentum. Numerical experiments indicate that both constraints can be satisfied when $a \approx 1/2$ fm or smaller, provided all lattice operators are tadpole improved.

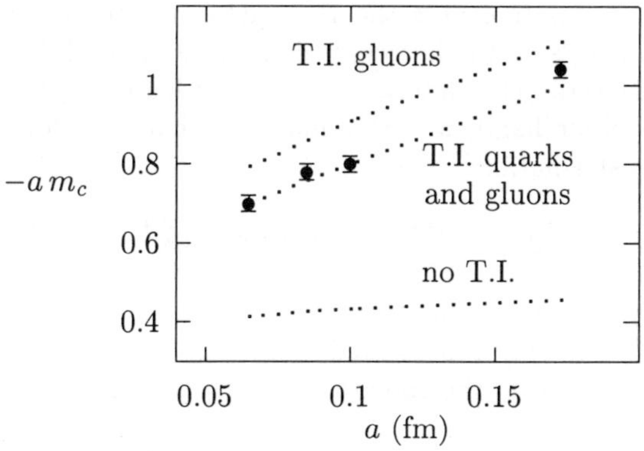

Figure 11: The critical bare quark mass for Wilson's lattice quark action versus lattice spacing. Monte Carlo data points are compared with perturbation theories in a theory with no tadpole improvement (T.I), tadpole-improved gluon dynamics, and tadpole-improved quark and gluon dynamics.

4 QCD on a Lattice

4.1 Classical Gluons[10]

The continuum action for QCD is

$$S = \int d^4x \, \tfrac{1}{2} \sum_{\mu,\nu} \operatorname{Tr} F_{\mu\nu}^2(x) \tag{81}$$

where

$$F_{\mu\nu} \equiv \partial_\mu A_\nu - \partial_\nu A_\mu + ig[A_\mu, A_\nu] \tag{82}$$

is the field tensor, a traceless 3×3 hermitian matrix. The defining characteristic of the theory is its invariance with respect to gauge transformations where

$$F_{\mu\nu} \to \Omega(x)\, F_{\mu\nu}\, \Omega(x)^\dagger \tag{83}$$

and $\Omega(x)$ is an arbitrary x-dependent SU_3 matrix.

The standard discretization of this theory seems perverse at first sight. Rather than specifying the gauge field by the values of $A_\mu(x)$ at the sites of the lattice, the field is specified by variables on the links joining the sites. In the classical theory, the "link variable" on the link joining a site at x to one at $x + a\hat{\mu}$ is determined by the line integral of A_μ along the link:

$$U_\mu(x) \equiv \mathcal{P} \exp\left(-i \int_x^{x+a\hat{\mu}} gA \cdot dy \right) \tag{84}$$

where the $\mathcal{P}$-operator path-orders the A_μ's along the integration path. We use U_μ's in place of A_μ's on the lattice, because it is impossible to formulate a lattice version of QCD directly in terms of A_μ's that has exact gauge invariance. The U_μ's, on the other hand, are SU_3 matrices that transform very simply under a gauge transformation:

$$U_\mu(x) \rightarrow \Omega(x)\, U_\mu(x)\, \Omega(x + a\hat{\mu})^\dagger . \tag{85}$$

This makes it easy to build a discrete theory with exact gauge invariance.

A link variable $U_\mu(x)$ is represented pictorially by a directed line from x to $x + \hat{\mu}$, where this line is the integration path for the line integral in the exponent of $U_\mu(x)$:

$U_\mu(x)$

In the conjugate matrix $U_\mu^\dagger(x)$ the direction of the line integral is flipped and so we represent $U_\mu^\dagger(x)$ by a line going backwards from $x + \hat{\mu}$ to x:

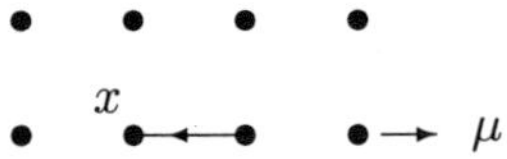

$U_\mu^\dagger(x)$

A Wilson loop function,

$$W(C) \equiv \tfrac{1}{3}\,\mathrm{Tr}\,\mathcal{P}\mathrm{e}^{-i\oint_C gA\cdot dx}, \tag{86}$$

for any closed path C built of links on the lattice can be computed from the path-ordered product of the U_μ's and $U_\mu^\dagger$'s associated with each link. For example, if C is the loop

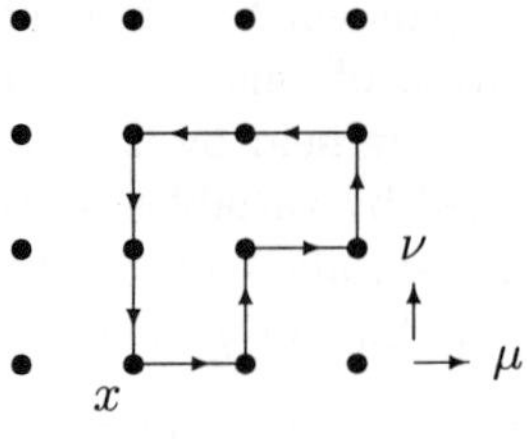

then

$$W(C) = \tfrac{1}{3}\,\mathrm{Tr}\left(U_\mu(x)U_\nu(x + a\hat{\mu}) \ldots U_\nu^\dagger(x)\right) . \tag{87}$$

Such quantities are obviously invariant under arbitrary gauge transformations Eq. (85).

One might wonder why we go to so much trouble to preserve gauge invariance when we quite willing give up Lorentz invariance, rotation invariance, etc. The reason is quite practical. With gauge invariance, the quark-gluon, three-gluon, and four-gluon couplings in QCD are all equal, and the bare gluon mass is zero. Without gauge invariance, each of these couplings must be tuned independently and a gluon mass introduced if one is to recover QCD. Tuning this many parameters in a numerical simulation is very expensive. This is not much of a problem in the classical theory, where approximate gauge invariance keeps the couplings approximately equal; but it is serious in the quantum theory because quantum fluctuations (loop-effects) renormalize the various couplings differently in the absence of exact gauge invariance. So while it is quite possible to formulate lattice QCD directly in terms of A_μ's, the resulting theory would have only approximate gauge invariance, and thus would be prohibitively expensive to simulate. Symmetries like Lorentz invariance can be given up with little cost because the symmetries of the lattice, though far less restrictive, are still sufficient to prevent the introduction of new interactions with new couplings (at least to lowest order in a).

We must now build a lattice lagrangian from the link operators. We require that the lagrangian be gauge invariant, local, and symmetric with respect to axis interchanges (which is all that is left of Lorentz invariance). The most local nontrivial gauge invariant object one can build from the link operators is the "plaquette operator," which involves the product of link variables around the smallest square at site x in the $\mu\nu$ plane:

$$P_{\mu\nu}(x) \equiv \tfrac{1}{3}\,\mathrm{Re}\,\mathrm{Tr}\left(U_\mu(x)U_\nu(x+a\hat{\mu})U_\mu^\dagger(x+a\hat{\mu}+a\hat{\nu})U_\nu^\dagger(x)\right). \tag{88}$$

To see what this object is, consider evaluating the plaquette centered about a point x_0 for a very smooth weak classical A_μ field. In this limit,

$$P_{\mu\nu} \approx 1 \tag{89}$$

since

$$U_\mu \approx \mathrm{e}^{-igaA_\mu} \approx 1. \tag{90}$$

Given that A_μ is slowly varying, its value anywhere on the plaquette should be accurately specified by its value and derivatives at x_0. Thus the corrections to Eq. (89) should be a polynomial in a with coefficients formed from gauge-invariant combinations of $A_\mu(x_0)$ and its derivatives: that is,

$$P_{\mu\nu} = 1 \quad - \quad c_1\, a^4\, \mathrm{Tr}\,(gF_{\mu\nu}(x_0))^2$$

$$
\begin{aligned}
&- \quad c_2\, a^6 \;\mathrm{Tr}\left(gF_{\mu\nu}(x_0)(\mathrm{D}_\mu^2 + \mathrm{D}_\nu^2)gF_{\mu\nu}(x_0)\right)\\
&+ \quad \mathcal{O}(a^8)
\end{aligned}
\tag{91}
$$

where c_1 and c_2 are constants, and D_μ is the gauge-covariant derivative. The leading correction is order a^4 because $F_{\mu\nu}^2$ is the lowest-dimension gauge-invariant combination of derivatives of A_μ, and it has dimension 4. (There are no F^3 terms because $P_{\mu\nu}$ is invariant under $U_\mu \to U_\mu^\dagger$ or, equivalently, $F \to -F$.)

It is straightforward to find the coefficients c_1 and c_2. We need only examine terms in the expansion of $P_{\mu\nu}$ that are quadratic in A_μ; the cubic and quartic parts of $F_{\mu\mu}^2$ then follow automatically, by gauge invariance. Because of the trace, the path ordering is irrelevant to this order. Thus

$$
\begin{aligned}
P_{\mu\nu} &= \tfrac{1}{3}\,\mathrm{Re}\;\mathrm{Tr}\,\mathcal{P}\mathrm{e}^{-i\oint_\Box gA\cdot dx}\\
&= \tfrac{1}{3}\,\mathrm{Re}\;\mathrm{Tr}\left[1 - i\oint_\Box gA\cdot dx - \tfrac{1}{2}\left(\oint_\Box gA\cdot dx\right)^2 + \mathcal{O}(A^3)\right]
\end{aligned}
\tag{92}
$$

where, by Stoke's Theorem,

$$
\begin{aligned}
\oint_\Box A\cdot dx &= \int_{-a/2}^{a/2} dx_\mu dx_\nu\,[\partial_\mu A_\nu(x_0+x) - \partial_\nu A_\mu(x_0+x)]\\
&= \int_{-a/2}^{a/2} dx_\mu dx_\nu\,[F_{\mu\nu}(x_0) + (x_\mu \mathrm{D}_\mu + x_\nu \mathrm{D}_\mu)F_{\mu\nu}(x_0) + \cdots]\\
&= a^2\,F_{\mu\nu}(x_0) + \frac{a^4}{24}\,(\mathrm{D}_\mu^2 + \mathrm{D}_\nu^2)F_{\mu\nu}(x_0) + \mathcal{O}(a^6, A^2).
\end{aligned}
\tag{93}
$$

Thus $c_1 = 1/6$ and $c_2 = 1/72$ in Eq. (91).

The expansion in Eq. (91) is the classical analogue of an operator product expansion. Using this expansion, we find that the traditional "Wilson action" for gluons on a lattice,

$$
S_{\mathrm{Wil}} = \beta \sum_{x,\mu>\nu} (1 - P_{\mu\nu}(x))
\tag{94}
$$

where $\beta = 6/g^2$, has the correct limit for small lattice spacing up to corrections of order a^2:

$$
S_{\mathrm{Wil}} = \int d^4x \sum_{\mu,\nu} \left\{ \tfrac{1}{2}\,\mathrm{Tr}\,F_{\mu\nu}^2 + \frac{a^2}{24}\,\mathrm{Tr}\,F_{\mu\nu}(\mathrm{D}_\mu^2 + \mathrm{D}_\nu^2)F_{\mu\nu} + \cdots \right\}.
\tag{95}
$$

We can cancel the a^2 error in the Wilson action by adding other Wilson loops. For example, the $2a \times a$ "rectangle operator"

$$R_{\mu\nu} = \tfrac{1}{3}\,\mathrm{Re\,Tr}\qquad (96)$$

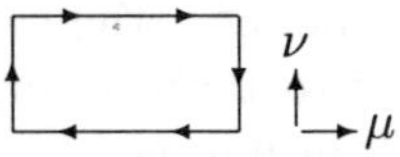

has expansion

$$R_{\mu\nu} = 1 - \frac{4}{6}\,a^4\,\mathrm{Tr}\,(gF_{\mu\nu})^2 - \frac{4}{72}\,a^6\,\mathrm{Tr}\,\left(gF_{\mu\nu}(4\,\mathrm{D}_\mu^2 + \mathrm{D}_\nu^2)gF_{\mu\nu}\right) - \cdots. \qquad (97)$$

The mix of a^4 terms and a^6 terms in the rectangle is different from that in the plaquette. Therefore we can combine the two operators to obtain an improved classical lattice action that is accurate up to $\mathcal{O}(a^4)$[12,13]:

$$S_{\mathrm{classical}} \equiv -\beta \sum_{x,\mu>\nu} \left\{ \frac{5P_{\mu\nu}}{3} - \frac{R_{\mu\nu} + R_{\nu\mu}}{12} \right\} + \mathrm{const} \qquad (98)$$

$$= \int d^4x \sum_{\mu,\nu} \tfrac{1}{2}\,\mathrm{Tr}\,F_{\mu\nu}^2 + \mathcal{O}(a^4). \qquad (99)$$

This process is the analogue of improving the derivatives in discretizations of non-gauge theories.[b]

Exercise: Defining the "twisted-rectangle operator"

$$T_{\mu\nu} = \tfrac{1}{3}\,\mathrm{Re\,Tr}\qquad . \qquad (100)$$

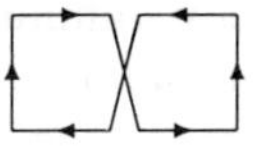

show that

$$S_{\mathrm{classical}} \equiv -\beta \sum_{x,\mu>\nu} \left\{ P_{\mu\nu} + \frac{T_{\mu\nu} + T_{\nu\mu}}{12} \right\} + \mathrm{const} \qquad (101)$$

$$= \int d^4x \sum_{\mu,\nu} \tfrac{1}{2}\,\mathrm{Tr}\,F_{\mu\nu}^2 + \mathcal{O}(a^4). \qquad (102)$$

This is an alternative to the improved gluon action derived in the previous exercise.

---

[b] An important step that I have not discussed is to show that the gluon action is positive for any configuration of link variables[13]. This guarantees that the classical ground state of the lattice action corresponds to $F_{\mu\nu} = 0$.

4.2 Quantum Gluons[c]

In the previous section we derived improved classical actions for gluons that are accurate through order a^4. We now turn these into quantum actions. The most important step is to tadpole improve the action by dividing each link operator U_μ by the mean link u_0: for example, the action built of plaquette and rectangle operators becomes

$$S = -\beta \sum_{x,\mu>\nu} \left\{ \frac{5}{3} \frac{P_{\mu\nu}}{u_0^4} - \frac{R_{\mu\nu} + R_{\nu\mu}}{12\, u_0^6} \right\}. \tag{103}$$

The u_0's cancel lattice tadpole contributions that otherwise would spoil weak-coupling perturbation theory in the lattice theory and undermine our procedure for improving the lattice discretization. Note that $u_0 \approx 3/4$ when $a = .4\,\mathrm{fm}$, and therefore the relative importance of the $R_{\mu\nu}$'s is larger by a factor of $1/u_0^2 \approx 2$ than without tadpole improvement. Without tadpole improvement, we cancel only half of the a^2 error.

The mean link u_0 is computed numerically by guessing a value for use in the action, measuring the mean link in a simulation, and then readjusting the value used in the action accordingly. This tuning cycle converges rapidly to selfconsistent values, and can be done very quickly using small lattice volumes. The u_0's depend only on lattice spacing, and become equal to one as the lattice spacing vanishes.

The expectation value of the link operator is gauge dependent. Thus to minimize gauge artifacts, u_0 is commonly defined as the Landau-gauge expectation value, $\langle 0|\frac{1}{3}\,\mathrm{Tr}\,U_\mu|0\rangle_{\mathrm{LG}}$. Landau gauge is the axis-symmetric gauge that maximizes u_0, thereby minimizing the tadpole contribution; any tadpole contribution that is left in Landau gauge cannot be a gauge artifact. An alternative procedure is to define u_0 as the fourth root of the plaquette expectation value,

$$u_0 = \langle 0|P_{\mu\nu}|0\rangle^{1/4}. \tag{104}$$

This definition gives almost identical results and is more convenient for numerical work since gauge fixing is unnecessary.

Tadpole improvement is the first step in a systematic procedure for improving the action. The next step is to add in renormalizations due to contributions from $k > \pi/a$ physics not already included in the tadpole improvement. These renormalizations induce $a^2\,\alpha_s(\pi/a)$ corrections,

$$\delta\mathcal{L} \;=\; \alpha_s\, r_1\, a^2 \sum_{\mu,\nu} \mathrm{Tr}(F_{\mu\nu} \mathrm{D}_\mu^2 F_{\mu\nu})$$

[c]This section is based on work with M. Alford, W. Dimm, G. Hockney and P. Mackenzie that is described in [14].

$$
\begin{aligned}
&+\quad \alpha_s\, r_2\, a^2 \sum_{\mu,\nu} \mathrm{Tr}(\mathrm{D}_\mu F_{\nu\sigma} \mathrm{D}_\mu F_{\nu\sigma}) \\
&+\quad \alpha_s\, r_3\, a^2 \sum_{\mu,\nu} \mathrm{Tr}(\mathrm{D}_\mu F_{\mu\sigma} \mathrm{D}_\nu F_{\nu\sigma}) \\
&+\quad \cdots, \qquad\qquad\qquad\qquad\qquad\qquad\qquad (105)
\end{aligned}
$$

that must be removed. The last term is harmless; its coefficient can be set to zero by a change of field variable (in the path integral) of the form

$$A_\mu \to A_\mu + a^2\, \alpha_s\, f(\alpha_s) \sum_\nu \mathrm{D}_\nu F_{\nu\mu}. \tag{106}$$

Since changing integration variables does not change the value of an integral, such field transformations must leave the physics unchanged.[d] Operators that can be removed by a field transformation are called "redundant." The other corrections are removed by renormalizing the coefficient of the rectangle operator $R_{\mu\nu}$ in the action, and by adding an additional operator. One choice for the extra operator is

$$C_{\mu\nu\sigma} \equiv \tfrac{1}{3}\,\mathrm{Re\,Tr}$$

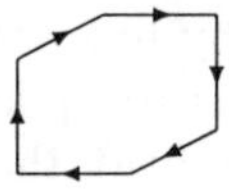

$$. \qquad (107)$$

Then the action, correct up to $\mathcal{O}(a^2\alpha_s^2, a^4)$, is[15]

$$S = -\beta \sum_{x,\mu>\nu} \left\{ \frac{5}{3}\frac{P_{\mu\nu}}{u_0^4} - r_{\mathrm{g}} \frac{R_{\mu\nu} + R_{\nu\mu}}{12\, u_0^6} \right\} + c_{\mathrm{g}}\, \beta \sum_{x,\mu>\nu>\sigma} \frac{C_{\mu\nu\sigma}}{u_0^6}, \tag{108}$$

where

$$
\begin{aligned}
r_{\mathrm{g}} &= 1 + .48\, \alpha_s(\pi/a) \qquad (109)\\
c_{\mathrm{g}} &= .055\, \alpha_s(\pi/a). \qquad (110)
\end{aligned}
$$

The coefficients r_{g} and c_{g} are computed by "matching" physical quantities, like low-energy scattering amplitudes, computed using perturbation theory in the lattice theory with the analogous quantity in the continuum theory. The lattice result depends upon r_{g} and c_{g}; these parameters are tuned until the lattice amplitude agrees with the continuum amplitude to the order in a and α_s required:

$$T_{\mathrm{lat}}(r_{\mathrm{g}}, c_{\mathrm{g}}) = T_{\mathrm{contin}}. \tag{111}$$

[d]One must, of course, include the jacobian for the transformation in the transformed path integral. This contributes only in one-loop order and higher; it has no effect on tree-level calculations.

Note that tadpole improvement has a big effect on these coefficients. Without tadpole improvement, $r_g = 1 + 2\alpha_s$; that is, the coefficient of the radiative correction is four times larger. Tadpole improvement automatically supplies 75% of the one-loop contribution needed without improvement. Since $\alpha_s \approx 0.3$, the unimproved expansion for r_g is not particularly convergent. However, with tadpole improvement, the one-loop correction is only about 10–20% of r_g. Indeed, for most current applications, one-loop corrections to tadpole-improved actions are negligible.

Exercise: Show that the gauge that maximizes $\langle 0|\frac{1}{3}\operatorname{Tr} U_\mu|0\rangle$ becomes Landau gauge ($\partial \cdot A = 0$) in the $a \to 0$ limit.

4.3 Monte Carlo Evaluation of Gluonic Path Integrals

A computer code for the Monte Carlo evaluation of gluonic path integrals can be designed in close analogy with our code for one-dimensional quantum mechanics. The Metropolis algorithm for generating random configurations must be adapted to work with SU_3 matrices. In our quantum mechanics example, the coordinate was updated by adding a random number. In QCD the gluon field is specified by link variables $U_\mu(x)$, which are exponentials of the fundamental field. Thus to update a link variable we must multiply by the exponential of a random field; that is we must multiply by a random SU_3 matrix M:

$$U_\mu \to M\,U_\mu \tag{112}$$

Typically the matrix M is chosen randomly from a set of 50 or 100 random SU_3 matrices that is generated once, at the start of the simulation; the only restrictions on this set are that it be large enough so that products of the various M's cover the entire space of SU_3 matrices, and that the inverse, $M^\dagger$, for each matrix M in the set also be included in the set. The M's can be generated by first creating a set of hermitian matrices H whose matrix elements are random numbers between -1 and 1. These are converted to SU_3 matrices by forming $1 + i\epsilon H$ and unitarizing it.[e] Parameter ϵ determines the size of the update; as before, it is adjusted so that roughly half of all trial updates are accepted.

A second modification of the Metropolis algorithm that is useful for QCD is to update each link variable several times (rather than just once) before moving on to the next variable in a single sweep through the lattice. This allows the

[e] To convert an arbitrary matrix $M = (m_1\, m_2\, m_3)$ into an SU_3 matrix: first normalize the first column m_1 to unity, then make the second column m_2 orthogonal to the (new) first column and normalize it, and replace the third column by the cross product of the (new) first two columns.

link variable to come into statistical equilibrium with its immediate neighbors on the lattice. The additional cost for these extra updates is relatively small because standard gluon actions are linear in the link variables. Thus the part of the action that must be computed when updating a particular $U_\mu(x)$ can be written

$$\Delta S(x,\mu) = \text{Re Tr}\,(U_\mu(x)\Gamma_\mu(x)) \tag{113}$$

where $\Gamma_\mu(x)$, which is a sum of products of the link variables, is independent of $U_\mu(x)$. Computing $\Gamma_\mu(x)$ is the most expensive part of the Metropolis update, but it need be computed only once for each set of successive updates of $U_\mu(x)$. Typically one does about 10 "hits" of the Metropolis algorithm before moving on the the next link variable.

Exercise: Design a computer code for evaluating gluonic path integrals using the Metropolis algorithm. Do this first for the simplest lattice action, the Wilson action (Eq. (94)):

$$S_{\text{Wil}} = -\tilde{\beta} \sum_{x,\mu>\nu} \frac{P_{\mu\nu}(x)}{u_0^4} \tag{114}$$

Run simulations at $\beta \equiv \tilde{\beta}/u_0^4 = 5.5$, which corresponds to a lattice spacing of around 0.25 fm. The lattice volume should be of order 2 fm on a side for typical QCD simulations; use $L/a = 8$ points on a side in your simulation. Set the Metropolis step size $\epsilon = 0.24$ and omit $N_{\text{cor}} = 50$ sweeps between Monte Carlo measurements. Compute Monte Carlo averages of $a \times a$ and $a \times 2a$ Wilson loops; you should obtain about 0.50 and 0.26, respectively.

Also try the improved action, Eq. (103). Use $\beta = 1.719$ and $u_0 = 0.797$ to again obtain $a \approx 0.25\,\text{fm}$[16]. The $a \times a$ and $a \times 2a$ Wilson loops have values of 0.54 and 0.28, respectively. (Wilson loops are unrenormalized and so these values need not agree with those from the Wilson action.)

4.4 A First Simulation

Perhaps the simplest physics calculation that involves just gluons is to compute the potential energy between a static quark and a static antiquark separated by a distance r. This "static potential" should be approximately Coulombic at short distances, but grow linearly at large distances, demonstrating quark confinement. It can be used in a Schrödinger equation to predict energy levels for the ψ/J and Υ families of mesons.

The euclidean Green's function or propagator G for a heavy nonrelativistic quark in a background gauge field A_μ satisfies the equation

$$\left(\mathrm{D}_t - \frac{\mathbf{D}^2}{2M}\right) G(x) = \delta^4(x) \tag{115}$$

where $\mathrm{D}_\mu = \partial_\mu - igA_\mu(x)$ is the gauge-covariant derivative. This equation is easily solved in the static-quark limit, where $M \to \infty$, to obtain

$$G_\infty(\mathbf{x},t) = \left[\mathcal{P}\mathrm{e}^{-i\int_0^t gA_0(\mathbf{x},t)dt}\right]^\dagger \delta^3(\mathbf{x}), \tag{116}$$

which on the lattice becomes

$$G_\infty(\mathbf{x},t) = U_t^\dagger(\mathbf{x},t-a)\, U_t^\dagger(\mathbf{x},t-2a)\, \ldots\, U_t^\dagger(\mathbf{x},0). \tag{117}$$

Propagation of a static antiquark is described by $G_\infty^\dagger$. Therefore we obtain the static potential $V(r)$, which is the energy of a static quark and antiquark a distance r apart, from expectation values of $r \times t$ Wilson loops:

$$W(r,t) \equiv \langle 0|\tfrac{1}{3}\,\mathrm{Tr}$$

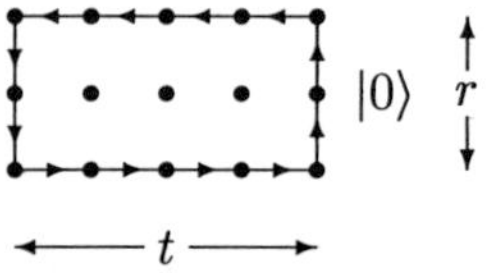

$$|0\rangle \tag{118}$$

where for large t

$$W(r,t) \to \text{const } \mathrm{e}^{-V(r)\,t}. \tag{119}$$

Thus, to calculate the static potential $V(r)$, we compute $W(r,t)$ for a variety of t's and then take the large-t limit

$$W(r,t)/W(r,t+a) \to a\,V(r). \tag{120}$$

There is one modification of this procedure that greatly improves the results. This is to replace the spatial link matrices, in the r direction, by "smeared" link variables $\tilde{U}_\mu(x)$:

$$\tilde{U}_\mu(x) \equiv \left(1 + \epsilon a^2 \Delta^{(2)}\right)^n U_\mu(x) \tag{121}$$

where

$$\Delta^{(2)} U_\mu(x) \equiv \sum_\rho \Delta_\rho^{(2)} U_\mu(x) \tag{122}$$

and $\Delta_\rho^{(2)}$ is a discretized, gauge-covariant derivative:

$$\begin{aligned}\Delta_\rho^{(2)}\, U_\mu(x) \quad &\equiv \quad \frac{1}{u_0^2\, a^2}\left(U_\rho(x)\, U_\mu(x+a\hat{\rho})\, U_\rho^\dagger(x+a\hat{\mu}) - 2\,u_0^2\, U_\mu(x)\right.\\ &\qquad \left. + \, U_\rho^\dagger(x-a\hat{\rho})\, U_\mu(x-a\hat{\rho})\, U_\rho(x-a\hat{\rho}+a\hat{\mu})\right). \end{aligned} \tag{123}$$

The low-momentum components of the smeared link variable are unchanged by the smearing, up to corrections of order a^2p^2; but at high momentum the smearing acts as an ultraviolet cutoff:

$$\left(1+\epsilon a^2\Delta^{(2)}\right)^n \rightarrow \left(1-\epsilon a^2 p^2\right)^n \approx \mathrm{e}^{-\epsilon a^2 p^2 n}. \tag{124}$$

By suppressing high-momentum gluons in the initial and final states of our matrix element, we suppress contributions from gluonic excitations and thereby hasten the convergence of $W(r,t)/W(r,t+a)$ to its asymptotic value. This significantly reduces the Monte Carlo statistical errors since these are generally much smaller at smaller t's.

Exercise: Add measurement code for $W(r,t)$ to your gluon Monte Carlo program. Run this code for the parameter sets given above and extract the static potential. Try this with unsmeared spatial U_μ's on the ends of the Wilson loop, and again with smeared links (try $n=4$ smearings with $\epsilon = 1/12$). In the smeared case asymptotic results appear at the first or second time step

My results for the Wilson and improved actions are shown in Fig. 12. Each run took about 3 hours on a 300 MHz personal computer (with 64 MB of RAM). The potential was computed for quarks separated along the lattice axes, as well as for separations along diagonal directions on the lattice (using nonplanar Wilson loops). The plots clearly show the linear rise of the potential at large r, which results in quark confinement; and careful examination shows the onset of Coulombic behavior at small r, which is due to asymptotic freedom. The Monte Carlo results are compared in the figure with global fits to the form

$$V(r) = \sigma r - b/r + c \tag{125}$$

where σ is the string tension. This parameterization works well for the range of r's we are considering (0.25 fm–1 fm). Note that the Monte Carlo data is not as smooth in the Wilson case as it is for the improved action. This is caused the $\mathcal{O}(a^2)$ errors in the Wilson action, coming from the second term in Eq. (95). This term breaks rotational invariance; it is the leading manifestation in the gluon dynamics of the cubic structure of our lattice. While the true potential is a function only of the magnitude of the quark separation, we expect differences on the lattice between potentials for separations that are along the grid axes and those for separations along diagonals. This is particularly apparent at $r=3a$ where there are two Monte Carlo points, one for separation $(3a,0,0)$ and the other for $(2a,2a,a)$. The two points are clearly separated on the plot for the Wilson action. The a^2 errors are removed from the improved action; the two $r=3a$ points merge on the plot for that action. The difference $a(V(2a,2a,a)-V(3a,0,0))$ is reduced from 0.065 ± 0.007 with the Wilson action to 0.003 ± 0.006 with the improved action in these particular simulations. The a^2 errors are larger at smaller r's, but more difficult to quantify there.

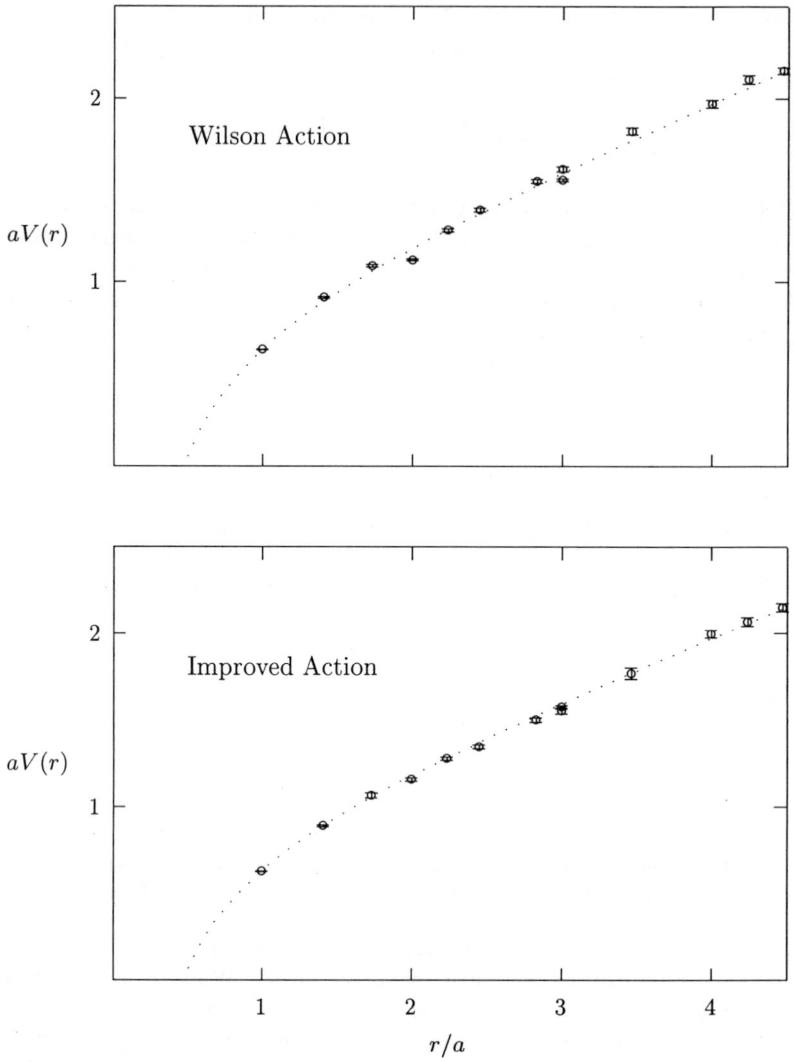

Figure 12: Static-quark potential computed using the Wilson gluon action and the $\mathcal{O}(a^2)$ improved action. The dotted line in each case is a fit of the Monte Carlo results to $\sigma r - b/r + c$. The lattice spacing $a \approx 0.25\,\text{fm}$ in each case. Each plot required about 3 hours of computer time using a 300 MHz personal computer. The improved action gives a smoother curve.

5 Conclusions

The QCD simulation discussed in the previous section is, of course, just a beginning. A reasonable next step would be to add heavy quarks to the simulation, using nonrelativistic QCD, and to compute spectra and wavefunctions for ψ/J and Υ mesons[10,17,18]. Far more costly to simulate, although no more subtle theoretically, are light quarks[10,16]. These are needed to analyze protons, neutrons and other standard hadrons. Finally, and most costly, one might include light-quark vacuum polarization, which is an essential step for high-precision calculations (better than 10–15%).

Acknowledgements

This work was supported by a grant from the National Science Foundation.

Appendix

The sample simulation code in Section 2 is written in Python, a simple but powerful computer language that is freely available for just about any type of computer from the web site www.python.org. These simulations are most efficient when the Numeric package is used with Python; this is included with some Python distributions but must downloaded separately (from www.python.org) for others. To use the code given in the text, collect it together in a single file as follows:

```
import Numeric
from whrandom import uniform
from math import *

# ... code from text goes here

# set parameters:
N = 20
N_cor = 20
N_cf = 100
a = 0.5
eps = 1.4

# create arrays:
x = Numeric.zeros((N,),Numeric.Float)
G = Numeric.zeros((N_cf,N),Numeric.Float)
```

```
# do the simulation:
MCaverage(x,G)
```

If the file is called `simulation.py`, it is run with the command `python simulation.`

To test the binning and bootstrap codes add the following to the the file:

```
def avg(G):                          # MC avg of G
    return Numeric.sum(G)/len(G)

def sdev(G):                         # std dev of G
    g = Numeric.asarray(G)
    return Numeric.absolute(avg(g**2)-avg(g)**2)**0.5

print 'avg G\n',avg(G)
print 'avg G (binned)\n',avg(bin(G,4))
print 'avg G (bootstrap)\n',avg(bootstrap(G))
```

The average of the binned copy of `G` should be the same as the average of `G` itself; the average of the bootstrap copy should be different by an amount of order the Monte Carlo error. Compute averages for several bootstrap copies to get a good feel for the errors.

Finally one wants to extract energies. This is done by adding code to compute $\Delta E(t)$:

```
def deltaE(G):                      # Delta E(t)
    avgG = avg(G)
    adE = Numeric.log(Numeric.absolute(avgG[:-1]/avgG[1:]))
    return adE/a

print 'Delta E\n',deltaE(G)
print 'Delta E (bootstrap)\n',deltaE(bootstrap(G))
```

Again repeating the evaluation for 50 or 100 bootstrap copies of `G` gives an estimate of the statistical errors in the energies. Additional code can be added to evaluate standard deviations from these copies:

```
def bootstrap_deltaE(G,nbstrap=100): # Delta E + errors
    avgE = deltaE(G)                 # avg deltaE
    bsE = []
```

```
    for i in range(nbstrap):      # bs copies of deltaE
        g = bootstrap(G)
        bsE.append(deltaE(g))
    bsE = Numeric.array(bsE)
    sdevE = sdev(bsE)             # spread of deltaE's
    print "\n%2s  %10s  %10s" % ("t","Delta E(t)","error")
    print 26*"-"
    for i in range(len(avgE)/2):
        print "%2d  %10g  %10g" % (i,avgE[i],sdevE[i])

bootstrap_deltaE(G)
```

The entire program should take only a few seconds to run on a modern personal computer.

This example almost completely ignores the powerful object-oriented features that distinguish Python from most other scripting languages.

References

1. See for example Chapter 3 in G. Baym's *Lectures on Quantum Mechanics* (Benjamin/Cummings, Menlo Park, 1973).
2. G.P. Lepage, J. Comp. Phys. **27** (1978) 192.
3. An extensive discussion of the systematics of Monte Carlo errors can be found in: G.P. Lepage, *The Analysis of Algorithms for Lattice Field Theory*, in *From Actions to Answers*, edited by T. DeGrand and D. Toussaint (World Scientific, Singapore, 1989).
4. See for example M. Creutz, *Quarks, gluons, and lattices* (Cambridge University Press, Cambridge, 1985).
5. Download python from www.python.org. See also M. Lutz and D. Ascher, *Learning Python* (O'Reilly and Associates, 1999).
6. G.P. Lepage and P.B. Mackenzie, *Phys. Rev.* **D48** (1993) 2250.
7. For an overview see K.G. Wilson, *Rev. Mod. Phys.* **55** (1983) 583.
8. G.P. Lepage, *Lattice QCD for Small Computers*, in *The Building Blocks of Creation,* edited by S. Raby and T. Walker (World Scientific Press, Singapore, 1994).
9. G.P. Lepage, Nucl. Phys. Proc. Suppl. **60A** (1998) 267
10. Parts of Section 3 and 4 are based on the following article; it is more detailed and also discusses simulations of light and heavy quarks: G.P. Lepage, *Redesigning Lattice QCD*, in *Perturbative and Nonperturbative Aspects of Quantum Field Theory* (Lecture Notes in Physics, 479), edited

by H. Latal and W. Schweiger (Springer-Verlag, Berlin, 1997).
11. See, for example, H. Kawai, R. Nakayama and K. Seo, *Nucl. Phys.* **B189** (1981) 40.
12. G. Curci, P. Menotti, and G. Paffuti, *Phys. Lett.* **130B** (1983) 205; Erratum: *ibid.* **135B** (1984) 516.
13. M. Lüscher and P. Weisz, *Comm. Math. Phys.* **97** (1985) 59.
14. M. Alford, W. Dimm, G.P. Lepage, G. Hockney and P. Mackenzie, *Phys. Lett.* **B361** (1995) 87.
15. M. Lüscher and P. Weisz, *Phys. Lett.* **158B**, (1985) 250, and references therein.
16. See for example: M. Alford, T.R. Klassen and G.P. Lepage, Phys. Rev. **D58** (1998) 034503.
17. G.P. Lepage and B. Thacker, *Nucl. Phys.* **B**(Proc. Suppl.)**4** (1988) 199; G.P. Lepage, K. Hornbostel, L. Magnea, U. Magnea and C. Nakhleh, *Phys. Rev.* **D46** (1992) 4052.
18. C. Davies et al., *Phys. Rev.* **D50** (1994) 6963; P. McCallum and J. Shigemitsu, *Nucl. Phys.* **B47** (Proc. Suppl.) (1996) 409; C. Davies et al., *Phys. Rev. Lett.* **73** (1994) 2654; C. Davies et al., *Phys. Lett.* **B345** (1995) 42; J. Shigemitsu's talk at *Lattice '96* (June 1996, St. Louis).

INTRODUCTION TO THE QCD SUM RULE APPROACH

A.V. RADYUSHKIN
Physics Department, Old Dominion University, Norfolk, VA 23529
and
Theory Group, Jefferson Lab, Newport News, VA 23606

In these lectures, I describe the techniques used within the QCD sum rule approach. The basic concepts of the approach are introduced using a simple model of quantum-mechanical oscillator in 2+1 dimensions. Then I discuss the field-theoretical extension and the construction of the operator product expansion for current correlators in QCD. The calculation of static parameters is illustrated with the example of sum rules in the vector and axial meson channels. Finally, the QCD sum rule calculation of the pion electromagnetic form factor is presented as an example of application of the method to dynamic characteristics of the hadrons.

1 Introduction

1.1 Introductory remarks

QCD sum rules, invented more than 20 years ago by Shifman, Vainshtein and Zakharov[1] are still a rather popular approach for theoretical study of the hadronic structure. Applied originally to the simplest static hadronic characteristics, like masses, leptonic widths, etc., they were also used to calculate much more complicated things like hadronic wave functions and form factors.

To begin with, I will briefly discuss the basic questions

- Why do we need QCD sum rules?
- What are QCD sum rules?
- How are the QCD sum rules working?

using for illustration the QCD calculation of the pion form factor. In subsequent sections, I will discuss in more detail the machinery of QCD sum rules in several cases: calculation of the properties of the lowest resonances in ρ-meson and π/A_1 channels, and analysis of the pion form factor in various kinematical situations.

1.2 Pion form factor in QCD

According to the factorization theorem[2,3], for sufficiently large momentum transfers, one can represent the pion form factor as a sum of terms of increasing complexity (see Fig.1). The first term (purely *soft* contribution), Fig.1a, contains no short-distance (SD) subprocesses. The second term, Fig.1b, contains a *hard* gluon exchange. There are also corrections to the hard term: higher

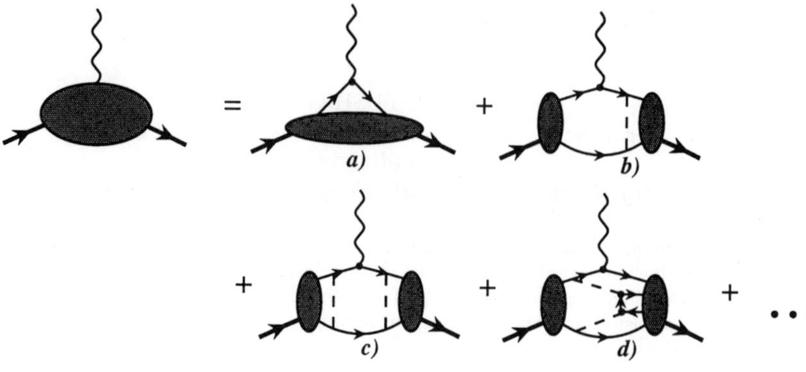

Figure 1: Structure of factorization for the pion form factor in QCD

order corrections, Fig.1c, containing extra α_s factors and higher twist corrections, Fig.1d. In perturbative QCD one can calculate only the hard terms of this expansion.

The first result of perturbative QCD for exclusive processes is the prediction for the asymptotic behavior of the pion electromagnetic form factor [4,2,5,6]:

$$F_\pi(Q^2) = \int_0^1 dx \int_0^1 t(xP,(1-x)P; yP',(1-y)P',q)\ \varphi(x)\ \varphi(y)\ dy\ , \quad (1)$$

where $\varphi(x)$ is the pion wave function giving the probability amplitude to find the pion in a state where quarks carry fractions xP and $(1-x)P$ of its longitudinal momentum P; $q = P' - P$ is the momentum transfer to the pion and $T(\ldots)$ is the perturbatively calculable short-distance amplitude. The wave function $\varphi(x)$ accumulates long-distance information about the pion structure and cannot be calculated within the pQCD approach. Normally, some phenomenological assumptions are used about the form of this function depending on a single parameter x. There is still a challenge to calculate it from first principles of QCD.

To calculate the soft contribution to the form factor one should know something like the light-cone wave function $\psi(x, k_\perp)$ depending also on the transverse momentum $k_\perp$. I say "something like", because in principle it is impossible to prove that the soft term reduces in QCD to the convolution

$$F(q^2) \sim \int \psi_P(x, k_\perp)\psi_P^*(x, k_\perp + xq)\ d^2k_\perp dx. \quad (2)$$

In fact, one should treat the soft term as a nonfactorizable amplitude and the problem is to calculate it in some reliable way.

The QCD sum rules are just giving the method to calculate the nonperturbative hadronic characteristics incorporating the asymptotic freedom property of QCD in a way similar to pQCD.

1.3 Basic ideas of the QCD sum rule approach

Among the existing approaches to the analysis of the nonperturbative effects in QCD the closest to perturbative QCD is the QCD sum rule method [1]. Let us formulate its basic ideas within the context of the pion form factor problem.

It is evident that one cannot directly study the soft contribution with the on-shell pions, because then only long distances are involved. But perturbative QCD can be applied in a situation when all relevant momenta q, p_1, p_2 are spacelike and sufficiently large: $|q^2|$, $|p_1^2|$, $|p_2^2| > 1\,\mathrm{GeV}^2$. To describe the virtual pions one should use some interpolating field, the usual (for the QCD sum rule practitioners) choice being the axial current $j_5^\alpha = \bar{d}\gamma_5\gamma^\alpha u$. Its projection onto the pion state $|P,\pi\rangle$ is proportional to f_π :

$$\langle 0|j_5^\alpha|P,\pi\rangle = if_\pi P^\alpha. \tag{3}$$

Via the dispersion relation

$$T(p_1^2,p_2^2,q^2) = \frac{1}{\pi^2}\int_0^\infty ds_1 \int_0^\infty ds_2\, \frac{\rho(s_1,s_2,q^2)}{(s_1-p_1^2)(s_2-p_2^2)} \tag{4}$$

one can relate the amplitude $T(p_1^2,p_2^2,q^2)$ to its time-like counterpart $\rho(s_1,s_2,q^2)$ containing the double pole term

$$\rho_\pi(s_1,s_2,q^2) = \pi^2 f_\pi^2 \delta(s_1-m_\pi^2)\delta(s_2-m_\pi^2)F_\pi(Q^2) \tag{5}$$

corresponding to the pion form factor. However, the axial current has nonzero projections onto other hadronic states (A_1-meson, say) as well, and the spectral density $\rho(s_1,s_2,q^2)$ contains also the part $\rho^{higher\ states}(s_1,s_2,q^2)$ related to other elastic and transition form factors. This is the price for going off the pion mass shell. The problem now is to pick out the F_π term from the whole mess.

Of course, calculating $T(p_1^2,p_2^2,q^2)$ in the lowest orders of perturbation theory one never observes something like the pion pole: one obtains a smooth function $\rho^{pert}(s_1,s_2,q^2)$ corresponding to transitions between the free-quark $\bar{u}d-$states with invariant masses s_1 and s_2, respectively. The difference between "exact" density $\rho(s_1,s_2,q^2)$ and its perturbative analog $\rho^{pert}(s_1,s_2,q^2)$ is reflected by additional nonperturbative contributions to $T(p_1^2,p_2^2,q^2)$. These

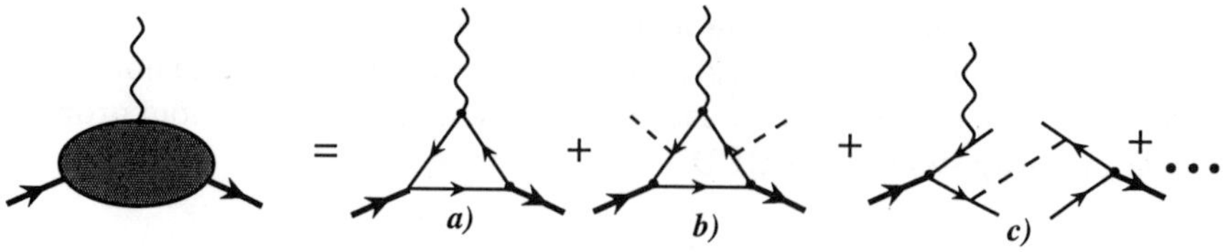

Figure 2: Structure of the operator product expansion for $T(p_1^2, p_2^2, q^2)$

contributions are due to quark and gluon condensates $\langle \bar{q}q \rangle$, $\langle GG \rangle$ etc., describing (and/or parameterizing) the nontrivial structure of the QCD vacuum state. Formally, these terms appear from the operator product expansion for the amplitude $T(p_1^2, p_2^2, q^2)$ (Fig.2):

$$T(p_1^2, p_2^2, q^2) = T^{pert}(p_1^2, p_2^2, q^2) + a\frac{\langle GG \rangle}{(p^2)^3} + b\frac{\alpha_s \langle \bar{q}q \rangle^2}{(p^2)^4} + \dots . \tag{6}$$

The problem now is to construct such a model of the spectral density $\rho(s_1, s_2, q^2)$ which gives the best agreement between two expressions for T, i.e., Eqs. (4) and (6). Naturally, having only a few first terms of the $1/p^2$-expansion one can hope to reproduce only the gross features of the hadronic spectrum in the relevant channel. Still, just using the simple fact that the condensate contributions die out for large p^2, one obtains the *global duality* relation between quark and hadronic densities

$$\int_0^\infty ds_1 \int_0^\infty ds_2 \left(\rho(s_1, s_2, q^2) - \rho^{pert}(s_1, s_2, q^2)\right) = 0. \tag{7}$$

In other words, integrally the two densities are rather similar. One can even expect that they are very close for high s to secure the convergence of the integral (7).

The standard ansatz is to approximate the higher states contribution into $\rho(s_1, s_2, q^2)$ by the free quark density (compare to the quasiclassical approximation for high levels in quantum mechanics):

$$\rho^{higher\ states}(s_1, s_2, q^2) = [1 - \theta(s_1 < s_0)\theta(s_2 < s_0)]\, \rho^{pert}(s_1, s_2, q^2), \tag{8}$$

with s_0 being the effective threshold for the higher states production.

1.4 Summary

To summarize, the basic ingredients of the QCD sum rule approach are

• Operator product expansion, or factorization of the short-distance, perturbatively calculable amplitudes, with the long-distance information accumulated by the quark and gluon condensates – the basic parameters of the approach – characterizing the properties of the QCD vacuum.

• Dispersion relations, relating the perturbatively calculated amplitude with the physical hadronic parameters.

• To extract the hadronic parameters one should construct an ansatz for the hadronic spectrum (like "first resonance plus continuum", etc.) and then perform fitting procedure.

Just from the above list it is clear that QCD sum rules cannot produce exact results. Their precision is unavoidably limited by the following facts :

• One can calculate normally only few first terms of the OPE.

• The model ansatz reproduces the real density only approximately.

• Furthermore, in some channels could appear nonperturbative short-distance contributions ("direct instantons").

As a result, using the QCD sum rules, one can normally expect only 10% - 20% accuracy.

2 QCD Sum Rules in Quantum Mechanics

We are going to apply the QCD sum rule method in the theory, where no exact results for hadrons were obtained so far. But, as formulated above, the QCD sum rule method is quite general. In fact, many features of the method can be demonstrated on the exactly solvable model of the quantum-mechanical oscillator. In the discussion below, we basically follow the pioneering paper[7], with some simplifying modifications[8,9].

2.1 Sum rules for the ground state of the two-dimensional oscillator.

Let us consider the simplest example of a system with confinement, the oscillator, with the confining potential being

$$V(\vec{x}) = \frac{m\omega^2}{2} r^2. \tag{9}$$

Formulas are the simplest (without square roots, complicated factorial terms, etc.) if we choose the oscillator in two spatial dimensions (+ time dimension). In this case the energy levels are given by

$$E_n = (2n+1)\omega, \tag{10}$$

and the values of the wave functions at the origin (quantum-mechanical analogues of f_π and other leptonic decay parameters) are the same for all levels corresponding to radial excitations with zero angular momentum:

$$|\psi_n(0)|^2 = \frac{m\omega}{\pi}. \tag{11}$$

To calculate E_0 and $\psi_0(0)$ using the sum rule approach one should proceed in a rather peculiar way: instead of concentrating on the lowest level, one should consider the weighted sum of all levels:

$$M(\epsilon) = \sum_{k=0}^{\infty} |\psi_k(0)|^2 e^{-E_k/\epsilon} \tag{12}$$

in which the lowest level dominates only for very small values of ϵ. For large ϵ all levels are essential. This combination (12) is directly related to the two-time Green function

$$G(x_1, t_1|x_2, t_2) = \sum_{k=0}^{\infty} \psi_k^*(x_2)\psi_k(x_1)e^{-iE_k(t_2-t_1)} \tag{13}$$

describing the probability amplitude for transition from the point x_1 at time t_1 to the point x_2 at time t_2. To get $M(\epsilon)$ one should take $x_2 = x_1 = 0$, $t_1 = 0$ and $t_2 = 1/i\epsilon$. In other words, the transition is taken in the imaginary time.

In our case, explicit form of $M(\epsilon)$ can be easily calculated using Eqs. (10) and (11):

$$M^{osc}(\epsilon) = \sum_{k=0}^{\infty} \left(\frac{m\omega}{\pi}\right) e^{-(2k+1)\omega/\epsilon} = \frac{m\omega}{\pi \sinh(\omega/\epsilon)} \tag{14}$$

In the large-ϵ limit one has

$$M^{osc}(\epsilon) = \frac{m\epsilon}{2\pi}\left(1 - \frac{\omega^2}{6\epsilon^2} + \frac{7}{360}\frac{\omega^4}{\epsilon^4} - \frac{31}{15120}\frac{\omega^6}{\epsilon^6} + \ldots\right) \tag{15}$$

i.e., the first term of the large-ϵ ($\equiv$small-time) expansion has no dependence on ω, the paramater characterizing the strength of oscillator potential. This simply means that at small time differences the Green function coincides with the free one:

$$G^{free}(2|1) = \frac{m}{2\pi i(t_2 - t_1)} \exp\left\{-i\frac{m(x_2 - x_1)^2}{2(t_2 - t_1)}\right\} \tag{16}$$

and

$$M^{free}(\epsilon) = \frac{m\epsilon}{2\pi} \tag{17}$$

The corrections to the free-particle behavior in Eq. (15) are in powers of ω^2. This corresponds to the perturbation expansion in the potential $V(r)$ (9):

$$M = M^{free} + M^{free} \otimes V \otimes M^{free} + M^{free} \otimes V \otimes M^{free} \otimes V \otimes M^{free} + \ldots \quad (18)$$

The function $M^{free}(\epsilon)$ can also be represented as a sum over the (free) states with $E_k = k^2/2m$:

$$M^{free}(\epsilon) = \int \frac{d^2k}{(2\pi)^2} e^{-E_k/\epsilon} = \frac{m}{2\pi} \int_0^\infty e^{-E/\epsilon} dE \equiv \frac{1}{\pi} \int_0^\infty e^{-E/\epsilon} \rho(E) dE \; . \quad (19)$$

We introduced here a new function

$$\rho^{free}(E) = \frac{m}{2} \theta(E > 0), \quad (20)$$

the spectral density corresponding to the free motion, which will play an important role in the subsequent discussion. The oscillator function $M^{osc}(\epsilon)$ also can be written in the spectral representation with the spectral density $\rho^{osc}(E)$ which is in this case a superposition of the delta-functions:

$$\rho^{osc}(E) = m\omega \sum_{k=0}^{\infty} \delta(E - (2k+1)\omega). \quad (21)$$

At first sight, the two densities are completely different. However, if one integrates them from zero to the midpoint between the first and second levels, one obtains the same result:

$$\frac{1}{m} \int_0^{2\omega} \rho^{free}(E) dE = \omega = \frac{1}{m} \int_0^{2\omega} \rho^{osc}(E) dE \; . \quad (22)$$

Moreover, similar relations hold for other levels:

$$\int_{2k\omega}^{2(k+1)\omega} \rho^{free}(E) dE = m\omega = \int_{2k\omega}^{2(k+1)\omega} \rho^{osc}(E) dE \; . \quad (23)$$

The equality is not violated even if one multiplies $\rho(E)$ by an extra power of the energy E:

$$\int_{2k\omega}^{2(k+1)\omega} \rho^{free}(E)\, E dE = m\omega^2(2k+1) = \int_{2k\omega}^{2(k+1)\omega} \rho^{osc}(E)\, E dE \; . \quad (24)$$

Thus, there exists the *local duality* between each resonance and the free states. It looks like each level "takes" its part of the free spectral density, leaving the integral unchanged.

Turning back to the ω-expansion for $M^{osc}(\epsilon)$ (Eq. (15))

$$M^{osc}(\epsilon) = M^{free}(\epsilon)\left\{1 - \frac{\omega^2}{6\epsilon^2} + \frac{7}{360}\frac{\omega^4}{\epsilon^4} - \frac{31}{15120}\frac{\omega^6}{\epsilon^6} + \dots\right\} \tag{25}$$

one can write

$$M^{osc}(\epsilon) - M^{free}(\epsilon) = O(\omega^2/\epsilon) + O(\omega^4/\epsilon^3) + \dots . \tag{26}$$

or, in terms of the spectral densities

$$\frac{1}{\pi}\int_0^\infty \left[\rho^{osc}(E) - \rho^{free}(E)\right] e^{-E/\epsilon} dE = \sum_{k=1} \frac{A_k}{\epsilon^k} . \tag{27}$$

The terms on the rhs, suppressed for large ϵ by powers of $1/\epsilon$ are normally referred to as "power corrections". Thus, the difference between the integrated spectral densities vanishes as $\epsilon \to \infty$:

$$\frac{1}{\pi}\int_0^\infty [\rho^{osc}(E) - \rho^{free}(E)] dE = 0\,, \tag{28}$$

since there is no $O(1)$ term in the rhs of Eq. (27). Eq. (28) is a standard *global duality* relation. In quantum mechanics one can obtain such a relation for any potential regular at the origin.

The lesson is that the two densities, despite the apparent absence of similarity, are rather close to each other in an integral sense. Now, the idea is to incorporate this similarity in the sum rule

$$|\psi_0(0)|^2 e^{-E_0/\epsilon} + \text{“}higher\ states\text{”} = \frac{m\epsilon}{2\pi}\left\{1 - \frac{\omega^2}{6\epsilon^2} + \frac{7}{360}\frac{\omega^4}{\epsilon^4} - \frac{31}{15120}\frac{\omega^6}{\epsilon^6} + \dots\right\} \tag{29}$$

by substituting the "higher states" contribution by that of the free states having energy above some threshold s_0. The sum rule then reads

$$|\tilde{\psi}_0(0)|^2 e^{-E_0/\epsilon} \approx \frac{\epsilon}{2}\left(1 - e^{-s_0/\epsilon}\right) - \frac{\omega^2}{12\epsilon} + \frac{7}{720}\frac{\omega^4}{\epsilon^3} - \frac{31}{30240}\frac{\omega^6}{\epsilon^5} + \dots \tag{30}$$

where

$$|\tilde{\psi}_0(0)|^2 = \frac{\pi}{m}|\psi_0(0)|^2$$

The parameters to be extracted from the sum rule are $E_0, \tilde{\psi}_0(0)$ and s_0. The first step is to extract E_0. To this end we differentiate Eq. (30) with respect to $(-1/\epsilon)$ to get

$$|\tilde{\psi}_0(0)|^2 E_0 e^{-E_0/\epsilon} \approx \frac{\epsilon^2}{2} - \frac{\epsilon^2}{2}\left(1 + \frac{s_0}{\epsilon}\right) e^{-s_0/\epsilon} + \frac{\omega^2}{12} - \frac{7}{240}\frac{\omega^4}{\epsilon^2} + \frac{31}{6048}\frac{\omega^6}{\epsilon^4} + \dots \tag{31}$$

The ratio of the two equations just gives the expression for E_0:

$$E_0 \approx \frac{\frac{\epsilon^2}{2} - \frac{\epsilon^2}{2}(1+\frac{s_0}{\epsilon})e^{-s_0/\epsilon} + \frac{\omega^2}{12} - \frac{7}{240}\frac{\omega^4}{\epsilon^2} + \frac{31}{6048}\frac{\omega^6}{\epsilon^4} + \ldots}{\frac{\epsilon}{2}(1-e^{-s_0/\epsilon}) - \frac{\omega^2}{12\epsilon} + \frac{7}{720}\frac{\omega^4}{\epsilon^3} - \frac{31}{30240}\frac{\omega^6}{\epsilon^5} + \ldots}\,. \tag{32}$$

If we would take the whole series in ω/ϵ and the exact expression for the higher states contribution, then the result for E_0 would be the ϵ-independent exact value $E_0 = \omega$. However, under approximations we made, the E_0-value extracted from Eq. (32) depends both on the auxiliary parameter ϵ and the threshold parameter s_0. The standard strategy is to take the value of s_0, for which E_0 has the minimal dependence on ϵ in the intermediate region $1 < \epsilon/\omega < 2$ where both the higher power corrections and the error due to the rough model of the higher states contribution are small ($\approx 5-10\%$). Truncating the series by terms of the second order in ω^2, one obtains[a] from such a fitting $s_0 = 1.6\omega$ and $E_0 = 0.9\omega$. Adding the third order terms improves the result: $s_0 = 1.75\omega$ and $E_0 = 0.95\omega$. In our case one can substitute into the sum rule the all-order result. The remaining ϵ-dependence is then entirely due to the use of an approximate model for the higher states. For sufficiently small values of ϵ all the curves give the exact value $E_0 = 2\omega$. The flattest curves are those with $s_0 \sim 2\omega$.

The second step is to extract the $\tilde{\psi}(0)$ value from the original sum rule (30)

$$|\tilde{\psi}_0(0)|^2 \approx e^{E_0/\epsilon}\left\{\frac{\epsilon}{2}(1-e^{-s_0/\epsilon}) - \frac{\omega^2}{12\epsilon} + \frac{7}{720}\frac{\omega^4}{\epsilon^3} - \frac{31}{30240}\frac{\omega^6}{\epsilon^5} + \ldots\right\}. \tag{33}$$

The analysis goes in a similar way. If one takes the third-order value $E_0 = 0.95\,\omega$, one obtains $|\tilde{\psi}_0(0)|^2 \sim 0.9\,\omega$. In the $\epsilon \to \infty$ limit one obtains the local duality relation

$$\{|\tilde{\psi}_0(0)|^2\}^{LD} = \frac{s_0}{2}. \tag{34}$$

So, the "exact value" for the duality interval s_0 is 2ω.

Summarizing, using the sum rule method one can determine the parameters of the lowest resonance with $\approx 10\%$ accuracy. The main source of the error is the use of a rather rough model for the spectral density of the higher states. It should be noted that in QCD situation is better: normally only the first resonance is narrow, while higher states are broader and broader, so that approximating them by free quark functions does not lead to large errors.

[a] Plotting E_0 for various values of s_0 is absolutely straightforward, and we encourage the readers perform the fitting as an exercise.

2.2 Sum rules for form factors of the oscillator states.

To calculate the form factors

$$F_{kl}(Q) = \int \psi_k^*(x)\psi_l(x)e^{iQx}\, d^2x \tag{35}$$

within the sum rule approach, one should construct a similar integral using the Green functions (13)

$$G(\,x,\, 1/i\epsilon \,|\, 0,\, 0\,) = \sum_{k=0}^{\infty} \psi_k^*(x_2)\psi_k(x_1)e^{-E_k/\epsilon}\,. \tag{36}$$

This gives the amplitude

$$\begin{aligned} M(\epsilon_1, \epsilon_2, Q^2) &= \int G(\,x,\, 1/i\epsilon_1 \,|\, 0,\, 0\,)\, G^*(\,x,\, 1/i\epsilon_2 \,|\, 0,\, 0\,)e^{iQx}\, d^2x \\ &= \sum_{k,l=0}^{\infty} \psi_l^*(0)\psi_k(0) \left[\int e^{iQx}\psi_k^*(x)\psi_l(x)d^2x\right] e^{-E_k/\epsilon_1 - E_l/\epsilon_2} \\ &= \sum_{k,l=0}^{\infty} \psi_l^*(0)\psi_k(0)F_{kl}(Q^2)e^{-E_k/\epsilon_1 - E_l/\epsilon_2} \end{aligned} \tag{37}$$

One can write this amplitude in the (double) spectral representation:

$$M(\epsilon_1, \epsilon_2, Q^2) = \frac{1}{\pi^2}\int_0^\infty \int_0^\infty e^{-s_1/\epsilon_1 - s_2/\epsilon_2}\rho(s_1, s_2, Q^2)ds_1 ds_2 \tag{38}$$

where the spectral function is the sum of double delta-functions:

$$\rho(s_1, s_2, Q^2) = \pi^2 \sum_{k,l=0}^{\infty} \psi_l^*(0)\psi_k(0)\, F_{kl}(Q)\, \delta(s_1 - E_k)\, \delta(s_2 - E_l). \tag{39}$$

In the free case one has

$$G^{free}(\,x,\, 1/i\epsilon \,|\, 0,\, 0) = \frac{m\epsilon}{2\pi}e^{-mx^2\epsilon/2} = \int \frac{d^2k}{(2\pi)^2}e^{ikx}e^{-k^2/2m\epsilon} \tag{40}$$

for the Green function and

$$\begin{aligned} M^{free}(\epsilon_1, \epsilon_2, Q^2) &= \int \frac{d^2k}{(2\pi)^2}e^{-k^2/2m\epsilon_1 - (k+Q)^2/2m\epsilon_2} \\ &= \left(\frac{m}{2\pi}\right)\frac{\epsilon_1\epsilon_2}{\epsilon_1 + \epsilon_2}\exp\left(-\frac{Q^2}{2m(\epsilon_1 + \epsilon_2)}\right) \end{aligned} \tag{41}$$

for the amplitude. The free spectral function $\rho^{free}(s_1, s_2, Q^2)$ can be easily extracted:

$$\rho^{free}(s_1, s_2, Q^2) = \pi^2 \int \frac{d^2k}{(2\pi)^2}\, \delta\left(s_1 - \frac{k^2}{2m}\right) \delta\left(s_2 - \frac{(k+Q)^2}{2m}\right)$$
$$= \frac{\theta(|\sqrt{2ms_1} + \sqrt{2ms_2}| \geq |Q| \geq |\sqrt{2ms_1} - \sqrt{2ms_2}|)}{\sqrt{4s_1s_2 - (s_1 + s_2 - Q^2/2m)^2}}. \quad (42)$$

The restriction on s_1, s_2 is in fact the triangle relation

$$|k_1| + |k_2| \geq |Q| \geq ||k_1| - |k_2||$$

between the initial momentum k_1 ($|k_1| = \sqrt{2ms_1}$), the momentum transfer Q and the final momentum $k_2 = (k_1 + Q)$ ($|k_2| = \sqrt{2ms_2}$).

The oscillator Green function is also known:

$$G^{osc}(x,\, 1/i\epsilon\,|\,0,\, 0) = \frac{m\omega}{2\pi \sinh(\omega/\epsilon)} \exp\left(-\frac{m\omega \vec{x}^2 \cosh(\omega/\epsilon)}{2\sinh(\omega/\epsilon)}\right) \quad (43)$$

and using it, one can calculate the amplitude

$$M(\epsilon_1, \epsilon_2, Q^2) = \frac{m\omega}{2\pi \sinh(\omega/\epsilon_1 + \omega/\epsilon_2)} \exp\left(-\frac{Q^2}{2m\omega} \frac{\sinh(\omega/\epsilon_1)\sinh(\omega/\epsilon_2)}{\sinh(\omega/\epsilon_1 + \omega/\epsilon_1)}\right). \quad (44)$$

Now, representing M as

$$M(\epsilon_1, \epsilon_2, Q^2) = \frac{m\omega}{\pi} \sum_{k=0}^{\infty} e^{-(2k+1)\omega(1/\epsilon_1 + 1/\epsilon_2)} \exp\left(-\frac{Q^2}{4m\omega}\right)$$
$$\exp\left(\frac{Q^2}{4m\omega}\left[1 - \frac{(1 - e^{-\omega/\epsilon_1})(1 - e^{-\omega/\epsilon_2})}{(1 - e^{-\omega/\epsilon_1} e^{-\omega/\epsilon_2})}\right]\right) \quad (45)$$

and expanding M in powers of $e^{-\omega/\epsilon_1} \equiv \mathcal{E}_1$ and $e^{-\omega/\epsilon_2} \equiv \mathcal{E}_2$ one can extract the form factors $F_{lk}(Q^2)$:

$$M(\epsilon_1,\, \epsilon_2,\, Q^2) = \frac{m\omega}{\pi} \sum_{k=0}^{\infty} \mathcal{E}_1^{2k+1} \mathcal{E}_2^{2k+1} \exp\left(-\frac{Q^2}{4m\omega}\right)$$
$$\left\{1 + \frac{Q^2}{4m\omega}\left[1 - \frac{(1-\mathcal{E}_1)(1-\mathcal{E}_2)}{1 - \mathcal{E}_1\mathcal{E}_2}\right] + \frac{1}{2}\left(\frac{Q^2}{4m\omega}\right)^2 [\ldots]^2 + \ldots\right\} \quad (46)$$

identifying them as the factors in front of the $\mathcal{E}_1^k \mathcal{E}_2^l$ terms:

$$F_{00}(Q^2) = \exp\left(-\frac{Q^2}{4m\omega}\right),$$

$$F_{01}(Q^2) = F_{10}(Q^2) = \frac{Q^2}{4m\omega}\exp\left(-\frac{Q^2}{4m\omega}\right),$$
$$F_{11}(Q^2) = \left(1-\frac{Q^2}{4m\omega}\right)^2\exp\left(-\frac{Q^2}{4m\omega}\right) \tag{47}$$

It is instructive to write down the amplitude $M(\epsilon_1, \epsilon_2, Q^2)$ for $Q^2 = 0$:

$$M(\epsilon_1, \epsilon_2, Q^2 = 0) = \frac{m\omega}{2\pi \sinh(\omega/\epsilon_1 + \omega/\epsilon_2)} = \frac{m\omega}{\pi}\sum_{k=0}^{\infty} \mathcal{E}_1^{2k+1}\mathcal{E}_2^{2k+1}\,. \tag{48}$$

It is easy to notice that
• There are no terms with $k \neq l$: only the diagonal transitions are allowed in this limit
• The numerical factor is just the same as in Eq. (14) for $M^{osc}(\epsilon)$. This means that $F_{kk}(0) = 1$ for all k, just as required by charge conservation.

Next step is to get the perturbative expansion in powers of ω :

$$\begin{aligned}
M(\epsilon_1, \epsilon_2, Q^2) &= \frac{m\epsilon_1\epsilon_2}{2\pi(\epsilon_1+\epsilon_2)}\left\{\left[1-\frac{\omega^2}{6}\left(\frac{1}{\epsilon_1}+\frac{1}{\epsilon_2}\right)^2+\ldots\right]\right. \\
&\quad \left.\times\exp\left(-\frac{Q^2}{2m}\frac{\frac{1}{\epsilon_1}(1+\frac{1}{6}\frac{\omega^2}{\epsilon_1^2}+\ldots)\frac{1}{\epsilon_2}(1+\frac{1}{6}\frac{\omega^2}{\epsilon_2^2}+\ldots)}{(\frac{1}{\epsilon_1}+\frac{1}{\epsilon_2})(1+\frac{\omega^2}{6}(\frac{1}{\epsilon_1}+\frac{1}{\epsilon_2})^2+\ldots)}\right)\right\} \\
&= \frac{m\epsilon_1\epsilon_2}{2\pi(\epsilon_1+\epsilon_2)}\left\{\left[1-\frac{\omega^2}{6}\left(\frac{1}{\epsilon_1}+\frac{1}{\epsilon_2}\right)^2+\frac{7\omega^4}{360}\left(\frac{1}{\epsilon_1}+\frac{1}{\epsilon_2}\right)^4+\ldots\right]\right. \\
&\quad \exp\left(-\frac{Q^2}{2m(\epsilon_1+\epsilon_2)}\right)\left[1+\frac{Q^2\omega^2}{6m\epsilon_1\epsilon_2(\epsilon_1+\epsilon_2)}\right. \\
&\quad \left.\left.-\frac{Q^2\omega^4[(\epsilon_1+\epsilon_2)^2+2\epsilon_1\epsilon_2]}{90m\epsilon_1^3\epsilon_2^3(\epsilon_1+\epsilon_2)}+\frac{Q^4\omega^4}{72m^2\epsilon_1^2\epsilon_2^2(\epsilon_1+\epsilon_2)^2}+\ldots\right]\right\} \\
&= M^{free}(\epsilon_1, \epsilon_2, Q^2)\left\{1+\sum_{k=1}^{\infty}(\omega^2)^k\varphi_k(\epsilon_1, \epsilon_2, Q^2)\right\}
\end{aligned} \tag{49}$$

The sum rule now reads

$$\begin{aligned}
&|\psi_0(0)|^2 F_{00}(Q^2)e^{-E_0/\epsilon_1 - E_0/\epsilon_2} + \text{“}higher\ states\text{”} \\
&\quad = M^{free}(\epsilon_1, \epsilon_2, Q^2)\left\{1+O\left(\frac{\omega^2}{\epsilon^2}\right)+O\left(\frac{\omega^4}{\epsilon^4}\right)+\ldots\right\}
\end{aligned} \tag{50}$$

To proceed further, one should take some ansatz for the higher states contribution. The ansatz similar to that used earlier is to assume that

$$\text{"}higher\ states\text{"} = \text{"}free\ states\text{"}$$

outside the square $(0, s_0) \otimes (0, s_0)$. Then the sum rule has the form

$$\begin{aligned} &|\psi_0(0)|^2 F_{00}(Q^2) e^{-E_0/\epsilon_1 - E_0/\epsilon_2} \\ &\quad = \frac{1}{\pi^2} \int_0^{s_0} ds_1 \int_0^{s_0} ds_2\, \rho^{free}(s_1, s_2, Q^2) e^{-s_1/\epsilon_1 - s_2/\epsilon_2} \\ &\qquad + O\left(\frac{\omega^2}{\epsilon}\right) + O\left(\frac{\omega^4}{\epsilon^3}\right) + \dots . \end{aligned} \tag{51}$$

In the $\epsilon_1, \epsilon_2 \to \infty$ limit one obtains the local duality relation:

$$|\psi_0(0)|^2 F_{00}(Q^2) = \frac{1}{\pi^2} \int_0^{s_0} \int_0^{s_0} \rho^{free}(s_1, s_2, Q^2) ds_1 ds_2 \tag{52}$$

with $\rho^{free}(s_1, s_2, Q^2)$ given by Eq. (42):

$$\rho^{free}(s_1, s_2, Q^2) = \frac{1}{4} \int d^2k\, \delta\left(s_1 - \frac{k^2}{2m}\right) \delta\left(s_2 - \frac{(k+Q)^2}{2m}\right) . \tag{53}$$

Using the two expressions given above, one can write

$$|\psi_0(0)|^2 F_{00}^{LD}(Q^2) = \frac{1}{(2\pi)^2} \int d^2k\, \theta\left(\frac{k^2}{2m} < s_0\right) \theta\left(\frac{(k+Q)^2}{2m} < s_0\right) . \tag{54}$$

This integral can be easily calculated if one notices that it defines the overlap area of two circles having the same radius $R^2 = 2ms_0$, with their centers separated by the distance equal to Q:

$$|\psi_0(0)|^2 F_{00}^{LD}(Q^2) = \frac{ms_0}{\pi^2} \left[\cos^{-1}(\kappa) - \kappa\sqrt{1-\kappa^2}\right] \theta(Q^2 < 8ms_0) , \tag{55}$$

where $\kappa = Q/\sqrt{8ms_0}$. Recall now that from the local duality relation for $M(\epsilon)$ we had

$$\{|\psi_0(0)|^2\}^{LD} = \frac{ms_0}{2\pi} \tag{56}$$

and this gives the result

$$F_{00}^{LD}(Q^2) = \frac{2}{\pi} \left[\cos^{-1}(\kappa) - \kappa\sqrt{1-\kappa^2}\right] \theta(Q^2 < 8ms_0) \tag{57}$$

having the correct normalization $F_{00}^{LD}(Q^2) = 1$.

The local duality prediction calculated for the "exact value" $s_0 = 2\omega$ of the duality interval

$$F_{00}^{LD}(Q^2)|_{s_0=2\omega} = \frac{2}{\pi}\left[\cos^{-1}\left(\frac{Q}{\sqrt{16m\omega}}\right)\right.$$
$$\left. - \frac{Q}{\sqrt{16m\omega}}\sqrt{1-\frac{Q^2}{16m\omega}}\right]\theta(Q^2 < 16m\omega) \tag{58}$$

is very close to the exact form factor

$$F_{00}^{exact}(Q^2) = \exp\left(-\frac{Q^2}{4m\omega}\right). \tag{59}$$

There is a very simple physics behind the local duality. One should just compare the local duality prescription for the form factor

$$F_{00}^{LD}(Q^2) = \frac{1}{|\psi_0(0)|^2}\int \frac{d^2k}{(2\pi)^2}\,\theta\left(\frac{k^2}{2m} < s_0\right)\theta\left(\frac{(k+Q)^2}{2m} < s_0\right) \tag{60}$$

and the exact formula

$$F_{00}(Q^2) = \int \frac{d^2k}{(2\pi)^2}\,\psi_0(k)\psi_0^*(k+Q) \tag{61}$$

to realize that the local duality, in fact, is equivalent to using a model form for the wave function in the momentum space:

$$\tilde{\psi}_0^{LD}(k) = \frac{1}{|\psi_0(0)|}\,\theta(k^2 < 2ms_0) \tag{62}$$

or, taking $s_0 = 2\omega$:

$$\tilde{\psi}_0^{LD}(k) = \sqrt{\frac{\pi}{m\omega}}\,\theta(k^2 < 4m\omega), \tag{63}$$

to be compared with the exact wave function

$$\tilde{\psi}_0(k) = 2\sqrt{\frac{\pi}{m\omega}}\exp\left(-\frac{k^2}{2m\omega}\right). \tag{64}$$

It is easy to establish that the model wave function correctly reproduces the simplest integral properties of the exact wave function:

$$\int \tilde{\psi}^{exact}(k)d^2k = \int \tilde{\psi}^{LD}(k)d^2k \tag{65}$$

$$\int |\tilde{\psi}^{exact}(k)|^2 d^2k = \int |\tilde{\psi}^{LD}(k)|^2 d^2k. \tag{66}$$

3 Condensates and Operator Product Expansion

3.1 Correlators

The basic object in the quantum mechanical example was the Green function $G(x_1, t_1 \,|\, x_2, t_2)$ describing the propagation of a particle from one point to another. In fact, it was sufficient to analyze the probability amplitude for the particle to remain at the same spatial point after some imaginary time elapsed:

$$M(\epsilon) = G(0, 0 \,|\, 0, 1/i\epsilon) \; . \tag{67}$$

The question is what should be its analog in the quantum field theory? Recall, that the motion in the external potential corresponds to a two-particle system with one of the particles being infinitely massive. If this particle moves, the other moves together with it. So, the Green function describes really a collective motion of two particles from one point to another. In quantum field theory this corresponds to the Green-like functions

$$\langle 0|T\,(j(x)\,j(y)\,)|\,0\rangle$$

normally referred to as correlators, with the currents $j(x)$ being the products of fields at the same point, *e.g.* $J^\mu(x) = \bar{q}(x)\gamma^\mu q(x)$. Due to the translation invariance

$$\langle 0|T\,(j(x)\,j(y)\,)|\,0\rangle \to \langle 0|T\,(j(x-y)\,j(0)\,)|\,0\rangle$$

it is always possible and convenient to take one point at the origin. In QFT, it is also more convenient to consider correlators in the momentum representation:

$$\Pi(q) = \int e^{iqx}\,\langle 0|T\,(j(x)\,j(0)\,)|\,0\rangle\, d^4x.$$

In the lowest order the correlator is given by the product of the relevant propagators, but there are also radiative corrections, which one can calculate in perturbative QCD, at least in the region where q is spacelike $q^2 < 0$ and large enough.

Next step is to relate $\Pi(q)$ to a sum over the physical states. This is performed *via* the dispersion relation

$$\Pi(q^2) = \frac{1}{\pi}\int_0^\infty \frac{\rho(s)}{s-q^2}ds + \text{“}\textit{subtractions}\text{”}.$$

The subtraction terms appear if the spectral density is such that the dispersion integral diverges. If, say, $\rho(s) \sim s^n$, one needs $(n+1)$ subtractions:

$$\Pi(q^2) = \frac{1}{\pi}\int_0^\infty \frac{\rho(s)}{s-q^2}ds$$

$$- \frac{1}{\pi}\int_0^\infty ds\,\rho(s)\Big\{\frac{1}{s-\lambda^2} + \frac{q^2-\lambda^2}{(s-\lambda^2)^2} + \ldots + \frac{(q^2-\lambda^2)^n}{(s-\lambda^2)^{n+1}}\}$$

$$+ \{\Pi(\lambda^2) + (q^2-\lambda^2)\Pi'(\lambda^2) + \ldots + \frac{(q^2-\lambda^2)^n}{n!}\Pi^{(n)}(\lambda^2)\Big\}. \quad (68)$$

After collecting all the terms containing $\rho(s)$, one observes that $\rho(s)$ is divided by $(s-q^2)(s-\lambda^2)^{n+1}$, and the integral converges. Note, that all the subtraction terms are polynomials of a finite order in Q^2 (from now on we use the notation $q^2 = -Q^2$, with $Q^2 > 0$).

3.2 Borel Transformation

Note, that in the oscillator studies, we dealt with the exponentially weighted sum:

$$M(\epsilon) = \frac{1}{\pi}\int_0^\infty e^{-s/\epsilon}\rho^{osc}(s)\,ds. \quad (69)$$

We observed then that it has a nice property that for ϵ values of an order of ω, both the contributions of the higher states were strongly damped and the perturbative expansion for $M(\epsilon)$ was converging fast enough. With the power weight implied by the dispersion relation, the suppression of the higher states contribution is not so effective.

However, it is rather easy to transform the original dispersion integral into the exponentially weighted one. To this end, one should apply the so-called Borel transformation, formally defined by the formula[1]

$$\Phi(M^2) = \lim_{\substack{Q^2,n\to\infty \\ Q^2/n=M^2}} \frac{1}{(n-1)!}(Q^2)^n\left[-\frac{d}{dQ^2}\right]^n \Pi(Q^2). \quad (70)$$

It is easy to check that applying the above procedure to $1/(s+Q^2)$ one really obtains an exponential. Indeed,

$$\left[-\frac{d}{dQ^2}\right]^n \frac{1}{s+Q^2} = \frac{n!}{(s+Q^2)^{n+1}}. \quad (71)$$

Multiplying the result by $\frac{(Q^2)^n}{(n-1)!}$ one obtains

$$n\,\frac{(Q^2)^n}{(s+Q^2)^{n+1}} = \frac{1}{Q^2/n}\frac{1}{(1+s/Q^2)^{n+1}} = \frac{1}{M^2}\frac{1}{(1+s/nM^2)^{n+1}}. \quad (72)$$

Finally, taking the $n\to\infty$ limit one obtains $\frac{1}{M^2}e^{-s/M^2}$. This means that

$$\Phi(M^2) \equiv B(Q^2\to M^2)\,\Pi(Q^2) = \frac{1}{\pi}\int_0^\infty \frac{ds}{M^2}\rho(s)e^{-s/M^2} + \text{“}nothing\text{”} \quad (73)$$

where "nothing" means that all the subtraction terms have disappeared, since no polynomial can survive the infinite number of differentiations. Thus, the function $\Phi(M^2)$ is the QFT analog of $M(\epsilon)$.

In many cases one can perform the Borel transformation using the formula[10]

$$B(Q^2 \to M^2)\{e^{-AQ^2}\} = \delta(1 - AM^2). \tag{74}$$

As an illustration, let us reproduce the result obtained above :

$$\begin{aligned} B\frac{1}{s+Q^2} &= B\int_0^\infty e^{-\alpha(s+Q^2)}\,d\alpha \\ &= \int_0^\infty e^{-\alpha s}\delta(1-\alpha M^2)\,d\alpha = \frac{1}{M^2}e^{-s/M^2}. \end{aligned} \tag{75}$$

Other functions, the Borel transforms of which we will need, are those having a power behavior in $1/Q^2$. In this case we have

$$\begin{aligned} B\frac{1}{(Q^2)^n} &= B\int_0^\infty e^{-\alpha Q^2}\frac{\alpha^{n-1}}{(n-1)!}\,d\alpha \\ &= \int_0^\infty \delta(1-\alpha M^2)\frac{\alpha^{n-1}}{(n-1)!}\,d\alpha = \frac{1}{(n-1)!(M^2)^n}. \end{aligned} \tag{76}$$

Thus, a power expansion for $\Phi(M^2)$ over $1/M^2$ converges much faster than the original expansion of $\Pi(Q^2)$ over $1/Q^2$, since each term is now suppressed factorially. This looks precisely like the Borel improvement of a series, and that is why Shifman, Vainshtein and Zakharov[1] referred to the transformation they introduced as "the Borel transformation". The representation for $\Phi(M^2)$ has the form of the Laplace transformation, and that is why some people call the sum rules obtained in this way "the Laplace transform sum rules". In fact, there is no contradiction: $\Phi(M^2)$ is the Borel transform of $\Pi(Q^2)$ and the Laplace transform of $\rho(s)$.

To summarize, the Borel transformation produces two improvements:

- higher state contributions in the spectral representation for $\Phi(M^2)$ are exponentially suppressed,
- its power series expansion in $1/M^2$ is factorially improved.

3.3 Heavy-light system and the $m \to \infty$ limit

Using the basic formula, it is rather easy to establish[11] that in the limit when one of the particles becomes infinitely massive, the QFT function $\Phi(M^2)$ converts into its quantum-mechanical analog. Consider the lowest-order diagram.

The propagator of the heavy particle is

$$\frac{1}{(q+k)^2 - m^2} = \frac{1}{q^2 - m^2 + 2qk + k^2} .$$

In the $m \to \infty$ limit all the bound state masses are equal to this mass plus some finite quantity. So, it makes sense to subtract this infinite constant. To this end one can represent the total momentum q as

$$q = mv + l$$

where v is the four-velocity. Then

$$q^2 - m^2 = 2m(vl) + l^2$$

The scalar product $(vl) \equiv E$ just has the meaning of the energy variable. There is only one more $O(m)$ term in the propagator:

$$2(qk) = 2m(vk)\{1 + O(1/m)\} .$$

With this accuracy, one can write down the relevant Feynman integral as

$$I(m, v, E) = \int \frac{d^4k}{2m(E + (vk))} S(k) ,$$

where $S(k)$ is the propagator of the light particle. Now, applying the Borel transformation $B(E \to \epsilon)$ which converts $1/(E + (vk))$ into $\frac{1}{\epsilon} e^{-(vk)/\epsilon}$ we get

$$M(m, v, \epsilon) \equiv B(E \to \epsilon) I(m, v, E) = \frac{1}{2m\epsilon} \int e^{-(vk)/\epsilon} S(k) d^4k.$$

It is easy to realize that the integral above is just the light particle propagator in the configuration space

$$M(m, v, \epsilon) = \frac{1}{2m\epsilon} \langle \varphi(0) \varphi(iv/\epsilon) \rangle$$

i.e., the Green function describing propagation in the imaginary time, since the velocity vector v^μ, in the rest frame of the heavy particle, is oriented exactly in the time direction.

Thus, we have demonstrated that the Borel transformed correlators are just the QFT generalization of the imaginary time Green functions we studied in the quantum-mechanical examples.

3.4 Perturbative vs nonperturbative contributions in QCD

From the technical point of view, the strategy now is
• to calculate the correlator $\Pi(Q^2)$ in the deep spacelike region where one can rely on the asymptotic freedom of QCD,
• apply the Borel transformation to the resulting expression to get a $1/M^2$ expansion for $\Phi(M^2)$,
• finally, using the dispersion representation for $\Phi(M^2)$, to construct a sum rule.

The question is, what are the contributions analogous to the ω^2/ϵ^2 corrections?

The standard idea about the QCD potential $V(r)$ is that it consists of a Coulomb-like part $\sim \alpha_s(1/r^2\Lambda^2)$ calculable perturbatively at short distances and a nonperturbative $\sim r$ long distance "confining" part:

$$V(r) = V^{\text{“}Coulomb\text{”}}(r) + V^{conf}(r)$$

The magnitude of the "Coulomb" effects is determined by the QCD running coupling constant

$$\alpha_s(Q^2) = \frac{4\pi}{9\log(Q^2/\Lambda^2)} + \ldots.$$

The standard modern estimate for the value of the QCD scale Λ is

$$\Lambda \sim 200\, MeV\ .$$

As we will see, the parameters that really appear in QCD sum rules have dimension of $(mass)^2$, and one should compare

$$\Lambda^2 \sim 0.04\,\text{GeV}^2$$

to the typical hadronic scales like ρ and nucleon masses (squared):

$$m_\rho^2 \approx 0.6\,\text{GeV}^2, \qquad\qquad m_N^2 \approx 0.9\,\text{GeV}^2.$$

To get such scales from Λ^2 one needs unnaturally large numerical factors of an order of 25 or even 100. So, it is most unlikely that the $O(\alpha_s)$-corrections can generate scales of an order of 1 GeV and , hence, the "Coulomb" part is practically irrelevant to the formation of the spectrum of light hadrons. Our problem now is to take into account the effects due to the long-range nonperturbative part of the QCD potential.

Our study of the quantum-mechanical oscillator (which is the simplest example of a system with confinement) has demonstrated that the propagation amplitude in the presence of the potential is different from the free Green function. Furthermore, we observed that the difference vanishes at short distances and that one can calculate exact Green function perturbatively, expanding in powers of the oscillator potential. In QCD, with two components of $V(r)$ one can imagine a double series, both in $V^{\text{"Coulomb"}}(r)$ and $V^{conf}(r)$. There is no problem to arrange an expansion in $V^{\text{"Coulomb"}}(r)$, – this will be just the standard perturbation theory in the QCD coupling constant. On the other hand, the confining potential $V^{conf}(r)$ in QCD is not even known. Moreover, the widespread belief is that it is determined by some essentially nonperturbative effects, *i.e.*, by those which cannot be seen in perturbation theory.

In such a situation, a possible way out is to proceed as follows:

• to construct perturbation expansion in terms of quark and gluon propagators (this amounts to taking into account only the Coulombic part of the potential),

• to postulate that quark and gluon propagators are modified by the long-range confinement part of the QCD potential; but the modification is soft in a sense that at short distances the difference between exact and perturbative (free-field) propagators vanishes.

To formalize this statement, one can write the exact propagator $\mathcal{D}^{exact}(x)$ as a vacuum average of a T-product of fields in the exact vacuum Ω (corresponding to a confining potential)

$$\mathcal{D}^{exact}(x) = \langle \Omega | T(\varphi(x)\varphi(0)) | \Omega \rangle.$$

According to the Wick theorem, one can write the T–product as the sum

$$T(\varphi(x)\varphi(0)) = \underbrace{\varphi(x)\varphi(0)} + :\varphi(x)\varphi(0):$$

of the "pairing" and the "normal" product. The "pairing" is just the expectation value of the T- product over the perturbative vacuum

$$\underbrace{\varphi(x)\varphi(0)} = \langle 0 | T(\varphi(x)\varphi(0)) | 0 \rangle.$$

i.e., the perturbative propagator. By this definition, the normal product $:\varphi(x)\varphi(0):$ vanishes if averaged over the perturbative vacuum:

$$\langle 0 | :\varphi(x)\varphi(0): | 0 \rangle = 0.$$

Thus, our assumption that $\mathcal{D}^{exact}(x) \neq \mathcal{D}^{pert}(x)$ is equivalent to the statement

$$\langle \Omega | :\varphi(x)\varphi(0): | \Omega \rangle \neq 0,$$

which is the starting point to calculating power corrections in QCD. To simplify the notation, in what follows we will write the lhs of the above equation as $\langle \varphi(x)\varphi(0) \rangle$.

3.5 Condensates

In the oscillator case the analog of $\langle\varphi(x)\varphi(0)\rangle$ is the difference between the exact Green function

$$G^{osc}(0,0\,|\,0,\tau/i) = \frac{m\omega}{2\pi\sinh(\omega\tau)}$$

and the free Green function

$$G^{free}(0,0\,|\,0,\tau/i) = \frac{m}{2\pi\tau}$$

where $\tau \equiv 1/\epsilon$ is just the imaginary time variable. Note, that both oscillator and free Green function are singular for $\tau \to 0$, but the difference

$$\begin{aligned} &G^{osc}(0,0\,|\,0,\tau/i) - G^{free}(0,0\,|\,0,\tau/i) \\ &= \frac{m}{2\pi}\left[-\frac{1}{6}\,\omega^2\tau + \frac{7}{360}\,\omega^4\tau^3 - \frac{31}{15120}\,\omega^6\tau^5 + \dots\right] \end{aligned} \tag{77}$$

is regular at that point, and one can even expand the difference into the Taylor series in $(\omega\tau)^2$. The expansion is possible essentially because the confining potential is regular (and even vanishing) at $\tau = 0$.

In the QCD sum rule approach it is also assumed that the confinement effects are sufficiently soft to allow for the Taylor expansion of $\langle\varphi(x)\varphi(0)\rangle$ at $x = 0$:

$$\begin{aligned} \langle\varphi(0)\varphi(x)\rangle &= \sum_{n=0}^{\infty}\frac{x^{\mu_1}\dots x^{\mu_n}}{n!}\langle\varphi(0)\partial_{\mu_1}\dots\partial_{\mu_n}\varphi(0)\rangle \\ &= \langle\varphi\varphi\rangle + x^{\mu}\langle\varphi\partial_\mu\varphi\rangle + \frac{x^{\mu_1}x^{\mu_2}}{2}\langle\varphi\partial_{\mu_1}\partial_{\mu_2}\varphi\rangle + \dots \end{aligned} \tag{78}$$

This is, in fact, the expansion of the nonlocal object $\langle\varphi(0)\varphi(x)\rangle$ over the vacuum matrix elements of the local composite operators. Not all operators really contribute to the expansion above. Using the Lorentz covariance, it is trivial to find that

$$\begin{aligned} \langle\varphi\partial_\mu\varphi\rangle &= 0, \\ \langle\varphi\partial_{\mu_1}\partial_{\mu_2}\varphi\rangle &= \frac{1}{4}g_{\mu_1\mu_2}\langle\varphi\partial^2\varphi\rangle, \end{aligned} \tag{79}$$

etc., so that finally one arrives at the expansion in x^2:

$$\langle\varphi(0)\varphi(x)\rangle = \sum_{n=0}^{\infty}\left(\frac{x^2}{4}\right)^n\frac{1}{n!(n+1)!}\langle\varphi(\partial^2)^n\varphi\rangle\,. \tag{80}$$

Thus, the modification of the propagator by the nonperturbative effects is now parametrized by the matrix elements of the composite operators like $\langle\varphi(\partial^2)^n\varphi\rangle$. The examples in QCD are

• $\langle\bar{q}q\rangle$ referred to as the quark condensate,
• $\langle\bar{q}D^2q\rangle$, characterizing the average virtuality of the vacuum quarks,
• the gluon condensate $\langle G^a_{\mu\nu}G^a_{\mu\nu}\rangle$, etc. Here $D_\mu \equiv \partial_\mu - igA_\mu$ is the covariant derivative and $G_{\mu\nu} = (i/g)[D_\mu, D_\nu]$ is the gluonic field strength. Note, that only gauge invariant composite operators should appear in QCD, *i.e.* each ∂_μ must be accompanied by the relevant A_μ. How this happens, will be discussed later.

3.6 Operator product expansion

It is instructive to analyze the above expansion in the momentum representation

$$\langle\varphi(0)\varphi(x)\rangle \equiv \int e^{ikx}\mathcal{D}(k)\,d^4k = \sum_{n=0}^{\infty}\left(\frac{x^2}{4}\right)^n \frac{1}{n!(n+1)!}\int(-k^2)^n\mathcal{D}(k)\,d^4k. \tag{81}$$

Note, that existence of $\langle\varphi(0)\varphi(0)\rangle$ implies that $\mathcal{D}(k)$, the nonperturbative part of the propagator, vanishes for large momenta faster than k^{-5}, while existence of $\langle\varphi(0)\partial^2\varphi(0)\rangle$ implies that $\mathcal{D}(k) < k^{-7}$, etc. All matrix elements $\langle\varphi(0)(\partial^2)^n\varphi(0)\rangle$ exist only if, for large momenta, the function $\mathcal{D}(k)$ vanishes faster than any power of $1/k^2$, say, like an exponential. Thus, the function $\mathcal{D}(k)$ is concentrated in the region of small momenta. This corresponds to the basic assumption that at large momenta one should enjoy the asymptotic freedom.

In other words, the exact propagator is represented as a sum of the perturbative $O(1/k^2)$ part and the nonperturbative ("condensate") part that vanishes fast when the momentum k increases. Using such a decomposition, we obtain a modified diagram technique for the correlator functions. Consider, *e.g.*, the lowest order diagram corresponding to the product of two propagators. After the decomposition it produces four diagrams (see Fig.3), where the dashed lines correspond to the nonperturbative part of the propagators.

The first term (Fig.3a) is just the ordinary purely perturbative diagram having $\sim \log(q^2)$ behavior for large q^2:

$$\int\frac{d^4k}{k^2(q-k)^2} \sim \log(q^2).$$

Two other diagrams (Figs.3b,c), as we will see below, behave like $1/q^2$. This behavior is completely determined by that of the perturbative propagator: the

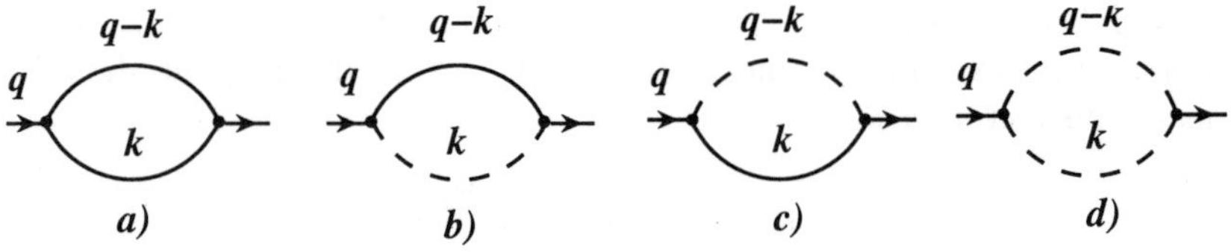

Figure 3: Operator product expansion for scalar one-loop diagram.

whole momentum flows through the perturbative line where its propagation is not suppressed. In the last diagram (Fig.3d), all possible momentum flows are strongly suppressed, *e.g.*, if $\mathcal{D}(k)$, the nonperturbative part of the propagator, vanishes exponentially at large k, so does this contribution also.

Now, let us analyze the contribution of the second diagram:

$$I^{\langle\varphi\varphi\rangle}(q^2) = \int \mathcal{D}(k^2)\frac{d^4k}{(q-k)^2}. \tag{82}$$

As we discussed earlier, the function $\mathcal{D}(k^2)$ is concentrated in the region of small momenta. So it makes sense to expand the perturbative propagator $1/(q-k)^2$ in k:

$$\frac{1}{(q-k)^2} = \frac{1}{q^2}\sum_{n=0}^{\infty}\frac{2^n\{qk\}^{(n)}}{(q^2)^n}\theta(k^2 < q^2) + \frac{1}{k^2}\sum_{n=0}^{\infty}\frac{2^n\{qk\}^{(n)}}{(k^2)^n}\theta(q^2 < k^2) \tag{83}$$

where

$$\{qk\}^{(n)} \equiv \{q_{\mu_1}\cdots q_{\mu_n}\}\{k^{\mu_1}\ldots k^{\mu_n}\}$$

is the product of two symmetric

$$O^{\cdots\mu_i\cdots\mu_j\cdots} = O^{\cdots\mu_j\cdots\mu_i\cdots}$$

and traceless

$$O^{\cdots\mu_i\cdots\mu_j\cdots}g_{\mu_i\mu_j} = 0$$

tensors formed from the vectors q and k, respectively. The symmetric-traceless tensors have a very useful property:

$$\int F(k^2)\{k^{\mu_1}\ldots k^{\mu_n}\}\,d^4k = 0$$

for $n \neq 0$, because it is impossible to construct a nontrivial symmetric-traceless combination using only the metric tensors $g_{\mu_i \mu_j}$. To do this, one should have at least one vector, *e.g.*,

$$\{p_{\mu_1} p_{\mu_2}\} = p_{\mu_1} p_{\mu_2} - \frac{1}{4} p^2 \, g_{\mu_1 \mu_2}.$$

Thus, in our case

$$\begin{aligned} I^{\langle\varphi\varphi\rangle}(p^2) &\equiv \int \mathcal{D}(k^2) \frac{d^4 k}{(q-k)^2} \\ &= \frac{1}{q^2} \int \mathcal{D}(k^2)\theta(k^2 < q^2) d^4 k + \int \mathcal{D}(k^2) \frac{d^4 k}{k^2} \theta(q^2 < k^2) \\ &= \frac{1}{q^2} \langle\varphi\varphi\rangle + \text{“}\textit{quickly vanishing contribution}\text{”}, \end{aligned} \tag{84}$$

where the “quickly vanishing” term is given by the integral over large momenta:

$$\text{“}\textit{quickly vanishing contribution}\text{”} = \int \left(\frac{1}{k^2} - \frac{1}{q^2}\right) \mathcal{D}(k^2)\, \theta(k^2 > q^2)\, d^4 k$$

where the function $\mathcal{D}(k^2)$ is strongly suppressed.

Thus, it is very easy to calculate this contribution: the whole external momentum flows through the perturbative line, so one should just multiply the perturbative propagator $1/q^2$ by the condensate $\langle\varphi\varphi\rangle$ factor. One can wonder, however, why the higher condensates (containing the ∂^2) did not appear? To answer this question, it is useful to analyze this contribution in the configuration representation. The original diagram decomposition now looks like

$$\begin{aligned} \mathcal{D}(x,0)\mathcal{D}(0,x) &= \mathcal{D}^{pert}(x,0)\mathcal{D}^{pert}(0,x) + \mathcal{D}^{pert}(x,0)\langle\, \varphi(0)\varphi(x)\rangle \\ &\quad + \mathcal{D}^{pert}(0,x)\langle\, \varphi(x)\varphi(0)\rangle + \langle\, \varphi(0)\varphi(x)\rangle\langle\, \varphi(x)\varphi(0)\rangle, \end{aligned} \tag{85}$$

and its second term can be written (modulo trivial factors like 2π) as

$$\langle\, \varphi(0)\varphi(x)\rangle / x^2.$$

Now, using the x^2-expansion for the $\langle\, \varphi(0)\varphi(x)\rangle$- factor, we obtain

$$\frac{1}{x^2}\langle\, \varphi(0)\varphi(x)\rangle = \frac{1}{x^2}\langle\varphi\varphi\rangle + \frac{1}{x^2}\frac{x^2}{8}\langle\, \varphi(0)\varphi(x)\rangle + O(x^2) + \ldots\,. \tag{86}$$

In the momentum representation, the first term just corresponds to the $\langle\, \varphi\varphi\rangle/q^2$ contribution, while the second is proportional to $\delta^4(q)$ and can be discarded since the external momentum q is assumed to be large. Higher terms are proportional to the derivatives of $\delta^4(q)$, and also can be ignored for large q. It is easy to realize, that the sum of all these terms just produces the "quickly vanishing contribution" discussed above. Only the terms singular for $x^2 = 0$ (in the configuration representation) can produce a power-like behavior in the momentum representation.

The expansion we discussed above is a simplified illustration of the operator product expansion[12]

$$T(j(x)j(0)) = \sum_{n=0} C_n(x^2) O_n(0). \tag{87}$$

It represents the T-product of two currents $j(x), j(0)$ as a sum of the local (composite) operators of increasing dimension (in units of mass). The coefficient functions $C_n(x^2)$ characterizing the contribution of a particular operator are ordered in their singularity at $x^2 = 0$. From a simple dimensional analysis it follows that there should be a simple relation between singularity of the coefficient function and the dimension of the relevant composite operator; *e.g.*, if

$$C_n(x^2) \sim \frac{1}{(x^2)^{N_0 - n}}$$

then

$$O_n \sim (\mu^2)^{d_0 + n},$$

so that after the Fourier transform one arrives at an expansion in powers of μ^2/q^2. There are, of course, more complicated contributions into the OPE, corresponding to more complicated original diagrams.

Similar decomposition of propagators into perturbative and nonperturbative parts can be also performed in the QCD case. At this level of discussion, the only difference is that in QCD one has to deal with two types of fields: quarks $q, \bar{q}$ and gluons A. Consider the decomposition of a two-loop diagram (see Fig.4).

• The first diagram (Fig.4a) corresponds to the purely perturbative contribution (unity operator in the OPE language),

• the second diagram (Fig.4b) (in fact there are two such diagrams) corresponds to the $\langle \bar{q} A q \rangle$-type contribution,

• the third (Fig.4c) is evidently proportional to the $\langle AA \rangle$ vacuum average, accompanied by the one-loop coefficient function,

• the fourth (Fig.4d) diagram (there is also another such diagram) brings in the four-quark vacuum average $\langle \bar{q} q \bar{q} q \rangle$.

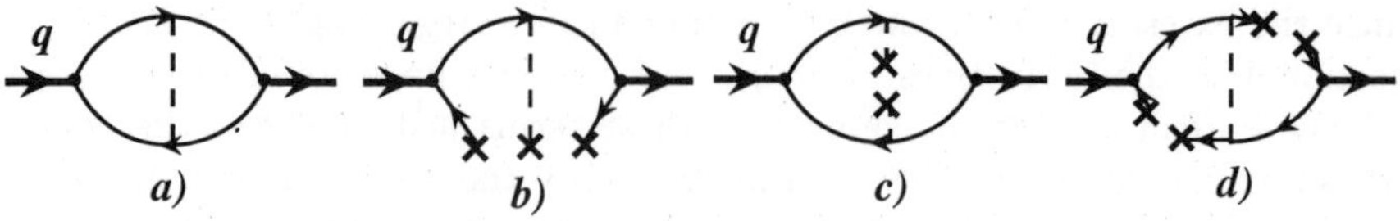

Figure 4: Operator product expansion for QCD two-loop diagram.

4 Operator Product Expansion: Gauge Invariance and Related Problems

Looking at the operators discussed in the previous section, one can notice that some of them, *e.g.*, $\langle \bar{q}\partial_\mu q\rangle$, $\langle \bar{q}A_\mu q\rangle$, $\langle A_\mu A_\nu\rangle$, etc., do not satisfy a very important requirement: they are not gauge invariant. And the product of gauge invariant operators in a gauge theory should have an expansion in gauge invariant operators like $\langle \bar{q}\hat{D}_\mu q\rangle$ and $\langle G^a_{\mu\nu}G^{a\,\mu\nu}\rangle$.

4.1 Notations

Let us introduce now the necessary notations. To begin with,

$$\hat{D}_\mu = \partial_\mu - ig\hat{A}_\mu$$

is the covariant derivative in the quark (fundamental) representation of the $SU(3)_c$ gauge group:

$$\hat{A}_\mu = A^a_\mu \tau^a,$$

with matrices $(\tau^a)_{AB}$ simply related to the Gell-Mann (3×3) λ-matrices

$$\tau^a \equiv \lambda^a/2.$$

The indices A, B correspond to 3 possible quark colors:$A, B = 1, 2, 3$. The gluons have 8 colors described by the a-indices: $a = 1, \ldots 8$. The covariant derivative

$$\tilde{D}_\mu = \partial_\mu - ig\tilde{A}_\mu$$

acting on the gluonic field contains the (8×8) σ-matrices

$$\tilde{A}_\mu = A^a_\mu \sigma^a$$

characteristic for the gluonic (adjoint) representation of the $SU(3)$ color group:

$$(\sigma^a)_{bc} = -if_{abc},$$

where f_{abc} are the structure constants of the group:

$$[\tau^a, \tau^b] = if^{abc}\tau^c.$$

The gluonic field strength tensor

$$G^a_{\mu\nu} = \partial_\mu A_\nu - \partial_\nu A_\mu + gf^{abc}A^b_\mu A^c_\nu$$

can be related to the commutator of the covariant derivatives:

$$\hat{G}_{\mu\nu} = \frac{1}{ig}[\hat{D}_\mu, \hat{D}_\nu] \tag{88}$$

$$\tilde{G}_{\mu\nu} = \frac{1}{ig}[\tilde{D}_\mu, \tilde{D}_\nu]. \tag{89}$$

4.2 External field method

A rather innocent-sounding statement that only gauge invariant operators should appear in the operator product expansion implies that there are numerous nontrivial correlations between the coefficient functions of non-gauge-invariant operators originating from completely different diagrams. For example, the operator $\langle \bar{q}\partial_\mu q\rangle$ appears when one expands the nonlocal operator $\langle \bar{q}(0)q(x)\rangle$ in the simplest combination (see Fig.5a)

$$\langle \bar{q}(0)S(-x)q(x)\rangle,$$

while the operator $\langle \bar{q}\hat{A}_\mu q\rangle$ necessary to form a gauge-invariant sum, results from a more complicated combination(see Fig.5b)

$$\int \langle \bar{q}(0)S(-y)\gamma^\mu(ig)\hat{A}_\mu(y)S(y-x)q(x)\rangle\, d^4y.$$

In both cases, the perturbative propagators (in configuration space) are written explicitly, while the nonperturbative parts, in accordance with our previous discussions, are denoted by vacuum averages of the relevant normal products of fields. By a straightforward integration one can establish that

$$\int S(-y)\gamma^\mu S(y-x)\, d^4y = -x^\mu S(-x)\,,$$

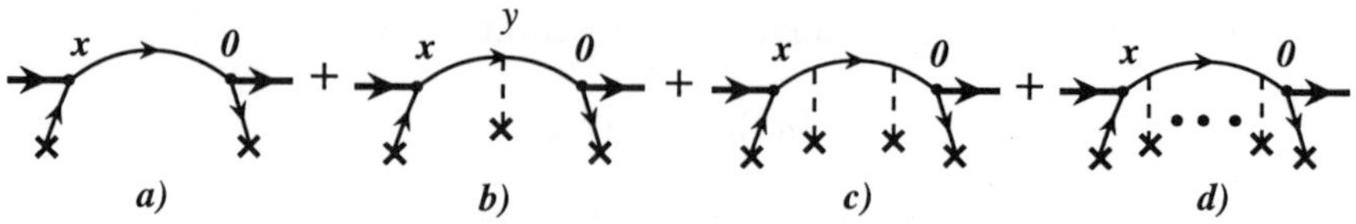

Figure 5: Summation over gluons inserted into hard quark propagator.

and, hence, the coefficient functions of the two operators are just those necessary to form the gauge invariant combination.

However, this is definitely not the method one would like to use in more complicated situations. In fact, it is much easier to analyze the sum of all the combinations differing only by the number of the gluons going into vacuum (see Fig.5).

One can write it as[13,14] $\langle \bar{q}(0)S(0,x;A)q(x)\rangle$, where $S(0,x;A)$ has the meaning of the perturbative quark propagator in the (external) vacuum gluonic field A. It satisfies the Dirac equation

$$\left[i\gamma^{\mu}(\frac{\partial}{\partial x^{\mu}} - ig\hat{A}_{\mu}(x)) - m\right] S(x,y;A) = -\delta^4(x-y). \tag{90}$$

Solving it by using perturbation theory in A, one gets back the original diagrammatic expansion.

The function $S(x,y;A)$ is not gauge invariant. From the good old electrodynamics it is known that propagation of a charged particle in the external field results in a phase factor given by the exponential of a line integral. In the QCD case it looks like

$$\hat{P}(x,y;A;C) = P\exp(ig\int_C \hat{A}_{\mu}(z)\, dz^{\mu}),$$

where P means that the color matrices should be ordered along the path C connecting the points x and y. Furthermore, it is known that the combination

$$\langle \bar{q}(y)\hat{P}(y,x;A;C)q(x)\rangle$$

is gauge invariant (though path-dependent). This gives the idea how one can construct the operator product expansion in an explicitly gauge invariant way. Let us try the ansatz

$$S(x,y;A) = \hat{E}(x,y;A)\mathcal{S}(x,y;A) \tag{91}$$

where $\hat{E}(x,y;A)$ is the P-exponential for the straight-line path connecting the points x and y:

$$\hat{E}(x,y;A) = P\exp\left[ig(x^\mu - y^\mu)\int_0^1 \hat{A}_\mu(z)\,dt\right]_{z=y+t(x-y)}.$$

It is straightforward to derive that the original Dirac equation is satisfied only if the function $\mathcal{S}(x,y;A)$ is a solution to the modified Dirac equation

$$\left[i\gamma^\mu\left(\frac{\partial}{\partial x^\mu} - ig\hat{\mathcal{A}}_\mu(x;y)\right) - m\right]\mathcal{S}(x,y;A) = -\delta^4(x-y) \tag{92}$$

that differs from the original one only by the change

$$A^a_\mu(x) \to \mathcal{A}^a_\mu(x;y) = (x^\nu - y^\nu)\int_0^1 G^b_{\nu\mu}(z)\,\tilde{E}^{ba}(z,y)\,t dt. \tag{93}$$

Here, $z = y + t(x-y)$ and $\tilde{E}$ is a straight-line-ordered exponential with the A-fields multiplied by the σ-matrices of the gluonic representation. It appears after one commutes the $\hat{E}$-exponentials through a τ-matrix:

$$(\tau^a)_{AB}\hat{E}_{BC}(z,y) = \hat{E}_{AB}(z,y)(\tau^b)_{BC}\tilde{E}^{ab}(z,y). \tag{94}$$

The commutation rule is based on the well-known formula

$$e^A B e^{-A} = B + [A,B] + \frac{1}{2!}[A,[A,B]] + \ldots \tag{95}$$

and the relation

$$[\tau_a, \tau_b] = -(\sigma_a)_{bc}\tau_c. \tag{96}$$

Surprisingly enough, the presence of the $\tilde{E}$-factor only simplifies the situation, because the derivatives in the Taylor expansion of the $G\tilde{E}$-combination are just the covariant ones:

$$G^b_{\nu\mu}(z)\,\tilde{E}^{ba}(z,y) = \sum_{n=0}^{\infty}\frac{1}{n!}\,G^a_{\nu\mu;\mu_1\ldots\mu_n}(y)(z-y)^{\mu_1}\ldots(z-y)^{\mu_n}. \tag{97}$$

As a result, the new field $\mathcal{A}^a_\mu(x;y)$ can be expressed in terms of the gluon field stress tensor $G_{\mu\nu}$ and its covariant derivatives (taken, of course, in the gluonic representation):

$$\mathcal{A}^a_\mu(x;y) = \sum_{n=0}^{\infty}\frac{1}{(n+2)n!}(x-y)^\nu(x-y)^{\mu_1}\ldots(x-y)^{\mu_n}\,G^a_{\nu\mu;\mu_1\ldots\mu_n}(y). \tag{98}$$

This means that the function $\mathcal{S}(x,y;A)$ depends on the gluonic field only through the gluon field stress tensor $G_{\mu\nu}$ and its covariant derivatives. Solving the equation for $\mathcal{S}(x,y;A)$ perturbatively, one obtains the $g\mathcal{A}$-expansion that has the same structure as the original gA-expansion. A very important difference is that the resulting operators have an explicitly gauge invariant form. For example, taking the lowest (zero) order term one obtains the contribution

$$\langle \bar{q}(y) S(y-x) \hat{E}(y,x;A) q(x) \rangle \tag{99}$$

containing the gauge-invariant operator

$$\langle \bar{q}(y) \hat{E}(y,x;A) q(x) \rangle = \sum_{n=0}^{\infty} \frac{1}{n!} (x-y)_{\mu_1} \dots (x-y)_{\mu_n} \langle \bar{q} \hat{D}^{\mu_1} \dots \hat{D}^{\mu_n} q \rangle |_y . \tag{100}$$

We used the direct straight-line path from x to y in our ansatz. In general, it is more convenient to use the path formed by two straight lines: from x to some fixed point z_0 (usually taken at the origin, $z_0 = 0$) and then from z_0 to y. The only change then will be that quark and gluon fields $q(\xi)$, $A(\xi)$ will be finally accompanied by the $E(\xi, z_0; A)$ exponentials, so that a covariant expansion will be obtained if one expands at the fixed point z_0. Using translation invariance, one can always substitute matrix elements $\langle \bar{q}(z_0) \dots G(z_0) \dots q(z_0) \rangle$, resulting from such a procedure, by $\langle \bar{q}(0) \dots G(0) \dots q(0) \rangle$. Less trivial is the disappearance of the z_0-dependence from the coefficient functions. One can enjoy it only if the calculations had no errors, and after summation over all diagrams (of the modified expansion) producing the same operator. The latter observation suggests that z_0 works as a gauge parameter. Indeed, it is easy to notice that the relevant P-exponentials

$$E(x, z_0; a) = P \exp[ig(x^\mu - z_0^\mu) \int_0^1 A_\mu(z)\, dt]|_{z=z_0+t(x-z_0)} \tag{101}$$

disappear (are equal to 1) if one imposes the gauge condition[15,16,17,18,14]

$$(x^\mu - z_0^\mu) A_\mu(x) = 0. \tag{102}$$

Furthermore, the "calligraphic" field $\mathcal{A}$ in this (Fock-Schwinger or fixed point) gauge coincides with the ordinary one $A = \mathcal{A}(\{G\})$, *i.e.*, in this gauge not only $G_{\mu\nu}$ can be expressed in terms of A_μ, but also the vector-potential A_μ can be represented as a (nonlocal) functional of the field strength tensor $G_{\mu\nu}$. This means that the Fock-Schwinger gauge is a physical gauge. Another example is the axial gauge[19]

$$n_\mu A^\mu(x) = 0 \tag{103}$$

with n_μ being a fixed vector. In this gauge one can also express[19] A_μ in terms of $G_{\mu\nu}$:

$$A_\mu(x) = n^\nu \int_0^\infty G_{\nu\mu}(x+tn)\,dt. \qquad (104)$$

Imposing the Fock-Schwinger gauge condition, one can forget about the exponentials. One should remember, however, that the "ordinary" derivatives in this gauge should be treated as the covariant ones, *e.g.*, they do not commute, etc. To illustrate the typical tricks and conventions, let us consider the simplest gauge invariant nonlocal condensate

$$\langle \bar{q}(0)\hat{E}(0,x;A)q(x)\rangle|_{(xA)=0} = \langle \bar{q}(0)q(x)\rangle$$
$$= \sum_{n=0}^{\infty} \frac{1}{n!} x_{\mu_1}\ldots x_{\mu_n}\langle \bar{q}\hat{D}^{\mu_1}\ldots\hat{D}^{\mu_n}q\rangle = \langle \bar{q}q\rangle + \frac{x^2}{8}\langle \bar{q}\hat{D}^2 q\rangle + \ldots . \qquad (105)$$

The ratio

$$\lambda^2 \equiv \frac{\langle \bar{q}\hat{D}^2 q\rangle}{\langle \bar{q}q\rangle} \qquad (106)$$

can be interpreted as the average virtuality of the vacuum quarks. It is possible, however, to represent the $\langle \bar{q}\hat{D}^2 q\rangle$ operator in a different form:

$$\langle \bar{q}\hat{D}^2 q\rangle \equiv \langle \bar{q}\hat{D}^\mu \hat{D}_\mu q\rangle = \langle \bar{q}\hat{D}^\mu \hat{D}^\nu q\rangle g_{\mu\nu} = \langle \bar{q}\hat{\not{D}}\hat{\not{D}}q\rangle - \langle \bar{q}\hat{D}^\mu \hat{D}^\nu \sigma_{\mu\nu}q\rangle$$
$$= -m_q^2\langle \bar{q}q\rangle - \frac{1}{2}\langle \bar{q}\,[\hat{D}^\mu,\hat{D}^\nu]\,\sigma_{\mu\nu}q\rangle = -m_q^2\langle \bar{q}q\rangle + \frac{1}{2}\langle \bar{q}ig\hat{G}^{\mu\nu}\sigma_{\mu\nu}q\rangle \qquad (107)$$

where we have used the identity

$$g_{\mu\nu} = \gamma_\mu\gamma_\nu - \frac{\gamma_\mu\gamma_\nu - \gamma_\nu\gamma_\mu}{2} = \gamma_\mu\gamma_\nu - \sigma_{\mu\nu},$$

the equation of motion

$$\hat{\not{D}}q(x) = -im_q q(x) \qquad (108)$$

and the definition of the field strength tensor

$$[\hat{D}_\mu,\hat{D}_\nu] = -ig\hat{G}_{\mu\nu}. \qquad (109)$$

Thus, the "average virtuality" of the vacuum quarks $\langle \bar{q}\hat{D}^2 q\rangle$ is directly related to the "average vacuum gluonic field strength" $\langle \bar{q}ig\hat{G}^{\mu\nu}\sigma_{\mu\nu}q\rangle$. In many papers one can find the notation

$$\langle \bar{q}ig\hat{G}^{\mu\nu}\sigma_{\mu\nu}q\rangle \equiv m_0^2\langle \bar{q}q\rangle. \qquad (110)$$

Using it, one can write the relation

$$\lambda^2 = \frac{m_0^2}{2} - m_q^2. \tag{111}$$

For light quarks, the m_q^2 term can be neglected.

4.3 Equations of motion and the current conservation

The equation of motion $\hat{\not{D}}q(x) = -im_q q(x)$ brings in correlations between matrix elements of different operators. These correlations, in fact, are very important to preserve the symmetry properties required by the structure of the currents $j(x)$ in the correlators. For example, one might expect that a correlator of the electromagnetic vector currents

$$\Pi_{\mu\nu}(p) = \int \langle T\, j_\mu(0) j_\nu(x)\rangle e^{ipx}\, d^4x \tag{112}$$

should be transverse:

$$\Pi_{\mu\nu}(p) = (p_\mu p_\nu - g_{\mu\nu}p^2)\,\Pi(p^2) \tag{113}$$

due to the current conservation condition $\partial^\mu j_\mu = 0$. Indeed, for the perturbative one-loop contribution, using the explicit form of the propagator in the configuration space

$$S(x) \propto -2i\frac{\not{x}}{x^2}\left[\frac{1}{x^2} + \frac{m^2}{4}\right] + \frac{m}{x^2} + O(m^3) \tag{114}$$

one finds that the integrand is

$$\frac{1}{(x^2)^4}\left[g_{\mu\nu}x^2 - 2x_\mu x_\nu\right] - \frac{m^2}{(x^2)^3}\left[g_{\mu\nu}x^2 - 4x_\mu x_\nu\right] + \dots\,. \tag{115}$$

It is easy to see that both terms satisfy the transversality condition

$$\frac{\partial}{\partial x^\mu}\{\ldots\} = 0,$$

and this means that

$$\Pi_{\mu\nu}^{pert}(p) = (p_\mu p_\nu - g_{\mu\nu}p^2)\Pi^{pert}(p^2)$$

at least up to the third order in m. One can, in fact, perform the calculations in the momentum space directly, to see that this property of $\Pi_{\mu\nu}^{pert}(p)$ is valid to all orders in m.

Now, let us consider the simplest nonperturbative contribution. In the momentum representation it looks like

$$\langle \bar{q}\gamma_\mu \frac{\not{p}+m}{p^2-m^2}\gamma_\nu q\rangle \sim \frac{m g_{\mu\nu}}{p^2}\langle\bar{q}q\rangle \ . \tag{116}$$

However, it seems that there is no way to get the $p_\mu p_\nu$ term, since there is only one $\not{p}$ term in the numerator. To solve this puzzle, let us consider this contribution in the configuration representation. Up to $O(m^2)$-terms one has

$$\begin{aligned}\langle\bar{q}(0)\gamma_\mu S(x)\gamma_\nu q(x)\rangle &\propto \langle\bar{q}(0)\gamma_\mu\left(-2i\frac{\not{x}}{(x^2)^2}+\frac{m}{x^2}\right)\gamma_\nu q(x)\rangle + O(m^2)\\ &= -\frac{2i}{(x^2)^2}\langle\bar{q}(0)\gamma_\mu\not{x}\gamma_\nu q(x)\rangle + \frac{m}{x^2}\langle\bar{q}(0)\gamma_\mu\gamma_\nu q(x)\rangle + O(m^2).\end{aligned} \tag{117}$$

Note, that vacuum averages of local operators with an odd number of indices vanish, and only those with an even number survive. Incorporating the Lorentz invariance, we obtain

$$\begin{aligned}\langle\bar{q}(0)\gamma_\mu\gamma_\nu q(0)\rangle &= g_{\mu\nu}\langle\bar{q}q\rangle,\\ \langle\bar{q}(0)\gamma_\mu\gamma_\alpha\gamma_\nu q(0)\rangle &= 0,\\ \langle\bar{q}(0)\gamma_\mu\gamma_\alpha\gamma_\nu D_\beta q(0)\rangle &= (g_{\mu\alpha}g_{\nu\sigma}+g_{\nu\alpha}g_{\mu\sigma}-g_{\mu\nu}g_{\alpha\sigma})\langle\bar{q}\gamma_\sigma D_\beta q\rangle,\\ \langle\bar{q}\gamma_\sigma D_\beta q\rangle &= \frac{g_{\sigma\beta}}{4}\langle\,\bar{q}\hat{\not{D}}q\rangle.\end{aligned} \tag{118}$$

Finally, incorporating the equation of motion

$$\langle\bar{q}\hat{\not{D}}q\rangle = -im\langle\bar{q}q\rangle \tag{119}$$

one obtains the transverse result

$$\langle\bar{q}(0)\gamma_\mu S(x)\gamma_\nu q(x)\rangle \propto \frac{m}{2(x^2)^2}\langle\bar{q}q\rangle[g_{\mu\nu}x^2+2x_\mu x_\nu]. \tag{120}$$

It is instructive to analyze also the $\langle GG\rangle$ contribution. First, using the exponential representation, it is easy to see that for the closed loop the exponentials corresponding to the upper and the lower lines cancel each other, and only the $O(G)$ terms remain. This general result can be illustrated on the example of one-loop diagrams with two vacuum gluons. To study, how the cancellation proceeds in this case, one should expand the exponential up to $O(A^2)$ terms:

$$\begin{aligned}S(x,A) &= S(x)\left\{1+ig(xA)+\frac{(ig)^2}{2}(xA)^2+\ldots\right\},\\ S(-x,A) &= S(-x)\left\{1-ig(xA)+\frac{(ig)^2}{2}(xA)^2+\ldots\right\}.\end{aligned} \tag{121}$$

Now, combining the contributions of the three relevant diagrams, one obtains

$$(ig)^2 \left\{ \frac{1}{2}(xA)^2 + \frac{1}{2}(xA)^2 - (xA)^2 \right\} = 0, \tag{122}$$

i.e., the $A-$dependence disappeared. To calculate the $O(G)$- terms, it is most convenient to use the Fock-Schwinger gauge $(xA) = 0$, in which

$$A_\mu(x) = x^\nu \int_0^1 G_{\nu\mu}(tx)\, tdt = \sum_{n=0}^{\infty} \frac{x^\nu x^{\mu_1} \ldots x^{\mu_n}}{(n+2)n!} G_{\nu\mu;\mu_1\ldots\mu_n}(0)$$
$$= \frac{1}{2} x^\nu G_{\nu\mu}(0) + \ldots . \tag{123}$$

Using this representation, one can construct the Feynman rules for the vertices corresponding to $G_{\mu\nu}$, $G_{\mu\nu;\mu_1}$, etc. For example, in the momentum representation

$$\frac{x_\nu}{2} G_{\nu\mu} \Rightarrow \frac{1}{2i} G_{\nu\mu} \frac{\partial}{\partial k_\nu}, \tag{124}$$

and the insertion of such a vertex into the quark propagator gives

$$\frac{1}{2i} \frac{\not{k}}{k^2} \gamma_\mu G_{\nu\mu} \frac{\partial}{\partial k_\nu} \frac{\not{k}}{k^2} = \frac{i}{2} G_{\nu\mu} \frac{\not{k}}{k^2} \gamma_\mu \frac{\not{k}}{k^2} \gamma_\nu \frac{\not{k}}{k^2} . \tag{125}$$

The result looks like an insertion of two A-fields, A_ν and A_μ, antisymmetrized with respect to the interchange of μ and ν. There is an important thing to remember while constructing the diagrammatic technique for the G-type vertices: in the momentum representation they contain derivatives acting onto all the propagators between the G-vertex and the Fock-Schwinger zero point. The G-inserted propagator looks simpler in the configuration space:

$$S(x,G) = -\frac{\not{x}}{2\pi^2 (x^2)^2} - \frac{g}{16\pi^2 x^2} x_\alpha \epsilon_{\alpha\beta\mu\nu} \hat{G}^{\mu\nu} \gamma_\beta \gamma_5 + O(z_0)\langle GG \rangle + \ldots \tag{126}$$

where z_0 is the Fock-Schwinger fixed point. If one takes $z_0 = 0$, then the $O(\langle GG \rangle)$ term disappears. This means that in the gauge $(xA) = 0$ only the diagram with the G-insertions into different lines survives. Its contribution in the configuration space is easily calculated by just multiplying the two $S(x,G)$ propagators. Taking the trace over the γ-matrices, one obtains the result

$$-\frac{2i}{3(16\pi^2)^2 (x^2)^2} (g_{\mu\nu} x^2 + 2x_\mu x_\nu) \langle g^2 G^2 \rangle \tag{127}$$

proportional to the same transverse structure as the contribution due to the $m_q\langle\bar{q}q\rangle$ operator. This is because both operators have the same dimension, equal to 4 in mass units.

The QCD sum rules for the "standard" mesons, like $\rho, \pi, K, etc.$, are based on the OPE including the operators of dimension 6, at least. One of these operators can be easily obtained from the Taylor expansion of the nonlocal quark condensate containing a γ-matrix

$$\langle\bar{q}(0)\gamma_\sigma q(x)\rangle = \langle\bar{q}\gamma_\sigma\hat{D}_\alpha q\rangle x^\alpha + \frac{1}{3!}\langle\bar{q}\gamma_\sigma\hat{D}_{\alpha_1}\hat{D}_{\alpha_2}\hat{D}_{\alpha_3}q\rangle x^{\alpha_1}x^{\alpha_2}x^{\alpha_3} + \dots . \quad (128)$$

We recall that the terms, containing an even number of derivatives, vanish since the total number of indices is odd in these cases. Incorporating the Lorentz invariance, we obtain

$$\begin{aligned}\langle\bar{q}(0)\gamma_\sigma q(x)\rangle &= \frac{x^\sigma}{4}\langle\bar{q}\hat{\not{D}}q\rangle + \frac{x^\sigma x^2}{144}(\langle\bar{q}\hat{\not{D}}\hat{D}^\alpha\hat{D}_\alpha q\rangle \\ &+\langle\bar{q}\,\hat{D}^\alpha\hat{\not{D}}\hat{D}_\alpha\rangle + \langle\bar{q}\,\hat{D}^\alpha\hat{D}_\alpha\hat{\not{D}}\rangle).\end{aligned} \quad (129)$$

Next step is to use the translation invariance $\langle\partial^\mu\bar{q}(0)\dots q(0)\rangle = 0$, the equation of motion $\hat{\not{D}}q = -im_q q$ and the commutation relation $[\hat{D}_\alpha, \hat{D}_\mu] = ig\hat{G}_{\alpha\mu}$ to get

$$\langle\bar{q}(0)\gamma_\sigma q(x)\rangle = \frac{-imx^\sigma}{4}\left(\langle\bar{q}q\rangle + \frac{x^2}{12}\langle\bar{q}\hat{D}^2q\rangle\right) + \frac{ig}{288}x^\sigma x^2\langle\bar{q}\gamma_\mu[\hat{D}_\alpha, \hat{G}^{\mu\alpha}]q\rangle. \quad (130)$$

The commutator $[\hat{D}_\alpha, \hat{G}^{\mu\alpha}]$ is equal to the covariant derivative of the gluonic field

$$[\hat{D}_\alpha, \hat{G}^{\mu\alpha}] = \hat{G}^{\mu\alpha}{}_{;\alpha} . \quad (131)$$

The final step is to use the gluonic equation of motion

$$\hat{G}^{\mu\alpha}{}_{;\alpha} = j^\mu_a\tau^a \equiv \sum_{i=u,d,s}\langle\bar{q}_i\gamma^\mu\tau_a q_i\rangle\tau^a \quad (132)$$

analogous to the Maxwell equation. Thus, due to the equations of motion, the apparently two-quark operator produces a four-quark term:

$$\langle\bar{q}(0)\gamma_\sigma q(x)\rangle = \frac{-imx^\sigma}{4}(\langle\bar{q}q\rangle + \frac{x^2}{12}\langle\bar{q}(0)\hat{D}^2q\rangle) + \frac{ig}{288}x^\sigma x^2\langle\bar{q}\gamma_\mu j^\mu_a\tau^a q\rangle. \quad (133)$$

5 QCD Sum Rules for ρ-meson

5.1 Dispersion relation

As pointed out by SVZ (whose analysis[1] we closely follow in this section) we closely to determine the basic parameters characterizing the ρ-meson, *e.g.*, its mass m_ρ and its e^+e^- coupling constant g_ρ, one should consider the correlator of two vector currents

$$j_\mu^{(\rho)} = \frac{1}{2}(\bar{u}\gamma_\mu u - \bar{d}\gamma_\mu d) \tag{134}$$

with the ρ-meson quantum numbers:

$$\Pi_{\mu\nu}^{(\rho)}(q) = (q_\mu q_\nu - q^2 g_{\mu\nu})\Pi^{(\rho)}(q^2) = i\int e^{iqx}\langle 0|T\,(j_\mu^{(\rho)}(x)\,j_\nu^{(\rho)}(0)\,)|\,0\rangle\, d^4x. \tag{135}$$

Via the dispersion relation

$$\Pi^{(\rho)}(q^2 = -Q^2) = \Pi^{(\rho)}(0) - \frac{Q^2}{12\pi^2}\int_0^\infty \frac{R^{I=1}(s)}{s(s+Q^2)}\, ds \tag{136}$$

this correlator is related to the total cross section of the e^+e^--annihilation to hadrons with the isospin $I = 1$, measured in units of the $e^+e^- \to \mu^+\mu^-$ cross section :

$$R^{I=1} = \frac{\sigma(e^+e^- \to hadrons, I = 1)}{\sigma(e^+e^- \to \mu^+\mu^-)}\,. \tag{137}$$

Indeed, the $j_\mu^{(\rho)}$ coincides with the isovector part of the electromagnetic current and is responsible, therefore, for the production of the hadrons with $I = 1$ in the e^+e^--annihilation. The correlator $\Pi_{\mu\nu}^{(\rho)}(q)$ contains, of course, only the hadronic part of the relevant factors, and that explains the division by $\sigma(e^+e^- \to \mu^+\mu^-)$.

5.2 Operator product expansion

On the other hand, one can construct the operator product expansion for the correlator $\Pi_{\mu\nu}^{(\rho)}(q)$ calculating the relevant Feynman diagrams. To begin with, one should calculate the perturbative diagrams shown in Fig.6 to extract the perturbative spectral density

$$\rho^{pert}(s) = \frac{3}{2}\,\theta(s)\,\left\{1 + \frac{\alpha_s}{\pi}\right\} \tag{138}$$

which is essentially the isovector ($I = 1$ isospin projection) quark charge squared $(\frac{1}{2})^2$ multiplied by 2 (u- and d- quarks) and multiplied by the number

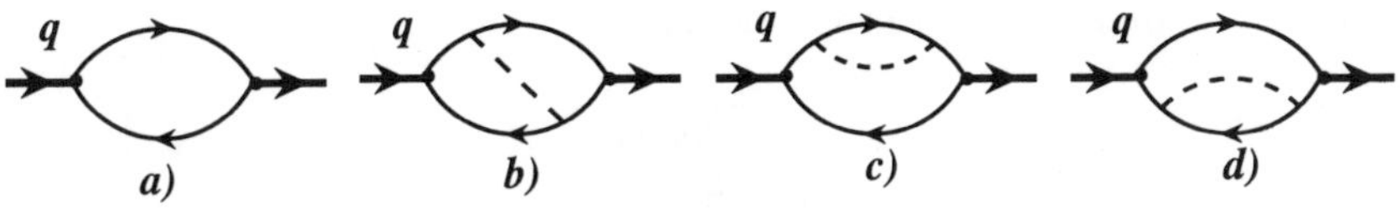

Figure 6: Perturbative contribution up to two loops.

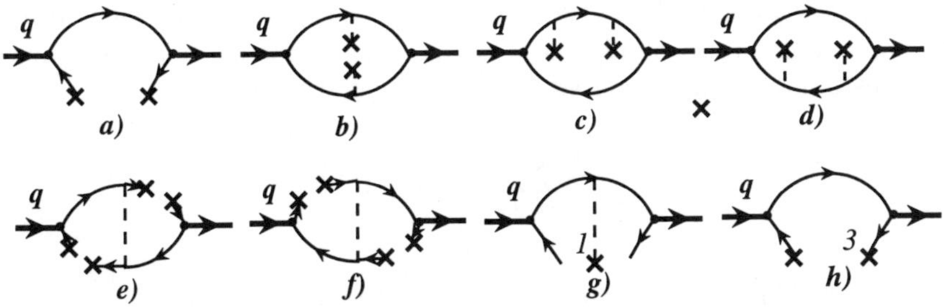

Figure 7: Nonperturbative contributions up to dimension six.

of colors (3). The simplest nonperturbative contribution is due to the quark condensate (see Fig.7a). It is proportional to the current quark masses:

$$\Pi^{\bar{q}q} \sim \frac{\langle 0|m_u \bar{u}u + m_d \bar{d}d|0\rangle}{(Q^2)^2},$$

with $m_u \sim 4MeV, m_d \sim 7MeV$. Hence, this contribution is rather small numerically. The contribution

$$\Pi^{GG} \sim g^2 \frac{\langle 0|GG|0\rangle}{(Q^2)^2}$$

due to the gluon condensate is given by three diagrams shown in Fig.7b-d. As mentioned at the end of the previous section, only the diagram 7b contributes in the Fock-Schwinger gauge $x_\mu A^\mu(x) = 0$. It is rather straightforward to calculate the diagrams 7e,f proportional to

$$\frac{g^2\langle\bar{q}q\bar{q}q\rangle}{(Q^2)^3}.$$

However, the contributions of the same four-quark type can be obtained also from the diagrams 7g,h which originally are bilinear in quark fields. Notice, however, the numbers 1 and 3 attached to the relevant lines going into the vacuum. They indicate how many covariant derivatives are acting on the relevant "vacuum" field. Thus, the matrix elements are $\langle\bar{q}\gamma_\mu\tilde{D}_\nu\hat{G}_{\alpha\beta}q\rangle$ and $\langle\bar{q}\gamma_\mu\hat{D}_{\nu_1}\hat{D}_{\nu_2}\hat{D}_{\nu_3}q\rangle$. As discussed at the end of Section 4, after vacuum averaging and commutations only the $\langle\bar{q}\gamma_\mu\tilde{D}_\nu\hat{G}_{\mu\nu}q\rangle$ component survives and, by equations of motion it is reduced to $\langle 0|\bar{q}\gamma_\mu\tau^a j^a_\mu q|0\rangle$ with

$$j^a_\mu = \sum_{i=u,d,s}\langle 0|\bar{q}_i\gamma_\mu\tau^a q_i|0\rangle.$$

5.3 Basic QCD sum rule for the ρ-meson channel

After the Borel transformation that modifies the dispersion integral

$$\int\frac{\rho(s)ds}{s+Q^2}\to\int e^{-s/M^2}\rho(s)\,\frac{ds}{M^2} \tag{139}$$

and the power corrections

$$\frac{1}{(Q^2)^n}\to\frac{1}{(n-1)!(M^2)^n} \tag{140}$$

one obtains the following QCD sum rule[1]:

$$\begin{aligned}\int_0^\infty e^{-s/M^2}R^{I=1}(s)ds &= \frac{3}{2}M^2\Big[1+\frac{\alpha_s(M)}{\pi}+\frac{4\pi^2}{M^4}(\langle m_u\bar{u}u\rangle+\langle m_d\bar{d}d\rangle)+\\ &+\frac{\pi^2}{3M^4}\langle\frac{\alpha_s}{\pi}GG\rangle-\frac{2\pi^3}{M^6}\langle\alpha_s(\bar{u}\gamma_\alpha\gamma_5\tau^a u-\bar{d}\gamma_\alpha\gamma_5\tau^a d)^2\rangle-\\ &-\frac{4\pi^3}{9M^6}\langle\alpha_s(\bar{u}\gamma_\alpha\tau^a u+\bar{d}\gamma_\alpha\tau^a d)\sum_{i=u,d,s}\langle 0|\bar{q}_i\gamma_\mu\tau^a q_i|0\rangle\rangle\Big]\end{aligned} \tag{141}$$

It is the starting point for further analysis.

5.4 *Fixing the condensates*

Next step is to specify the numerical values of the matrix elements on the r.h.s. of this sum rule (the same values will be used in all other sum rules).

• The first matrix element is in fact fixed by the current algebra:

$$\langle m_u \bar{u}u \rangle + \langle m_d \bar{d}d \rangle = -\frac{1}{2} f_\pi^2 m_\pi^2 = -1.7 \cdot 10^{-4} \mathrm{GeV}^4 \tag{142}$$

• The value of the gluonic condensate was extracted from the analysis of the QCD charmonium sum rules[1]

$$\langle \frac{\alpha_s}{\pi} GG \rangle = 1.2 \cdot 10^{-2} \mathrm{GeV}^4. \tag{143}$$

• The four-quark operators are first approximated by the value dictated by the vacuum dominance (or factorization) hypothesis:

$$\langle \bar{\psi}_{\{A\}} \psi_{\{B\}} \bar{\psi}_{\{C\}} \psi_{\{D\}} \rangle = \langle \bar{\psi}_{\{A\}} \psi_{\{B\}} \rangle \langle \bar{\psi}_{\{C\}} \psi_{\{D\}} \rangle - \langle \bar{\psi}_{\{A\}} \psi_{\{D\}} \rangle \langle \bar{\psi}_{\{C\}} \psi_{\{B\}} \rangle \tag{144}$$

where the subscripts A, B, C, D include spin, color and flavour. Using the orthogonality relation

$$\langle \bar{\psi}_{\{A\}} \psi_{\{B\}} \rangle = \frac{\delta_{AB}}{N} \langle \bar{\psi} \psi \rangle \tag{145}$$

one can reduce then any four quark operator to the $\langle \bar{q}q \rangle^2$ form. In our case

$$\begin{aligned} &\alpha_s \langle (\bar{u}\gamma_\alpha \gamma_5 \tau^a u - \bar{d}\gamma_\alpha \gamma_5 \tau^a d)^2 \rangle |_{vacuum\ dominance} = \\ &\qquad = \frac{32}{9} \alpha_s \langle \bar{q}q \rangle^2 \approx 6.5 \cdot 10^{-4}\, \mathrm{GeV}^6 \end{aligned} \tag{146}$$

$$\begin{aligned} &\alpha_s \langle (\bar{d}\gamma_\alpha \tau^a d) \sum_{i=u,d,s} \langle \bar{q}_i \gamma_\alpha \tau^a q_i \rangle |_{vacuum\ dominance} = \\ &\qquad = -\frac{32}{9} \alpha_s \langle \bar{q}q \rangle^2 \approx -6.5 \cdot 10^{-4}\, \mathrm{GeV}^6, \end{aligned} \tag{147}$$

with the numerical values following from the estimate

$$\alpha_s \langle \bar{q}q \rangle^2 \approx -1.8 \cdot 10^{-4} \mathrm{GeV}^6 \tag{148}$$

which can be justified using the above relation between the quark condensate, quark masses ($m_u + m_d = 11$ MeV, experimental values of the pion mass and pion decay constant f_π, under the assumption that the relevant normalization point corresponds to $\alpha_s = 0.7$. In fact, this estimate amounts to fixing one of the most important input parameters of the QCD sum rule method and the quoted value is supported by successful predictions for many different resonances.

5.5 *Qualitative analysis of the QCD sum rule*

Using the vacuum dominance hypothesis, we can rewrite the sum rule in a more compact form:

$$\int_0^\infty e^{-s/M^2} R^{I=1}(s)ds = \frac{3}{2}M^2\left[1+\frac{\alpha_s(M)}{\pi} - \frac{2\pi^2 f_\pi^2 m_\pi^2}{M^4} + \right.$$
$$\left. + \frac{\pi^2}{3M^4}\langle\frac{\alpha_s}{\pi}GG\rangle - \frac{448\pi^3}{M^6}\alpha_s\langle\bar{q}q\rangle^2\right]. \quad (149)$$

Substituting the numerical values of the condensate parameters gives

$$\int_0^\infty e^{-s/M^2} R^{I=1}(s)ds = \frac{3}{2}M^2\left[1+\frac{\alpha_s(M)}{\pi}+0.1\left(\frac{0.6\text{GeV}^2}{M^2}\right)^2-0.14\left(\frac{0.6\text{GeV}^2}{M^2}\right)^3\right] \quad (150)$$

Our experience with the sum rules for the quantum-mechanical oscillator (see Section 2) suggests that to calculate the ρ-meson mass we should incorporate also the "daughter" sum rule resulting from the original one after the differentiation with respect to $1/M^2$:

$$\int_0^\infty e^{-s/M^2} sR^{I=1}(s)ds = \frac{3}{2}M^4\left[1+\frac{\alpha_s(M)}{\pi}-0.1(\frac{0.6\text{GeV}^2}{M^2})^2+0.28(\frac{0.6\text{GeV}^2}{M^2})^3\right]. \quad (151)$$

Note now that power corrections in the sum rules above are small for the values of the Borel parameter M as low as $M^2 = m_\rho^2 = 0.6\,\text{GeV}^2$:

$$\int_0^\infty e^{-s/m_\rho^2} R^{I=1}(s)ds = \frac{3}{2}m_\rho^2[1+0.1+0.1-0.14], \quad (152)$$

$$\int_0^\infty e^{-s/m_\rho^2} sR^{I=1}(s)ds = \frac{3}{2}m_\rho^4[1+0.1-0.1+0.28]. \quad (153)$$

At such a low value of M^2 one should expect that the integral over the physical cross section is dominated by the lowest resonance, *i.e.*, by the ρ-meson. Its contribution to $R^{I=1}(s)$, in the limit of the vanishing width, can be written as

$$R_\rho^{I=1}(s) = \frac{12\pi^2 m_\rho^2}{g_\rho^2}\delta(s-m_\rho^2). \quad (154)$$

These observations allow one to make a rough estimate of the predictions following from the QCD sum rule. To this end we, following SVZ[1]

- neglect, for $M^2 = m_\rho^2$, the power corrections in the r.h.s. of the sum rules,

• neglect, for $M^2 = m_\rho^2$, the higher state contributions in the l.h.s. This gives the following relation

$$\frac{12\pi^2 m_\rho^2}{g_\rho^2} e^{-1} \approx \frac{3}{2} m_\rho^2. \tag{155}$$

Simplifying it results in a famous SVZ[1] expression for the g_ρ constant in terms of fundamental constants π and e:

$$\frac{g_\rho^2}{4\pi} \cong \frac{2\pi}{e} \approx 2.3, \tag{156}$$

to be compared with the experimental value 2.36 ± 0.18.

5.6 *Fitting the spectrum parameters*

The standard procedure is to extract the parameters of the hadronic spectrum requiring the best agreement between the two sides of the sum rule. Of course, having only a few first terms of the operator product expansion one should not hope to reproduce all the details of the resonance structure in a particular channel. Only the gross features of the spectrum can be extracted from the QCD sum rules. So, let us take the ansatz "first narrow resonance plus continuum" which has been discussed in the oscillator case. In fact, one should expect that in QCD this ansatz should work better, since the higher hadronic resonances are very broad, and the total cross section of their production is known to be well reproduced by the parton model (*i.e.*, perturbative) calculation. Specifically, we take

$$R_\rho^{I=1}(s) = \frac{12\pi^2 m_\rho^2}{g_\rho^2}\delta(s - m_\rho^2) + \frac{3}{2}\left(1 + \frac{\alpha_s}{\pi}\right)\theta(s - s_0). \tag{157}$$

The parameters to fit are the coupling constant g_ρ^2, the ρ-meson mass m_ρ^2 and the effective continuum threshold s_0.

The strategy will be just the same as used for the oscillator states:
• transfer the continuum contribution to the right hand side,
• write the original sum rule in the form

$$f_\rho^2 e^{-m_\rho^2/M^2} = \frac{3}{2} M^2 \left(1 - e^{-s_0/M^2}\right) + \text{“}\textit{power corrections}\text{”}, \tag{158}$$

where, for brevity, we introduced the notation $f_\rho^2 = \frac{12\pi^2 m_\rho^2}{g_\rho^2}$,
• write the daughter sum rule in the form

$$f_\rho^2 m_\rho^2 e^{-m_\rho^2/M^2} = \frac{3}{2} M^4 \left(1 - (1 + \frac{s_0}{M^2})e^{-s_0/M^2}\right) + \text{“}\textit{power corrections}\text{”}, \tag{159}$$

• divide the second sum rule by the first to get the expression for m_ρ^2 as a function of s_0 and M^2 (for fixed values of the condensates),
• look for the value of the effective threshold s_0 that produces the most stable curve for m_ρ^2 (recall that with all power corrections taken into account, and the exact ansatz for the higher states, one should obtain an M^2-independent result for m_ρ^2).

In this way one would obtain[b] $s_0 \approx 1.5\,\mathrm{GeV}^2$ and $m_\rho^2 \approx 0.58\,\mathrm{GeV}^2$. Next step is to extract g_ρ^2 from the first sum rule, using these values of s_0 and m_ρ^2. The value $s_0 \approx 1.5\,\mathrm{GeV}^2$ provides the most stable curve for g_ρ^2, corresponding to $g_\rho^2 \approx 2.17$. The result is rather sensitive to the m_ρ^2-value. If one takes the experimental value $m_\rho^2 \approx 0.6\,\mathrm{GeV}^2$, then the most stable curve still is that for $s_0 \approx 1.5\mathrm{GeV}^2$, but the g_ρ^2-value is a little higher $g_\rho^2 \approx 2.40$ and closer to the experimental one $g_\rho^2 \approx 2.36 \pm 0.18$. Anyway, it should be emphasized that all the g_ρ^2-values extracted from the QCD sum rule differ from the experimental value by less than 10%, the claimed precision of the method.

6 QCD Sum Rules for Pion

6.1 *Axial vector current*

Vector currents are a special case, because the relevant correlators can be related to directly measurable cross sections. From purely theoretical point of view, other currents such as scalar, pseudoscalar, axial vector etc., are not worse, but experimental information in these cases is normally limited to the masses of the lowest states. Still, the axial vector current is in a better situation. First, its projection onto the pion state is just the pion decay constant f_π:

$$\langle 0|\bar{d}\gamma_5\gamma_\mu u|P\rangle = if_\pi P_\mu. \tag{160}$$

Second, measuring the decays of the heavy lepton $\tau \to \nu_\tau + X$, one can measure the coupling of a hadronic state X to the axial current. Our goal now is to show how one can use the QCD sum rules to study the hadronic spectrum in the axial vector channel. Again, our presentation here closely follows the classic SVZ analysis[1].

Since the axial vector current $a_\mu(x) = \bar{d}\gamma_5\gamma_\mu u$ is not conserved, the relevant

---

[b]With all the numbers explicitly given, the readers are encouraged to perform the fitting as an excercise.

correlator $\Pi^{(ax)}_{\mu\nu}(q)$ is a sum of two independent functions:

$$\Pi^{(ax)}_{\mu\nu}(q) = i\int e^{iqx}\langle 0|T\left(a^+_\mu(x)\,a^-_\nu(0)\right)|\,0\rangle\, d^4x = -g_{\mu\nu}\Pi_1(Q^2) + q_\mu q_\nu \Pi_2(Q^2), \tag{161}$$

or, in terms of the transverse and longitudinal components

$$\Pi^{(ax)}_{\mu\nu}(q) = -\left\{g_{\mu\nu} - \frac{q_\mu q_\nu}{q^2}\right\}\Pi_1(Q^2) - \frac{q_\mu q_\nu}{q^2}\{\Pi_1(Q^2) + Q^2\Pi_2(Q^2)\}. \tag{162}$$

6.2 Massless pion

The common belief is that the pion mass vanishes in the limit of exact chiral symmetry so that the pion is a Goldstone particle. It can be shown, that the operator product expansion in this case really requires the presence of a massless particle in the limit $m_q \to 0$. In this limit the imaginary part of the longitudinal function $\Pi_1(Q^2) + Q^2\Pi_2(Q^2) \equiv \Pi_{||}$ vanishes, and $\Pi_{||}$ is a polynomial in Q^2. Now, let us switch on a small quark mass and keep terms linear in this mass. The statement is that the function $\Pi_1(Q^2) + Q^2\Pi_2(Q^2)$ is exactly calculable in this approximation:

$$\Pi_1(Q^2) + Q^2\Pi_2(Q^2) = \frac{(m_u + m_d)\langle \bar{u}u + \bar{d}d\rangle}{Q^2} + O(m_q^2). \tag{163}$$

There are no terms linear in m_q of higher orders in $1/Q^2$. Let us show that this equation really implies the existence of a nearly massless pion. First we should express the Π-functions in terms of the hadronic contributions, using the dispersion relations:

$$\Pi_1(Q^2) = \frac{1}{\pi}\int_0^\infty \frac{\mathrm{Im}\,\Pi_1}{s+Q^2}ds + \text{“}subtr.\text{”} \tag{164}$$

$$\Pi_2(Q^2) = \frac{1}{\pi}\int_0^\infty \frac{\mathrm{Im}\,\Pi_2}{s+Q^2}ds + \text{“}subtr.\text{”} \tag{165}$$

$$\Pi_1(Q^2) + Q^2\Pi_2(Q^2) = \frac{1}{\pi}\int_0^\infty \frac{\mathrm{Im}\,\Pi_1 - s\mathrm{Im}\,\Pi_2}{s+Q^2}ds + \text{“}subtr.\text{”} \tag{166}$$

where $\mathrm{Im}\,\Pi_1$ contains contributions only from pseudovector (spin-1) states and their contribution should be transverse, i.e., proportional to $(g_{\mu\nu} - q_\mu q_\nu/q^2)$:

$$\mathrm{Im}\,\Pi_1 = \sum_A \pi f_A^2 \delta(s - M^2), \tag{167}$$

while Im Π_2 contains contributions both from pseudovector (spin-1) states and from the pseudoscalar (spin-0) ones and their contribution into $\Pi^{(ax)}_{\mu\nu}(q)$ should be longitudinal *i.e.*, proportional to $q_\mu q_\nu$:

$$s\mathrm{Im}\Pi_2 = \sum_A \pi f_A^2 \delta(s - M_A^2) + \sum_P \pi f_P^2 m_P^2 \delta(s - M_P^2). \tag{168}$$

Hence, only pseudoscalar states contribute to the longitudinal component of $\Pi^{(ax)}_{\mu\nu}$. For small quark masses we have

$$\begin{aligned}\frac{(m_u + m_d)\langle \bar{u}u + \bar{d}d\rangle}{Q^2} + O(m_q^2) &= -\sum_P \frac{f_P^2 m_P^2}{Q^2 + m_P^2} = \\ &= -\sum_P \left(\frac{f_P^2 m_P^2}{Q^2} - \frac{f_P^2 m_P^4}{Q^4} + \ldots \right) .\end{aligned} \tag{169}$$

All $O(Q^{-4})$ terms should be $O(m_q^2)$. This means that

$$f_P^2 m_P^4 = O(m_q^2) \tag{170}$$

for all states. Hence, the particles with masses that remain constant as $m_q \to 0$, decouple in the chiral limit:

$$m_P^2 = O(m_q^0), \tag{171}$$

$$f_P^2 = O(m_q^2). \tag{172}$$

The last relation also means that they do not contribute to the term linear in m_q. The only way to get this term is to assume that there exists a state for which the $O(m_q)$-term is brought by the m_P^2-factor:

$$m_P^2 = O(m_q^1), \tag{173}$$

$$f_P = O(m_q^0). \tag{174}$$

Such a state is naturally identified with the pion. In addition, we get the well-known relation

$$f_\pi^2 m_\pi^2 = -(m_u + m_d)\langle \bar{u}u + \bar{d}d\rangle \tag{175}$$

demonstrating explicitly the vanishing of the pion mass in the chiral limit.

To derive the starting statement, one should consider $q_\mu q_\nu \Pi^{(ax)}_{\mu\nu}(q)$. By equations of motion it is related to the correlator of the pseudoscalar densities:

$$\begin{aligned} q_\mu q_\nu \Pi^{(ax)}_{\mu\nu}(q) &= Q^2(\Pi_1(Q^2) + Q^2\Pi_2(Q^2)) = \\ = -i \int e^{iqx} \langle 0| T \left((\bar{u}(x)\gamma_5 d(x)) \; (\bar{d}(0)\gamma_5 u(0)) \right) &|0\rangle \, d^4x + \text{const.}\end{aligned} \tag{176}$$

The constant on the right-hand side is due to contact terms which normally arise if one differentiates a T-product. This constant corresponds to the $1/Q^2$-term in the combination $\Pi_1(Q^2) + Q^2\Pi_2(Q^2)$. Direct calculation of this term gives the result stated in the beginning of this subsection.

Thus, the operator product expansion implies that $m_\pi^2 \to 0$ in the $m_q \to 0$ limit, and , hence, the pion mass should not be fitted from the sum rule. Within our accuracy one can set $m_\pi^2 = 0$ from the start.

6.3 QCD sum rule for the axial channel

The operator product expansion for the correlator of two axial currents is completely analogous to that of two vector currents, the only difference being the presence of the extra γ_5-matrices in the vertices corresponding to the currents. For massless quarks, however, this produces no changes both in perturbative and $O(\langle GG\rangle)$ contributions. However, the four-quark terms shown in Fig.7e,f change sign, because one of the γ-matrix type terms is substituted by the condensate factor $\langle\bar{q}q\rangle$ which has the structure of **1** with respect to the Dirac indices.

After the Borel transformation the sum rule looks as follows:

$$\int_0^\infty e^{-s/M^2}\,\mathrm{Im}\,\Pi_2(s)ds = \frac{M^2}{4\pi}\Bigg[1 + \frac{\alpha_s(M)}{\pi} + \frac{\pi^2}{3M^4}\langle\frac{\alpha_s}{\pi}GG\rangle + \\ + \frac{4\pi^3}{M^6}\langle\alpha_s(\bar{u}\gamma_\alpha\gamma_5\tau^a d\,\bar{d}\gamma_\alpha\gamma_5\tau^a u)\rangle + \\ + \frac{4\pi^3}{9M^6}\langle\alpha_s(\bar{u}\gamma_\alpha\tau^a u + \bar{d}\gamma_\alpha\tau^a d)\sum_{q=u,d,s}\langle 0|\bar{q}\gamma_\mu\tau^a q|0\rangle\Bigg]. \quad (177)$$

In the axial channel we have two states that are the lowest among other states with the same quantum numbers: a pseudoscalar (the pion) and an axial vector (A_1-meson) ones. Thus, one can try the ansatz with two low-lying states ("pion+A_1+continuum"):

$$\mathrm{Im}\,\Pi_2(s) = \pi f_\pi^2\delta(s) + \pi f_{A_1}^2\delta(s - m_{A_1}^2) + \frac{1}{4\pi}\left(1 + \frac{\alpha_s}{\pi}\right)\theta(s > s_0^{(A_1)}). \quad (178)$$

One can also treat the A_1 as a part of the continuum. This, more crude approximation, corresponds to the simplified ansatz

$$\mathrm{Im}\,\Pi_2(s) = \pi f_\pi^2\delta(s) + \frac{1}{4\pi}\left(1 + \frac{\alpha_s}{\pi}\right)\theta(s > s_0^{(\pi)}). \quad (179)$$

The constants f_π, f_{A_1} are defined as follows:

$$\langle 0|\bar{d}\gamma_5\gamma_\mu u|\pi\rangle = if_\pi P_\mu, \qquad \langle 0|\bar{d}\gamma_5\gamma_\mu u|A_1\rangle = if_{A_1}\epsilon_\mu, \quad (180)$$

where ϵ_μ is the A_1 polarization vector.

6.4 Fitting the sum rule

Substituting the numerical values of the condensates into the basic sum rule gives

$$\int_0^\infty e^{-s/M^2}\mathrm{Im}\,\Pi_2(s)ds = \frac{M^2}{4\pi}\left[1+\frac{\alpha_s(M)}{\pi}\right.$$
$$\left.+0.1\left(\frac{0.6\mathrm{GeV}^2}{M^2}\right)^2 + 0.22\left(\frac{0.6\mathrm{GeV}^2}{M^2}\right)^3\right]. \tag{181}$$

The "daughter" sum rule is

$$\int_0^\infty e^{-s/M^2}\mathrm{Im}\,\Pi_2(s)sds = \frac{M^2}{4\pi}\left[1+\frac{\alpha_s(M)}{\pi}\right.$$
$$\left.-0.1\left(\frac{0.6\mathrm{GeV}^2}{M^2}\right)^2 - 0.44\left(\frac{0.6\mathrm{GeV}^2}{M^2}\right)^3\right]. \tag{182}$$

Note, that in the $m_\pi^2 = 0$ approximation the pion does not contribute to the daughter sum rule, and the spectrum structure for $s\mathrm{Im}\,\Pi_2(s) = \mathrm{Im}\,\Pi_1(s)$ is more like in the vector channel: first non-zero mass resonance + continuum. This is reflected in the structure of the relevant sum rule: power corrections are negative. They are considerably larger than those in the ρ-channel, and one should expect that the A_1-mass (squared) is much larger than m_ρ^2.

The fitting procedure [c] is most straightforward in the case of the simplified ansatz. One should transfer the continuum contribution to the right hand side to get f_π^2 as a function of $s_0^{(\pi)}$ and the Borel parameter M^2:

$$4\pi^2 f_\pi^2 = M^2\left(1-e^{-s_0^{(\pi)}/M^2}\right)\left[1+\frac{\alpha_s(M)}{\pi}\right] + \frac{\pi^2}{3M^2}\langle\frac{\alpha_s}{\pi}GG\rangle + \frac{704\pi^3}{M^4}\alpha_s\langle\bar{q}q\rangle^2\,. \tag{183}$$

The most stable curve corresponds to $s_0 \approx 0.7\,\mathrm{GeV}^2$ and $4\pi^2 f_\pi^2 \approx 0.65\,\mathrm{GeV}^2$, the experimental value being $4\pi^2 f_\pi^2 = 0.67\,\mathrm{GeV}^2$.

One can also try the ansatz in which A_1 is treated as a (narrow) resonance. The strategy is just the same as in the ρ-case. To this end, one should

- differentiate the daughter (second) sum rule with respect to $1/M^2$ to get the third sum rule,
- transfer the continuum contribution to the right hand side in all sum rules,

[c] Again, we encourage the readers to obtain the curves themselves.

• extract $m^2_{A_1}$ and $s_0^{(A_1)}$ from the ratio of the third sum rule to the second,
• using the values $m^2_{A_1} \approx 1.6\,\mathrm{GeV}^2$ and $s_0^{(A_1)} \approx 2.4\,\mathrm{GeV}^2$ obtained from the previous fitting, find $f^2_{A_1}$ from the second sum rule,
• using the value $4\pi^2 f^2_{A_1} \approx 1.95\,\mathrm{GeV}^2$ obtained from the previous fitting, find f_π^2 from the first sum rule. The value $f_\pi^2 \approx 0.69\,\mathrm{GeV}^2$ obtained in this way is very close to the experimental one $f_\pi^2|^{exp} = 0.67\,\mathrm{GeV}^2$. The A_1 mass value $m^2_{A_1} \approx 1.6\,\mathrm{GeV}^2$ extracted from the QCD sum rule also is in a very good agreement with experiment.

7 QCD Sum Rules for Pion Form Factor

7.1 Three-point function

To analyze the pion form factor, we consider the three-point function, *i.e.*, the correlator of the three currents[20,21]:

$$T^\mu_{\alpha\beta}(p_1,p_2) = i\int e^{-ip_1x+ip_2y}\langle\{j_\beta(y)J^\mu(0)j^+_\alpha(x)\}\rangle d^4x d^4y, \tag{184}$$

where

$$J^\mu = \frac{2}{3}\bar{u}\gamma^\mu u - \frac{1}{3}\bar{d}\gamma^\mu d \tag{185}$$

is the electromagnetic current and

$$j_\alpha = \bar{d}\gamma_5\gamma_\alpha u \tag{186}$$

is the axial current used in the preceding section to study the static properties of the pion. The pion contribution

$$\langle 0|j_\beta(y)|p_2\rangle\langle p_2|J^\mu(0)|p_1\rangle\langle p_1|j^+_\alpha(x)|0\rangle \tag{187}$$

into this three-point correlator can be obtained by inserting the pion state(s) between the currents. The form factor $F_\pi(q^2)$ appears in the middle matrix element:

$$\langle p_2|J^\mu(0)|p_1\rangle = (p_1^\mu + p_2^\mu)F_\pi(q^2), \tag{188}$$

while the other two are projections of the axial current onto the initial and final pion states

$$\langle 0|j_\beta(0)|p_2\rangle = if_\pi(p_2)_\beta \tag{189}$$

$$\langle p_1|j^+_\alpha(0)|0\rangle = if_\pi(p_1)_\alpha. \tag{190}$$

The three-point function $T^\mu_{\alpha\beta}(p_1,p_2)$ is the sum of different structures each characterized by the relevant invariant amplitude $t_i(p_1^2,p_2^2,q^2)$. The first idea is to study the pion form factor analyzing the invariant amplitude corresponding to the structure $(p_2)_\beta(p_1)_\alpha(p_1^\mu + p_2^\mu)$. However, there are other structures $((p_1)_\beta(p_2)_\alpha, (p_1)_\beta(p_1)_\alpha, (p_2)_\beta(p_2)_\alpha$ that coincide with $(p_2)_\beta(p_1)_\alpha$ in the limit $q \to 0$. This complication disappears if all the basic structures are expanded in $P = (p_1 + p_2)/2$ and $q = p_2 - p_1$. Then the pion contribution is

$$(p_2)_\beta(p_1)_\alpha(p_1^\mu + p_2^\mu) = 2\left(P_\beta + \frac{q_\beta}{2}\right)\left(P_\alpha - \frac{q_\alpha}{2}\right)P^\mu = 2P_\beta P_\alpha P^\mu + O(q),$$

and the "best" structure (not changing when q varies) is $P_\alpha P_\beta P^\mu$. The simplest way to extract the relevant amplitude (hereafter referred to as T) is to multiply $T^\mu_{\alpha\beta}(p_1, p_2)$ by $n_\mu n^\alpha n^\beta$, where n is a lightlike vector orthogonal to q. The property $n^2 = 0$ kills all the structures containing $g_{\alpha\beta}$ and other g-terms, while the property $(nq) = 0$ kills all the structures containing any q-factor. Thus, the amplitude we are going to analyze is

$$T(p_1^2, p_2^2, q^2) = \frac{n_\mu n^\alpha n^\beta}{2(nP)^3} T^\mu_{\alpha\beta}(p_1, p_2). \tag{191}$$

7.2 Dispersion representation

To extract information about the form factors of physical states, one should incorporate the double dispersion relation

$$T(p_1^2, p_2^2, q^2) = \frac{1}{\pi^2}\int_0^\infty ds_1 \int_0^\infty ds_2 \,\frac{\rho(s_1, s_2, q^2)}{(s_1 - p_1^2)(s_2 - p_2^2)} + \text{“}subtractions\text{”}. \tag{192}$$

The subtraction terms, in general, are polynomial in p_1^2 and/or p_2^2. They disappear after one applies $B(p_1^2 \to M_1^2)B(p_2^2 \to M_2^2)$, the Borel transformation in both variables [d]. The double Borel transform $\Phi = B_1 B_2 T$ is related to the double spectral density $\rho(s_1, s_2, q^2 = -Q^2)$ by

$$\Phi(M_1^2, M_2^2, q^2 = -Q^2) = \frac{1}{\pi^2}\int_0^\infty \frac{ds_1}{M_1^2}\int_0^\infty \frac{ds_2}{M_2^2}\exp\left(-\frac{s_1}{M_1^2} - \frac{s_2}{M_2^2}\right)\rho(s_1, s_2, Q^2). \tag{193}$$

7.3 Operator product expansion

In contrast to the two-point correlators $\Pi(Q^2)$ analyzed in the previous Section, the three-point amplitudes $T(p_1^2, p_2^2, q^2)$ have three arguments. To incorporate

[d] In our specific case, the subtractions are exhausted by a constant, and a single differentiation with respect to any variable is sufficient to remove it.

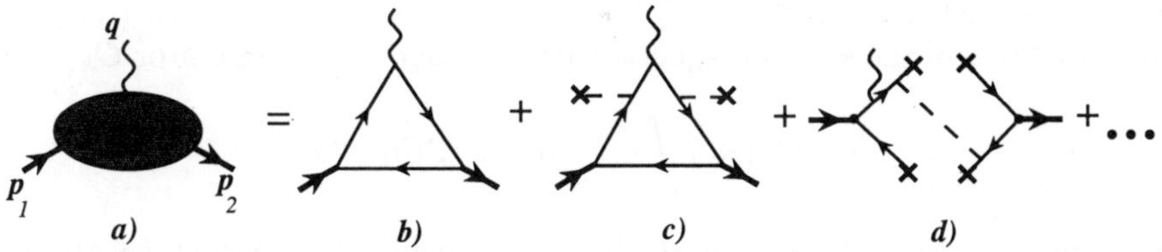

Figure 8: Structure of OPE in case of symmetric kinematics.

asymptotic freedom, one should take the invariants p_1^2, p_2^2 corresponding to the "hadronized" channels to be negative and sufficiently large, say, $|p_1^2|, |p_2^2| > 1\,\mathrm{GeV}^2$. Depending on the value of the third invariant q^2 that works as an external parameter (it also should be negative, of course), there are three essentially different situations.

• The simplest is the symmetric case when Q^2 is of the same order of magnitude as $|p_1^2|, |p_2^2|$:

$$|p_1^2| \sim |p_2^2| \sim Q^2 \sim \mu^2.$$

The operator product expansion for T in this case has the following structure:

$$T(p_1^2, p_2^2, q^2) = \sum_i \frac{1}{(\mu^2)^{d_i}} C_i\left(\frac{\mu^2}{p_1^2}, \frac{\mu^2}{p_2^2}, \frac{\mu^2}{q^2}\right) \langle \mathcal{O}_i \rangle \tag{194}$$

where $C_i/(\mu^2)^{d_i}$ is a perturbatively calculable short-distance coefficient function, $\langle \mathcal{O}_i \rangle$ is the vacuum matrix element of a local operator , *i.e.*, some condensate term and d_i is the dimension of $\mathcal{O}_i$ in mass units (see Fig.8).

• The case when Q^2 is small is more complicated:

$$|p_1^2| \sim |p_2^2| \sim \mu^2, \qquad Q^2 \ll \mu^2.$$

The coefficient functions C_i calculated in the symmetric kinematics, might become singular in this limit because of $(\mu^2/Q^2)^n$- or $\log(\mu^2/Q^2)$-terms. The operator product expansion in such a situation has a modified form:

$$T(p_1^2, p_2^2, q^2) = \sum_i \frac{1}{(\mu^2)^{d_i}} \tilde{C}_i\left(\frac{\mu^2}{p_1^2}, \frac{\mu^2}{p_2^2}, \frac{\mu^2}{q^2}\right) \langle \mathcal{O}_i \rangle$$
$$+ \sum_j \frac{1}{(\mu^2)^{d_j}} E_k\left(\frac{\mu^2}{p_1^2}, \frac{\mu^2}{p_2^2}\right) \Pi_k(Q^2) \tag{195}$$

where the coefficient functions $\tilde{C}_i$ are regular in the $Q^2 \to 0$ limit, $\Pi_k(Q^2)$ is the correlator of the electromagnetic current J and a local operator $\mathcal{O}_k$

$$\Pi_k(Q^2) = i \int e^{-iqx} \langle \mathcal{O}_k(x)\, J(0)\rangle\, d^4x,$$

d_k is the dimension of $\mathcal{O}_k$ in mass units. The coefficient functions E_k are (the Fourier transforms of) those for the operator product expansion of two axial currents

$$T(j\,j^+) = \sum_k E_k\, \mathcal{O}_k.$$

In the modified OPE for the three-point function, the contributions of the first type correspond to a situation when large momentum flow connects all three external vertices, while those of the second type describe the situation when large momentum flows only between the vertices x and y related to the axial currents. The momenta in the J-channel are of the order of Q, *i.e.*, small, and the correlator $\Pi_j(Q^2)$ is a nonperturbative object accumulating long-distance information.

• In the opposite limit of large Q^2

$$|p_1^2| \sim |p_2^2| \sim \mu^2, \qquad Q^2 \gg \mu^2,$$

a simple expansion in powers of $1/|p_1^2|, 1/|p_2^2$ (*i.e.*,in $1/\mu^2$) might be destabilized by large $(Q^2/\mu^2)^n$-terms. Appearance of terms growing with Q^2 is a bit surprising, since form factors should decrease as $Q^2 \to \infty$. But the appearance of such terms is just the artifact of the $1/\mu^2$ expansion: even a very reasonable function like $1/(Q^2+\mu^2)$, vanishing as $Q^2 \to \infty$, produces a divergent series when expanded in $1/\mu^2$. Thus, a possible way out in the large-Q^2 limit is a resummation of the parametrically enhanced contributions of $(Q^2/\mu^2)^n$-type. In fact, it can be established, that such contributions result from the Taylor expansion of the original nonlocal combinations $\langle\varphi(0)\varphi(x)\rangle$ (nonlocal condensates), and the idea is to construct sum rules without expanding the nonlocal condensates, treating them as some phenomenological functions describing the propagation of quark and gluons in the physical vacuum.

7.4 Perturbative contribution

The structure of the power corrections in the operator product expansion, as we discussed above, depends on the particular Q^2-region under study. However, the starting point of any expansion is the perturbative contribution, in our case corresponding to the triangle diagram plus radiative corrections to it.

Using Feynman parametrization, one can rather easily obtain the following representation[21] for the double Borel transform of the lowest-order perturbative triangle diagram (see Fig.8a)

$$\Phi^{pert}(M_1^2, M_2^2, Q^2) = \frac{3}{2\pi^2(M_1^2+M_2^2)} \int_0^1 x(1-x)$$
$$\exp\left\{-\frac{xQ^2}{(1-x)(M_1^2+M_2^2)}\right\} I(x, M_1^2, M_2^2, m_q^2, e_q)\, dx \tag{196}$$

where $I(x, M_1^2, M_2^2, m_q^2, e_q)$ is the factor containing the dependence on the quark masses

$$I(x, M_1^2, M_2^2, m_q^2, e_q) = e_u \exp\left\{-\left(\frac{m_u^2}{1-x} + \frac{m_d^2}{x}\right)\left(\frac{1}{M_1^2} + \frac{1}{M_2^2}\right)\right\} + \{u \leftrightarrow d\}. \tag{197}$$

In view of extreme smallness of the u- and d-quark masses, this factor can be safely set to unity. The x-variable may be interpreted as the fraction of the total momentum P carried by the passive quark (active is the quark interacting with the EM current). This integral representation is very convenient to study the behavior of $\Phi^{pert}(M_1^2, M_2^2, Q^2)$ for small and large Q^2. In the low-Q^2 limit we get

$$\Phi^{pert}(M_1^2, M_2^2, Q^2) = \frac{1}{4\pi^2(M_1^2+M_2^2)}\left\{1 - 2\frac{Q^2}{M_1^2+M_2^2}\right. \tag{198}$$
$$\left. - \sum_{N=0}^{\infty}\left(\frac{-Q^2}{M_1^2+M_2^2}\right)^{N+2} \frac{N+3}{N!}\left(\log\left[\frac{Q^2}{M_1^2+M_2^2}\right] + \frac{1}{N+3} - \psi(N+1)\right)\right\}.$$

The first two terms of the expansion can be obtained by simply expanding the exponential factor in Q^2 under the integral sign. However, the third term of such an expansion produces a logarithmically divergent integral

$$\ldots \int_0^1 \frac{1}{2!}\left(\frac{Q^2}{M_1^2+M_2^2}\right)^2 \frac{x^3}{1-x}\, dx,$$

and higher terms have power divergences when integrated in the region $x \to 1$. This region corresponds to a situation when the whole large momentum P is carried by the passive quark, and the active quark carries small momentum proportional to q. This is precisely the situation producing additional terms in the operator product expansion at small Q^2. The logarithmic factor $\log(Q^2)$, singular when $Q^2 \to 0$, just signalizes that there are long-distance effects (quarks propagate over distances $\sim 1/Q$) limiting the applicability of perturbation theory.

In the large-Q^2 limit one should expand $\Phi^{pert}(M_1^2, M_2^2, Q^2)$ in $1/Q^2$ rather than in Q^2:

$$\Phi^{pert}(M_1^2, M_2^2, Q^2) = \frac{3}{2\pi^2}\frac{M_1^2+M_2^2}{(Q^2)^2}\left\{1 - 8\frac{M_1^2+M_2^2}{Q^2} + \right.$$
$$\left. +\frac{1}{6}\sum_{N=2}^{\infty}(-)^N(N+1)(N+3)!\left(\frac{M_1^2+M_2^2}{Q^2}\right)^N\right\}. \tag{199}$$

Thus, the perturbative contribution vanishes like $1/Q^4$ as $Q^2 \to \infty$. Combining the spectral representation for $\Phi(M_1^2, M_2^2, Q^2)$ and the explicit expression for it, we get the equation for the perturbative spectral density $\rho^{pert}(s_1, s_2, Q^2)$

$$\frac{1}{\pi^2}\int_0^\infty \frac{ds_1}{M_1^2}\int_0^\infty \frac{ds_2}{M_2^2}\exp\left(-\frac{s_1}{M_1^2}-\frac{s_2}{M_2^2}\right)\rho^{pert}(s_1, s_2, Q^2) =$$
$$= \frac{3}{2\pi^2(M_1^2+M_2^2)}\int_0^1 x(1-x)\exp\left(-\frac{xQ^2}{(1-x)(M_1^2+M_2^2)}\right)dx. \tag{200}$$

Solving it (this amounts to a double inverse Laplace transformation) gives the perturbative spectral density:

$$\rho^{pert}(s_1, s_2, Q^2) = \frac{3}{2}\left[\frac{(Q^2)^2}{2!}+\frac{(Q^2)^3}{3!}\right]\frac{1}{\sqrt{(s_1+s_2+Q^2)^2-4s_1s_2}}. \tag{201}$$

In the low-Q^2 limit the perturbative spectral density can be expanded in Q^2:

$$\rho^{pert}(s_1, s_2, Q^2)|_{Q^2\to 0} = \frac{1}{4}\delta(s_1-s_2)\theta(s_1)\theta(s_2) +$$
$$+\frac{Q^2}{4}(s_1+s_2)\delta''(s_1-s_2)\theta(s_1)\theta(s_2) + \ldots. \tag{202}$$

For large Q^2 one can expand it in $1/Q^2$:

$$\rho^{pert}(s_1, s_2, Q^2)|_{Q^2\to\infty} = \frac{3}{2}\theta(s_1)\,\theta(s_2)\left\{\frac{s_1+s_2}{Q^4} - 4\frac{s_1^2+s_2^2+4s_1s_2}{Q^6} + \ldots\right\}. \tag{203}$$

Thus, for small Q^2, the perturbative spectral density is concentrated on the line $s_1 = s_2$: there are no transitions between states with different masses. Along this line, the spectral density does not vary: it has no dependence on $s_1 + s_2$. In the opposite limit, when $Q^2 \to \infty$, the situation is, in a sense, reversed: the density is nonzero in the whole region $s_1 > 0, s_2 > 0$, there is no dependence on $(s_1 - s_2)$, and the density grows like $s_1 + s_2$. The last observation means that higher states are relatively more important for large Q^2 than for the small ones.

7.5 Sum rule for the pion form factor at moderate Q^2

The physical spectral density $\rho(s_1, s_2, Q^2)$, of course, differs from its perturbative analog. It has a rather complicated structure on the s_1, s_2-plane. In particular, $\rho(s_1, s_2, Q^2)$ contains the term corresponding to the pion form factor

$$\rho_{\pi\pi}(s_1, s_2, Q^2) = \pi^2 f_\pi^2 F_\pi(Q^2)\delta(s_1 - m_\pi^2)\delta(s_2 - m_\pi^2). \tag{204}$$

In addition, it contains the contributions corresponding to transitions between the pion and higher resonances, and also the terms related to elastic and transition form factors of the higher resonances. Constructing the two-dimensional (s_1, s_2) analog of the simplest ansatz "lowest state plus continuum" we can treat all the contributions, except for the $\rho_{\pi\pi}$, as "continuum", *i.e.*, assume that $\rho(s_1, s_2, Q^2) = \rho^{pert}(s_1, s_2, Q^2)$ outside the square $(0, s_0) \times (0, s_0)$:

$$\rho(s_1, s_2, Q^2) = \rho_{\pi\pi}(s_1, s_2, Q^2) + (1 - \theta(s_1 < s_0)\theta(s_2 < s_0))\, \rho^{pert}(s_1, s_2, Q^2). \tag{205}$$

Now, transferring the continuum contribution to the right-hand side of the sum rule and using the operator product expansion in the symmetric kinematics $(Q^2 \sim M_1^2, M_2^2)$, we obtain the QCD sum rule for the pion form factor:

$$f_\pi^2 F_\pi(Q^2) = \frac{1}{\pi^2}\int_0^{s_0} ds_1 \int_0^{s_0} ds_2 \exp\left(-\frac{s_1 + s_2}{M^2}\right) \rho^{pert}(s_1, s_2, Q^2) + \\ + \frac{\alpha_s\langle GG\rangle}{12\pi M^2} + \frac{16}{81}\frac{\pi\alpha_s\langle\bar{q}q\rangle^2}{M^4}\left(13 + 2\frac{Q^2}{M^2}\right). \tag{206}$$

To treat initial and final states on equal footing and to simplify the analysis, we took $M_1^2 = M_2^2$. Unfortunately, the double integral in this sum rule cannot be calculated explicitly for arbitrary Q^2. However, it is instructive to consider the formal limit $Q^2 = 0$:

$$f_\pi^2 F_\pi(0) = \frac{1}{4\pi^2}\frac{M^2}{2}\left(1 - e^{-2s_0/M^2}\right) + \frac{\alpha_s\langle GG\rangle}{12\pi M^2} + \frac{208}{81}\frac{\pi\alpha_s\langle\bar{q}q\rangle^2}{M^4}. \tag{207}$$

This should be compared to the f_π^2 sum rule:

$$f_\pi^2 = \frac{m^2}{4\pi^2}\left(1 - e^{-s_0/m^2}\right) + \frac{\alpha_s\langle GG\rangle}{12\pi m^2} + \frac{176}{81}\frac{\pi\alpha_s\langle\bar{q}q\rangle^2}{m^4}. \tag{208}$$

The condensate terms look almost identical, but comparing the perturbative terms we observe that M^2, the Borel parameter of the form factor sum rule, should be larger by factor 2 than m^2, the Borel parameter of the f_π^2 sum rule: $M^2 = 2m^2$. After this change, the $\langle GG \rangle$-term in the form factor sum rule is multiplied by $\frac{1}{2}$ and the $\langle \bar{q}q \rangle^2$-term is multiplied by $\frac{1}{4}$

$$f_\pi^2 F_\pi(0) = \frac{m^2}{4\pi^2}\left(1 - e^{-s_0/m^2}\right) + \frac{\alpha_s \langle GG \rangle}{24\pi m^2} + \frac{52}{81}\frac{\pi \alpha_s \langle \bar{q}q \rangle^2}{m^4}. \tag{209}$$

Thus, we obtained a sum rule, which looks like that for f_π^2, but with the gluonic condensate term smaller by factor $2 \approx (1.4)^2$, and the quark condensate term smaller by factor $\sim 3.5 \approx (1.5)^3$. Since it is the values of the condensates that (after the fitting procedure) determine all the hadronic scales, we conclude that all the hadronic parameters having the dimension of $(mass)^2$, *i.e.*, s_0 and the combination $f_\pi^2 F_\pi(0)$, extracted from this sum rule are by a factor $1.4 - 1.5$ smaller than s_0^π and f_π^2, respectively[e]. To get s_0 closer to s_0^π, one should restore the matching between the perturbative and the condensate terms, *i.e.*, decrease the perturbative term by a factor 2 to 4. This allows us to make an educated guess that QCD sum rule for the pion form factor will most effectively work in the intermediate region where the form factor varies between 0.5 and 0.25, *i.e.*, for Q^2 between $0.5\,\text{GeV}^2$ and $2\,\text{GeV}^2$.

The readers can check these guesses by explicit fitting procedure. To this end, one should plot the combination $f_\pi^2 F_\pi(Q^2)$ as a function of the Borel parameter M^2 for different values of the effective threshold s_0. As a "true" value of s_0 one can take the value for which the curve is constant as $M^2 \to \infty$. One would observe that the value of s_0 obtained in this way slowly grows with Q^2, from $0.6\,\text{GeV}^2$ for $Q^2 = 0.5\,\text{GeV}^2$ to $1.0\,\text{GeV}^2$ for $Q^2 = 3\,\text{GeV}^2$. The growth of s_0 reflects the fact that, when the Q^2 value increases, the condensate contributions remain constant (with the quark term even slowly growing) whereas the perturbative term decreases. The constancy of the condensate contributions is an artifact of our approximation. If one includes operators of higher dimensions in the OPE, there appear terms diminishing the total condensate contribution for large Q^2. These terms, however, have the structure $(Q^2/M^2)^n$, and one should resum them to get a reasonable result. However, if one restricts the analysis to the region $0.5\,\text{GeV}^2 < Q^2 < 3\,\text{GeV}^2$, the predictions of our sum rule agree with existing pion form factor data within $10\% - 20\%$ accuracy.

[e] One should not worry about the fact that a straightforward extrapolation of the form factor sum rule to the point $Q^2 = 0$, gives $F_\pi(0) \approx 0.7$. This sum rule is valid only for sufficiently high Q^2, and, as we emphasized, there appear additional terms in the OPE which will increase the condensate contributions to make them equal to those in the f_π^2 sum rule and recover the $F_\pi(0) = 1$ normalization condition.

7.6 Local quark-hadron duality for the pion form factor

The approach described above allows to obtain the form factor at different values of Q^2 by independent fitting procedure, point by point. But it is tempting to see the Q^2-curve as a whole. This can be done by looking at the $F_\pi(Q^2)$-curves at fixed s_0 for different values of the Borel parameter M^2. For $M^2 > 1.5\,\text{GeV}^2$, all the curves are close to each other, approaching the limiting (s_0-dependent) form as $M^2 \to \infty$. The nature of this convergence is very simple: the condensate terms disappear from the sum rule in the $M^2 \to \infty$ limit, and the limiting curve corresponds to the local duality relation:

$$f_\pi^2 F_\pi(Q^2) = \frac{1}{\pi^2} \int_0^{s_0} ds_1 \int_0^{s_0} \rho^{pert}(s_1, s_2, Q^2)\, ds_2. \tag{210}$$

Note, that the function $\rho^{pert}(s_1, s_2, Q^2)$ describes the transition from a free-quark ($\bar{q}q$) state with invariant mass s_1 to a similar state with mass s_2. The local duality relation states, essentially, that one can calculate the pion form factor by averaging the form factors of such transitions over the appropriate duality interval s_0. In the two-point case, the local duality produces the relation between the pion decay constant f_π^2 and the duality interval s_0

$$s_0 = 4\pi^2 f_\pi^2 \approx 0.7\,\text{GeV}^2. \tag{211}$$

We recall that the duality interval is the effective threshold for production of higher resonances, and is very natural that it is just in the middle between $m_\pi^2 \approx 0$ and $m_{A_1}^2 \approx 1.6\,\text{GeV}^2$. Using the explicit form for $\rho^{pert}(s_1, s_2, Q^2)$, we obtain[21]

$$F_\pi(Q^2)|^{local\ duality} = 1 - \frac{1 + 6s_0/Q^2}{(1 + 4s_0/Q^2)^{3/2}}. \tag{212}$$

This formula gives a correct value for the form factor at $Q^2 = 0$, but the predicted slope is wrong:

$$F_\pi(Q^2)|^{local\ duality} = 1 - \frac{3}{4}\frac{Q}{\sqrt{s_0}} + \ldots. \tag{213}$$

This is just the same situation we encountered applying the local duality to the form factor of the lowest oscillator state. The reason is that there is a serious mismatch between perturbative and physical spectral densities: the perturbative density is concentrated in the narrow stripe $|s_1 - s_2| < Q$ near the $s_1 = s_2$ line, while physical density, in addition to the elastic form factors (points at the $s_1 = s_2$ line), contains also transition form factors (line $s_1 = 0, s_2 > s^{(3\pi\, threshold)}$, *etc.*) corresponding to the regions very far from the $s_1 =$

s_2 line. In other words, our model for the hadronic spectrum is too rough at small nonzero Q^2. The perturbative density in this region can be approximated by

$$\rho^{pert}(s_1, s_2, Q^2) \sim \frac{1}{Q}\theta(|s_1 - s_2| < Q). \tag{214}$$

Integrating it over any line $s_1 + s_2$ =const., one obtains a result analytic in Q^2. In our model, however, there are small triangles not included into the integration region. Their area is $\sim Q^2$, and the density is $\sim 1/Q$, so the resulting contribution (to be subtracted from 1) is $O(Q) = O(\sqrt{Q^2})$. To avoid this, one can try a "triangle" model for the higher states: "pion + (continuum outside the triangle formed by the line $(s_1 + s_2 = S_0)$". The local duality relation in this case gives[21]

$$F_\pi(Q^2)^{triangle\, l.d.} = \frac{S_0}{8\pi^2 f_\pi^2 (1 + Q^2/2S_0)^2}. \tag{215}$$

To get the correct normalization $F_\pi(0) = 1$ one should take $S_0 = 8\pi^2 f_\pi^2 = 2s_0$. However, the slope in this case is too small $F_\pi^{tr.l.d.}(Q^2)|_{S_0=2s_0} = 1 - Q^2/2s_0$, by factor 2 smaller than the experimental one, and the curve goes well above both the experimental points and the curve corresponding to the "squared" form of the local duality. To match with the latter, one should take $S_0 = \sqrt{2}s_0$. In this case the area of the local duality region is the same. Outside the small-Q^2 region, there is good agreement between the two local duality curves and the data approximated by the fit $F_\pi^{fit}(Q^2) \approx 1/(1 + Q^2/0.47)$.

7.7 One-gluon exchange contribution

The contribution into the pion form factor we calculated using the lowest order $O(\alpha_s^0)$ term in the perturbative spectral density corresponds to the so-called "soft diagram". Since this contribution decreases as $1/Q^4$ for large Q^2, it is normally ignored in the perturbative QCD analysis concentrated on the study of the "hard gluon exchange diagram" that has the $1/Q^2$ asymptotic behavior dictated by the quark counting rules.

However, the soft contribution as given, e.g., by the local quark-hadron duality, is very close to the experimental data leaving not much place for any other contribution. To estimate the contribution due to the "hard gluon" exchange diagrams, one should include the $O(\alpha_s(Q^2)$-terms in the perturbative spectral density. To simplify the analysis, one can use the local duality approximation. Still, the two-loop calculation is rather complicated, but the results,

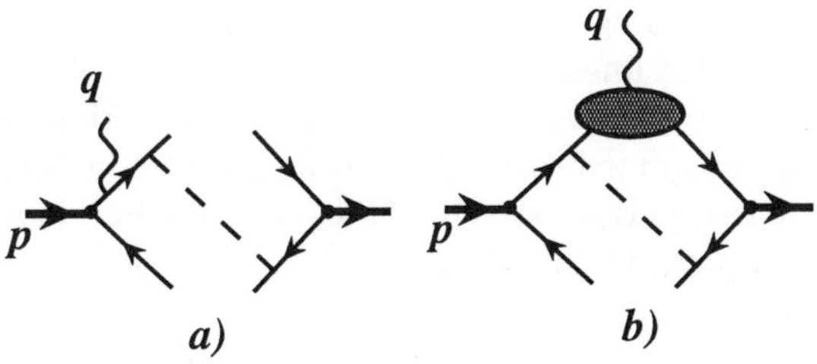

Figure 9: Gauge-invariant set of $\langle \bar{q}q \rangle^2$ diagrams.

with a rather high accuracy, can be approximated by a simple interpolation

$$\frac{1}{\pi^2 f_\pi^2}\int_0^{s_0} ds_1 \int_0^{s_0} ds_2 \Delta\rho^{pert}(s_1, s_2, q^2) \approx \frac{\alpha_s}{\pi}\frac{1}{1+Q^2/2s_0} \tag{216}$$

between the $Q^2 = 0$ value (related by the Ward identity to the $O(\alpha_s)$ term of the 2-point correlator) and the asymptotic behavior $F_\pi^{as}(Q^2) = 8\pi\alpha_s(Q^2)f_\pi^2/Q^2$.

7.8 Pion form factor at small Q^2

As we discussed earlier, if the QCD sum rule derived in the moderate-Q^2 region is simple-mindedly extrapolated to the point $Q^2 = 0$, one obtains the expression

$$f_\pi^2 F_\pi(0) \stackrel{?}{=} \frac{m^2}{4\pi^2}\left(1 - e^{-s_0/m^2}\right) + \frac{\alpha_s\langle GG\rangle}{24\pi m^2} + \frac{52}{81}\frac{\pi\alpha_s\langle\bar{q}q\rangle^2}{m^4} \tag{217}$$

that differs from the sum rule for f_π^2

$$f_\pi^2 = \frac{m^2}{4\pi^2}\left(1 - e^{-s_0/m^2}\right) + \frac{\alpha_s\langle GG\rangle}{12\pi m^2} + \frac{176}{81}\frac{\pi\alpha_s\langle\bar{q}q\rangle^2}{m^4}. \tag{218}$$

The condensate terms have the coefficients essentially smaller than it is necessary to enjoy the simple Ward identity prediction $F_\pi(Q^2) = 1$. The latter requires that the QCD sum rule for $f_\pi^2 F_\pi(0)$ should coincide with that for f_π^2. The Ward identity, being just the statement that the total electric charge of the pion is equal to that of the electron, should not be violated. Hence there should exist the additional condensate contributions "visible" at $Q^2 = 0$ (to satisfy the Ward identity) and dying rapidly for $Q^2 > m_\rho^2$ (to reproduce the intermediate-Q^2 sum rule).

To get a feeling about how these terms might look like, consider the $\alpha_s\langle\bar{q}q\rangle^2$ diagrams. To enjoy the consequences of the Ward identity, one should have a gauge invariant set of diagrams. This can be achieved by inserting the photon vertex into all possible lines, including also the "external" lines corresponding to quarks going into vacuum. However, in perturbation theory, this brings in terms $\sim 1/Q^2$ singular in the $Q^2 \to 0$ limit. Such a "coefficient function" cannot be calculated perturbatively. Thus, one should separate the large- and small-momentum parts of the diagram

$$T(p^2, q^2) = C(p^2) \otimes \Pi(q^2) \tag{219}$$

and treat the Q^2-dependent part as a correlator of the original electromagnetic current J and some local operator constructed from the quark fields going "into vacuum" (see Fig.9b):

$$\Pi(q^2) \sim \int e^{iqx} \langle T(J_\mu(0)\mathcal{O}(x))\rangle d^4x. \tag{220}$$

The question now is whether it is possible to calculate contributions like the correlators $\Pi(q^2)$ at small q? There are essentially two types of these correlators depending on the type of the operator $\mathcal{O}$. First, one should take into account that under the T-product sign, one cannot throw out the operators apparently vanishing due to the equations of motion, *e.g.*, those containing $\not{D}$ acting on the quark field, since there appear contact terms. In this case $\Pi(q^2)$ is a constant proportional to the quark condensate $\langle \bar{q}q \rangle$. In all other cases one can (approximately) calculate $\Pi(q^2)$ using the following QCD sum rule strategy:

• Calculate $\Pi(q^2)$ for large Q^2 using the operator product expansion. This normally gives

$$\Pi(q^2 = -Q^2) \sim \frac{\langle \bar{q}q \rangle}{Q^2} + k\frac{\langle \bar{q}D^2 q \rangle}{Q^4} + \dots.$$

• Use the dispersion relation

$$\Pi(Q^2) = \frac{1}{\pi} \int_0^\infty \frac{\rho(s)\, ds}{s + Q^2}.$$

• Assume an appropriate ansatz for the spectral density:

$$\rho(s) = \pi A \delta(s - m_\rho^2) + \pi B \delta(s - m_R^2).$$

Since the perturbative density is zero in this case, the higher states (lying above the ρ-meson) are approximated by an effective resonance R.

• Requiring the best agreement between the two representations for $\Pi(Q^2)$, extract the values of the constants A, B and m_R^2.

• The result is

$$\Pi(Q^2) = \frac{A}{Q^2 + m_\rho^2} + \frac{B}{Q^2 + m_R^2},$$

and one can use it for small Q^2.

In this way one can obtain the QCD sum rule for the pion form factor valid in the low-Q^2 region[22]:

$$f_\pi^2 F_\pi(Q^2) = \text{“}perturbative\, part\text{”} + \frac{\alpha_s \langle GG \rangle}{12\pi m^2}\left[\frac{1}{2} + \frac{m_\rho^2}{2(m_\rho^2 + Q^2)}\right] +$$
$$+\frac{16}{81}\frac{\pi \alpha_s \langle \bar{q} q \rangle^2}{m^4}\left[4 + 6\left(\frac{1.6\, m_\rho^2}{m_\rho^2 + Q^2} - \frac{0.6\, m_R^2}{m_R^2 + Q^2}\right) + \frac{m_\rho^2}{m_\rho^2 + Q^2} + \frac{Q^2}{m^2}\right]. \quad (221)$$

This sum rule possesses all the required properties:

• For $Q^2 = 0$ its left hand side exactly coincides with that of the f_π^2 sum rule.

• additional terms die out when Q^2 gets large.

• The curve for the pion form factor calculated with the help of this sum rule in the region of small Q^2, at the matching point $Q^2 = m_\rho^2$, agrees within 10% with the curve obtained from the moderate-Q^2 sum rule.

• the pion charge radius, extracted from the low-Q^2 sum rule,

$$\langle r_\pi^2 \rangle^{1/2}|_{SR} = 0.66 \pm 0.03\, fm$$

is in good agreement with the experimental value

$$\langle r_\pi^2 \rangle^{1/2}|_{exp.} = 0.636 \pm 0.036\, fm\,.$$

8 Conclusions

Thus, the QCD sum rules give an accurate description of the pion form factor in the whole region where the data exist. They allow one to trace its behavior from the normalization point $Q^2 = 0$, through the low-Q^2 region, to the region of moderately large Q^2. Everywhere the QCD sum rule results are in agreement with experiment within the accuracy of the method. The QCD sum rules unambigously demonstrate that, in the experimentally accessible region, the form factor is dominated by the soft contribution, which is not calculable within a purely perturbative approach. They show that the one-gluon-exchange mechanism, though dominant in the asymptotic limit, is of minor importance in the region $Q^2 < 4\,\text{GeV}^2$ and, maybe, till even larger values of Q^2.

Acknowledgements

This work was supported by the US Department of Energy under contract DE-AC05-84ER40150.

References

1. M.A. Shifman, A.I. Vainshtein and V.I. Zakharov: *Nucl. Phys.* B **147**, 385, 448 (1979).
2. A.V. Efremov and A.V. Radyushkin, *Phys. Lett.* B **94**, 245 (1980); *Theor. Mat. Fiz.* **42**, 147 (1980).
3. G.P. Lepage and S.J. Brodsky, *Phys. Lett.* B **87**, 359 (1979); *Phys. Rev.* D **22**, 2157 (1980).
4. V.L. Chernyak, A.R. Zhitnitsky and V.G. Serbo, *JETP Lett.* **26**, 594 (1977); V.L. Chernyak and A.R. Zhitnitsky: *JETP Lett.* **25**, 510 (1977); Yad. Fiz. **31**, 1053 (1980); *Phys. Rep.* **112**, 173 (1984).
5. G.R. Farrar and D.R. Jackson, *Phys. Rev. Lett.* **43**, 246 (1979).
6. G.P. Lepage and S.J. Brodsky, *Phys. Lett.* B **87**, 359 (1979)
7. A.I. Vainshtein, V.I. Zakharov, V.A. Novikov, M.A. Shifman, *Sov.J.Nucl.Phys.* **32**, 840 (1980).
8. V.A. Beilin, A.V. Radyushkin, Form-factors of oscillator states from QCD sum rules, report JINR-P2-87-216, Dubna (1987).
9. A.V. Radyushkin, Habilitation dissertation, JINR, Dubna (1987), unpublished.
10. V.A. Nesterenko and A.V. Radyushkin, *Sov.J.Nucl.Phys.* **39**, 811 (1984).
11. A.V. Radyushkin, *Phys. Lett.* B **271**, 218 (1991).
12. K. Wilson, *Phys. Rev.* **179**, 1499 (1969).
13. A.V. Efremov and A.V. Radyushkin, *Theor. Mat. Fiz.* **44**, 774 (1981).
14. S. N. Nikolaev and A.V. Radyushkin, *Phys.Lett.* **110B**, 476 (1982); *Nucl. Phys.* B **213**, 285 (1983).
15. V.A. Fock, *Sov. Phys.*, 12 (1937).
16. J. Schwinger, *Phys. Rev.* **82**, 664 (1951).
17. M.S. Dubovikov and A.V. Smilga, *Nucl.Phys.* B **185**, 109 (1981).
18. M.A. Shifman, *Nucl. Phys.* B **173**, 13 (1980).
19. S.V. Ivanov, G.P. Korchemsky and A.V. Radyushkin, *Sov.J.Nucl.Phys.* **44**, 145 (1986).
20. B.L. Ioffe and A.V. Smilga, *Phys.Lett.* B **114**, 53 (1982); *Nucl. Phys.* B **216**, 373 (1983).
21. V.A. Nesterenko and A.V. Radyushkin, *Phys.Lett.* B **115**, 410 (1982).
22. V.A. Nesterenko and A.V. Radyushkin, *JETP Lett.* **39**, 707 (1984).

QUARK CONFINEMENT AND THE HADRON SPECTRUM

Nora Brambilla[a]
Institute for Theoretical Physics, University Vienna
Boltzmanngasse 5, A-1090 Vienna, Austria
E-mail: brambill@doppler.thp.univie.ac.at nora.brambilla@mi.infn.it

Antonio Vairo
High Energy Institute, Austrian Academy of Science
Nikolsdorfergasse 18, A-1050 Vienna, Austria
E-mail: vairo@hephy.oeaw.ac.at antonio.vairo@cern.ch

Dedicated to the 60th birthday of Dieter Gromes

These lectures contain an introduction to the following topics:

- Phenomenology of the hadron spectrum.
- The static Wilson loop in perturbative and in lattice QCD. Confinement and the flux tube formation.
- Non static properties: effective field theories and relativistic corrections to the quarkonium potential.
- The QCD vacuum: minimal area law, Abelian projection and dual Meissner effect, stochastic vacuum.

1 Introduction

Quarks appear to be confined in nature. This means that free quarks have not been detected so far but only hadrons, their bound states. Therefore, predictions of the hadron spectrum are an explicit and direct test of our understanding of the confinement mechanism as a result of the low energy dynamics of QCD. In these lectures we will focus on heavy quark bound states and for simplicity we will only treat the quark-antiquark case i.e. the mesons. Indeed, also in order to extract the Cabibbo–Kobayashi–Maskawa matrix elements and to study CP violation from the experimental decay rate of heavy mesons, hadronic matrix elements are needed. It is reasonable to expect that

[a]Lectures given by Nora Brambilla at HUGS at CEBAF, 13th Annual Hampton University Graduate Studies at the Continuous Electron Beam Facility, May 26-June 12, 1998.

to this aim the somewhat simpler matching of the theoretical prediction for the spectrum to experiment has to be previously established. Moreover, we need to achieve some understanding of the bound state dynamics in QCD in order to make reliable identifications of the gluonic degrees of freedom in the spectrum (hybrids, glueballs). All this is relevant to the programme of most of the accelerators machines, Godfrey et al. (1998). From a more general point of view, the issue about the consequences of a nontrivial vacuum structure and the nonperturbative definition of a field theory are questions that overlap with the domain of supersymmetric and string theories.

These lectures are quite pedagogical and introductory and contain several illustrative exercises. The interested reader is referred for details to the quoted references.

The plan of the lectures is the following one. In Secs. 2 and 3 we give a brief overview on the hadron spectrum and on the phenomenological models devised to explain it. The nonperturbative phenomenological parameter σ is introduced and connected to the Regge trajectories as well as to the string models. In Sec. 4 we evaluate perturbatively and nonperturbatively (via lattice simulations) the static Wilson loop. We discuss the area law behaviour and the flux tube formation as signals of confinement. In Sec. 5 we summarize some existing model independent results on the heavy quark interaction. QCD effective field theories are introduced. In Sec. 6 we connect confinement to the structure of the QCD vacuum. We study with some detail the Minimal Area Law model. We discuss the Abelian Higgs model and the Dual Meissner effect and introduce the idea of 't Hooft Abelian projection. We list some of the results obtained on the lattice with partial gauge fixing. Finally we briefly review two models of the QCD vacuum: Dual QCD and the Stochastic Vacuum Model. Each section is supplemented with some exercises. We tried to be as self-contained as possible reporting all the relevant definitions and the basic concepts.

2 The Hadron Spectrum

The meson and the baryon resonances together with an introduction to the quark model have been discussed at this school by Jim Napolitano. Since these lectures are mainly concerned with the quark confinement mechanism, they contain only a general overview on the spectrum pointing out its relevant features.

Let us concentrate on the meson spectrum as given in Figs. 1-3. In principle, one should be able to explain and predict it only by means of the

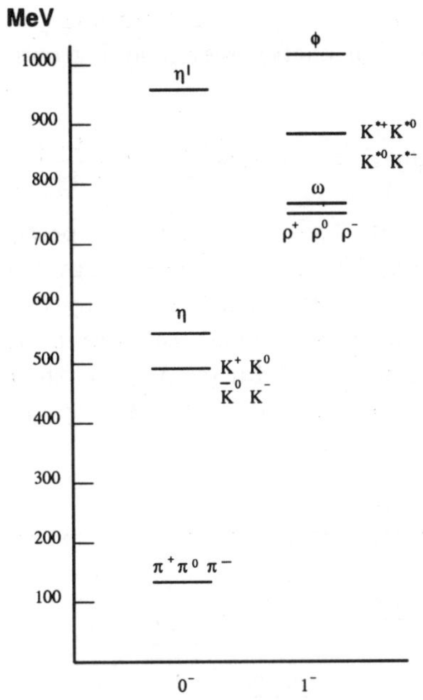

Figure 1: *The spectrum of the lightest mesons labeled by* $(\text{spin})^{\text{parity}}$.

QCD Lagrangian

$$L = -\frac{1}{4}F^{(a)}_{\mu\nu}F^{(a)\mu\nu} + i\sum_{q=1}^{N_f}\bar{\psi}^i_q\gamma^\mu(D_\mu)_{ij}\psi^j_q - \sum_{q=1}^{N_f} m_q\bar{\psi}^i_q\psi^i_q \tag{1}$$

where

$$\begin{aligned} F_{\mu\nu} &\equiv \partial_\mu A_\nu - \partial_\nu A_\mu + ig[A_\mu, A_\nu] \\ (D_\mu)_{ij} &\equiv \delta_{ij}\partial_\mu + ig\frac{\lambda^a_{ij}}{2}A^a_\mu\,; \qquad\qquad A_\mu \equiv A^a_\mu\frac{\lambda_a}{2} \quad a = 1,\ldots,8. \end{aligned} \tag{2}$$

ψ^j_q are the quark fields of flavour $q = 1, \ldots, 6 = N_f$ and colour $j = 1, \ldots, 3$, m_q the (current) quark masses and λ^a the Gell-Mann matrices of $SU(3)$. However, if we proceed to calculate the spectrum from this Lagrangian using the familiar tool of a perturbative expansion in the coupling constant g, we get no match with the experimental data presented in Figs. 1-3. This is a consequence of the most relevant feature of the Lagrangian (1): asymptotic freedom (Gross and Wilczek (1973), Politzer (1973)). The coupling constant ($\alpha_s \equiv g^2/4\pi$) vanishes in the infinitely high energy region and grows uncontrolled in the infrared energy region:

$$\frac{d\,\alpha_s(\mu)}{d\log\mu^2} \equiv \beta(\alpha_s) = -\alpha_s\left(\beta_0\frac{\alpha_s}{4\pi} + \beta_1\left(\frac{\alpha_s}{4\pi}\right)^2 + \cdots\right) \tag{3}$$

where $\beta_0 = 11 - 2/3N_f > 0$. As a consequence, a perturbative treatment is expected to be reliable in QCD only when the energy μ is large compared to Λ_{QCD}, which is the infrared energy scale defined by Eq. (3)[b]. The point is that the relevant energies in a quark bound state are in most cases of the order of the naturally occurring scale of 1 fm ($\simeq \Lambda^{-1}_{QCD}$), the average hadron size. The only exception are the mesons made up with top quarks. Unfortunately, they have not enough time to exist![c]

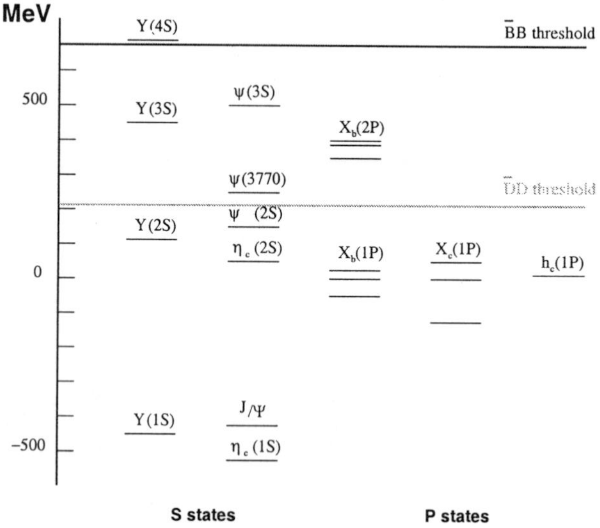

Figure 2: *The experimental heavy meson spectrum ($b\bar{b}$ and $c\bar{c}$) relative to the spin-average of the $\chi_b(1P)$ and $\chi_c(1P)$ states.*

The masses appearing in the Lagrangian are the so-called "current masses" and fall into two categories: light quark masses $m_u = 1.5 \div 5$ MeV, $m_d = 3 \div 9$

[b] $\Lambda^{N_f=4}_{\overline{\rm MS}} = 305 \pm 25 \pm 50$ MeV; this value corresponds to $\alpha_s(M_Z) = 0.117 \pm 0.002 \pm 0.004$, Particle Data Group (PDG) (1998).

[c] The top quark decays into a real W and a b with a large width. The toponium lifetime would be even smaller than the revolution time, thereby precluding the formation of a mesonic bound state, Quigg (1997), Bigi et al. (1986).

MeV, $m_s = 60 \div 170$ MeV (m_u, $m_d \ll \Lambda_{QCD}$ and $m_s \sim \Lambda_{QCD}$) and heavy quark masses $m_c = 1.1 \div 1.4$ GeV, $m_b = 4.1 \div 4.4$ GeV, $m_t = 173.8 \pm 5.2$ GeV ($m_c, m_b, m_t \gg \Lambda_{QCD}$)[d]. The spectrum should be obtained from the QCD Lagrangian with these values of the masses. However, in the light quark sector, spontaneous chiral symmetry breaking and non-linear strongly coupled effects cooperate in a highly non trivial way. Indeed, it is peculiar of QCD that, due to confinement, the quark masses are not physical, i.e. directly measurable quantities. Therefore, it turns out to be useful, in order to make phenomenological predictions, to introduce the so-called constituent quark masses, containing the current masses as well as mass corrections also due to confinement effects. These constituent masses can be defined, for example, using the additivity of the quark magnetic moments inside a hadron or using phenomenological potential models to fit the spectrum. For quarks heavier than Λ_{QCD} the difference between current and constituent masses is not quite relevant.

In the framework of the constituent quark model the meson states are classified as follows. For equal masses the quark and the antiquark spins combine to give the total spin $\mathbf{S} = \mathbf{S}_1 + \mathbf{S}_2$ which combines with the orbital angular momentum $\mathbf{L}$ to give the total angular momentum $\mathbf{J}$. The resulting state is denoted by $n^{2S+1}L_J$ where $n-1$ is the number of radial nodes. As usual, to $L=0$ is given the name S, to $L=1$ the name P, to $L=2$ the name D and so on. The resonances are classified via the J^{PC} quantum numbers, $P = (-1)^{L+1}$ being the parity number and $C = (-1)^{L+S}$ the C-parity.

Light mesons (as well as baryons) of a given internal symmetry quantum number but with different spins obey a simple spin (J)-mass (M) relation. They lie on a Regge trajectory

$$J(M^2) = \alpha_0 + \alpha' M^2 \tag{4}$$

with $\alpha' \simeq 0.8 - 0.9\,\mathrm{GeV}^{-2}$, see Fig. 3. Up to now, free quarks have not been detected. The upper limit on the cosmic abundance of relic quarks, n_q, is $n_q/n_p < 10^{-27}$, n_p being the abundance of nucleons, while cosmological models predict $n_q/n_p < 10^{-12}$ for unconfined quarks. The fact that no free

[d]The masses are scale dependent objects. For what concerns the experimental values quoted above (PDG (1998)), the u, d, s quark masses are estimates of the so-called current quark masses in a mass independent subtraction scheme such as $\overline{\mathrm{MS}}$ at a scale $\mu \simeq 2$ GeV. The c and b quark masses are estimated from charmonium, bottomonium, D and B masses. They are the running masses in the $\overline{\mathrm{MS}}$ scheme. These can be different from the constituent masses obtained in potential models, see below. We remark that it exists a definition of the mass, the pole quark mass (appropriate only for very heavy quarks) which does not depend of the renormalization scale μ. The quark masses are calculated in lattice QCD, QCD sum rules, chiral perturbation theory. For further explanations see Dosch and Narison (1998), Jamin et al. (1998), Kenway (1998), Leutwyler (1996).

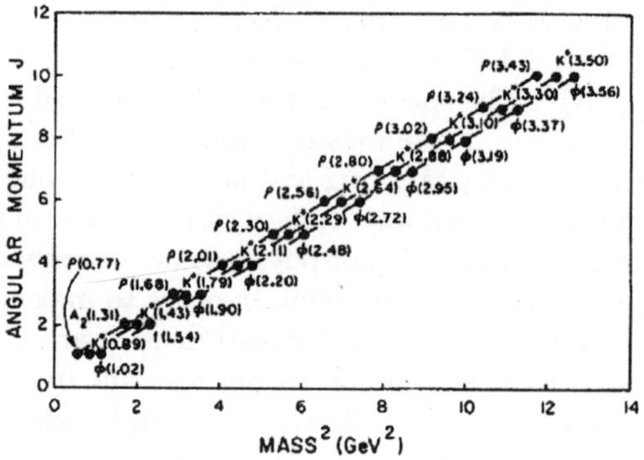

Figure 3: *The Regge trajectories for the ρ, K^* and ϕ. From Godfrey et al. (1985).*

quarks have been ever detected hints to the property of quark confinement. Hence, the interaction among quarks has to be so strong at large distances that a $q\bar{q}$ pair is always created when the quarks are widely separated. From the data it is reasonable to expect that a quark typically comes accompanied by an antiquark in a hadron of mass 1 GeV at a separation of 1 fm ($\simeq \Lambda_{QCD}^{-1}$). This suggests that between the quark and the antiquark there is a linear energy density (called string tension) of order

$$\sigma = \frac{\Delta E}{\Delta r} \simeq 1 \frac{\text{GeV}}{\text{fm}} \simeq 0.2\,\text{GeV}^2. \tag{5}$$

The evidence for linear Regge trajectories (see Fig. 3) supports this picture. A theoretical framework is provided by the string model, Nambu (1974). In this model the hadron is represented as a rotating string with the two quarks at the ends. The string is formed by the chromoelectric field responsible for the flux tube configuration and for the quark confinement (see Fig. 4), Buchmüller (1982). Upon solution of the Exercise 2.1, the reader can verify that it is possible to establish the relation

$$\alpha' = \frac{1}{2\pi\sigma} \tag{6}$$

between the slope of the Regge trajectories and the string tension. The string tension σ emerges as a key phenomenological parameter of the confinement physics.

From the light mesons spectrum of Fig. 1 it is evident that

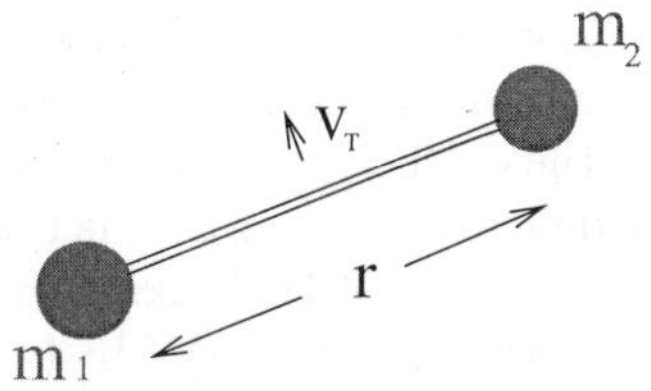

Figure 4: *Picture of the quark-antiquark bound state in the string model.*

- the separations between levels are considerably larger than the mesons masses $\Longrightarrow$ they are truly relativistic bound states;
- the splitting between the pseudoscalar π and the vector ρ mesons is so large to be anomalous $\Longrightarrow$ it is due to the Goldstone boson nature of the π.

Therefore, understanding the light meson spectrum means to solve a relativistic many-body bound state problem where confinement is strongly related with the spontaneous breaking of chiral symmetry. Since the target of these lectures is to gain some understanding of the confinement mechanism in relation to the spectrum, we will try to separate problems and to consider first the spectrum of mesons built by heavy valence quarks only. In this case we have still bound states of confined quarks, however, due to the large mass of the quarks involved, we can hope to treat relativistic and many-body (i.e. quark pair creation effects) contributions as corrections. The problem simplifies remarkably if we consider mesons made up by two valence heavy quarks ($b\bar{b}$, $c\bar{c}$, $b\bar{c}$, ...) i.e. quarkonium.

2. Exercises

2.1 Consider two massless and spinless quarks connected by a string of length R rotating with the endpoints at the speed of light, so that each point at distance r from the centre has the local velocity $v/c = 2r/R$. Recalling that the string tension σ is the linear energy density of the string between the quarks, calculate the total mass and the total angular momentum and demonstrate that they lie on a Regge trajectory with $\alpha' = 1/(2\pi\sigma)$.

2.2 Consider the spin-independent Lagrangian

$$\begin{aligned} L &= -m_1\sqrt{1-\mathbf{v}_1^2} - m_2\sqrt{1-\mathbf{v}_2^2} - U_0(r) \\ &\quad -\frac{U_+(r)}{4}(\mathbf{v}_1-\mathbf{v}_2)^2 - \frac{U_-(r)}{4}(\mathbf{v}_1+\mathbf{v}_2)^2 \end{aligned}$$

where $\mathbf{x}_1, \mathbf{v}_1$ and $\mathbf{x}_2, \mathbf{v}_2$ are the positions and velocities respectively, of the quark and the antiquark and $\mathbf{r} = \mathbf{x}_1 - \mathbf{x}_2$. U_0 is the static potential and U_+ and U_- are the coefficients of the velocity dependent terms in the potential. Velocity dependent terms of this type are obtained in Baker et al. (1995), Brambilla et al. (1994). Making some simplifications (circular orbits, $v_j \equiv \omega \times \mathbf{r}_j$ with $\mathbf{r}_j = (-1)^{j+1} r_j \hat{\mathbf{r}}$, $r_1 + r_2 = r$, $m_1 = m_2$), calculate the energy E and the angular momentum $\mathbf{J}$. Determine the moment of inertia of the colour field produced by the rotating quarks and establish the physical meaning of U_+. Then, obtain the slope of the Regge trajectories in the case in which $U_0 = \sigma r$ and $U_+ = -Ar$.

3 Quarkonium and Confining Phenomenological Potentials

The relevant features of the quarkonium spectrum are: the pattern of the levels, the spin separation between pseudoscalar mesons $n^1S_0(0^{-+})$ and vector mesons $n^3S_1(1^{--})$ (called hyperfine splitting), the spin separations between states within the same $L \neq 0$ and S multiplets (e.g. the splitting in the 1^3P_J multiplet $\chi_c(1P)$ in charmonium cf. Fig. 2) (called fine splitting), and the transition and decay rates, see PDG. To separate the sub-structure from the radial and orbital splittings it is convenient to work in terms of spin-averaged splittings. Spin-averaged states are obtained summing over masses of given L and n, and weighting by $2J+1$. The hyperfine spin splittings appear to scale roughly with a $1/m_Q$ dependence. We note that states below threshold are considerably narrow since they can decay only by annihilation.

The fact that all the splittings are considerably smaller than the masses implies that all the dynamical scales of the bound state, such as the kinetic energy or the momentum of the heavy quarks, are considerably smaller than the quark masses. Therefore, the quark velocities are nonrelativistic: $v \ll 1$. The energy scales in quarkonium are the typical scales of a nonrelativistic bound state: the momentum scale $m_Q v$ and the energy scale $m_Q v^2$. Being the time scale $T_g \sim 1/m_Q v$ associated with the binding gluons smaller than the time scale $T_Q \sim 1/m_Q v^2$ associated with the quark motion, the gluon interaction between heavy quarks appears "instantaneous". Therefore, it can be modelled with a potential and the energy can be obtained solving the corresponding Schrödinger equation. In the extreme nonrelativistic limit of very heavy quarks the spin splittings vanish and the spin-averaged spectrum is described by a single static central potential.

A lot of work has been done to find the phenomenological form of the static potential (see the report of Gromes, Lucha and Schöberl (1991)). Lowest order perturbation theory for QCD gives a flavour-independent central potential

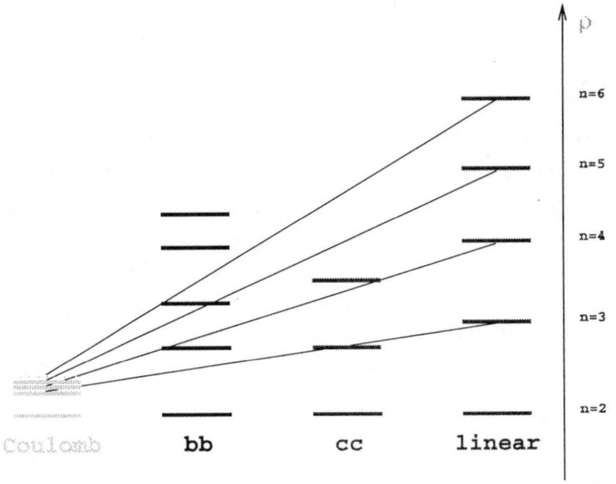

Figure 5: $\rho = (E_n - E_1)/(E_2 - E_1)$, *where* E_n *is either the nth energy level of the physical system or the nth eigenvalue of the Schrödinger equation corresponding to the indicated potential.*

based on the one-gluon exchange, which has a Coulomb-like form (see Exercise 3.1)

$$V_0(r) = -\frac{4}{3}\frac{\alpha_s}{r} \tag{7}$$

where r is the distance between the two quarks and α_s is the strong coupling constant [e] [f]. This cannot be the final answer since it does not confine quarks and gives a spectrum incompatible with the data (see Fig. 5). Nevertheless we can regard the one-gluon exchange formula to be valid for $r \simeq 0.1$ fm.

It was found that the addition of some positive power of r to Eq. (7) rescues the phenomenology. The intuitive argument of a constant energy density (string tension), exposed in Sec. 2, led to the flavor-independent Cornell

[e] To compute Eq. (7), cf. Exercise 3.1, it is necessary to evaluate products of the type $\lambda^{(1)} \cdot \lambda^{(2)}$ in the representation of interest. It turns out that, out of all two-body channels, the colour singlet $(q\bar{q})$ is the most attractive. This hints to the fact that coloured mesons should not exist.

[f] Of course, α_s in Eq. (7) depends on a scale. If we work in a physical gauge, e.g. in a lightlike gauge, the quarks in the bound state interact through gluon exchange and these interactions are renormalized by loop corrections. In the extreme nonrelativistic limit, only the corrections to the gluon propagator survive and this give α_s evaluated at the square momentum of the exchanged gluon $Q^2 \simeq \mathbf{Q}^2$.

potential (Eichten et al. (1978))

$$V_0(r) = -\frac{4}{3}\frac{\alpha_s}{r} + \sigma r + \text{const.} \tag{8}$$

Here, α_s and σ are regarded as free parameters to be fitted on the spectrum. The Schrödinger equation with the potential (8) and parameters $\alpha_s = 0.39$ and $\sigma = 0.182$ GeV^2 gives quite a satisfactory agreement with the data. In Eichten et' al. (1980), the coupling to charmed meson decay-channels was also taken into account. It was found that the mass shifts due to the coupled channel effects are indeed large also below threshold and yet they do not spoil the predictions of the naive potential model. Indeed these effects can essentially be absorbed into a redefinition of the effective parameters. Since then several

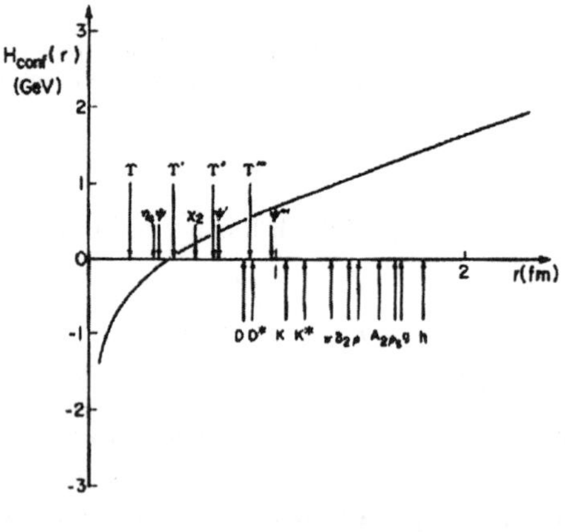

Figure 6: *The rms $q\bar{q}$ separations in some representative mesons is shown with respect to the Cornell potential. All the phenomenological potentials agree in the range of* 0.1 – 1 *fm which is the physical range for quarkonia. From Godfrey and Isgur (1985).*

different phenomenological forms of the static potential have been exploited, e.g. the Richardson potential,

$$V_0(r) = \int \frac{d^3\mathbf{Q}}{(2\pi)^3} \exp\{i\mathbf{Q}\cdot\mathbf{r}\} \frac{\text{const.}}{Q^2\log(1+Q^2/\Lambda^2)}$$

(Richardson (1979)), the logarithmic potential, $V_0(r) = A\log(r/r_0)$ (Quigg and Rosner (1977)), the Martin potential, $V_0(r) = A(r/r_0)^\alpha$ (Martin (1980)). By

nL	$\langle v_b^2 \rangle$	$\langle v_c^2 \rangle$	$\sqrt{\langle r_b^2 \rangle}$/fm	$\sqrt{\langle r_c^2 \rangle}$/fm
$1S$	0.080	0.27	0.24	0.43
$2S$	0.081	0.35	0.51	0.85
$3S$	0.096	0.44	0.73	1.18
$1P$	0.068	0.29	0.41	0.67
$2P$	0.085	0.39	0.65	1.04
$1D$	0.075	0.34	0.54	0.87

Table 1: *From Bali et al. (1997). Notice that for a Coulombic system $v \sim \alpha/n$ and for a confined system v grows with n.*

fitting the parameters, all these potentials can reproduce the spectrum. This is not surprising if we look at Fig. 6: the potentials essentially agree in the region $r \sim 0.1 \div 0.8$ fm in which the $\sqrt{\langle r^2 \rangle}$ for quarkonia sits (Buchmüller and Tye (1981)). On the other hand, it is a considerable limit of the potential approach that the connection with the true QCD parameters of Eq. (1) remains totally hidden and mysterious.

However, the existence of a spin substructure in the spectrum indicates that relativistic corrections to the static central potential V_0 have to be taken into account. From the radial level splitting (see Exercise 3.2) as well as from the fits with the phenomenological potential, we find for

$$\begin{array}{lll} c & \text{in} & \psi,\ v^2 \simeq 0.3 \\ b & \text{in} & \Upsilon,\ v^2 \simeq 0.1 \quad , \end{array} \tag{9}$$

and therefore we expect relativistic corrections of order $20 \div 30\%$ for the charmonium spectrum and up to 10% for the bottomonium spectrum. This determination of the heavy quark velocity in the bound states is confirmed by lattice calculation, see Tab.1. Notice that relativistic corrections are of critical importance for the observables sensitive to the details of the wave functions (e.g. the radiative transitions) (Mc Clary et al. (1983)).

The phenomenological potential model predictions of the relativistic corrections are calculated by means of a Breit–Fermi Hamiltonian of the type

$$H = \sum_{j=1,2} \left(m_j + \frac{p_j^2}{2m_j} - \frac{p_j^4}{8m_j^3} \right) + V_0 + V_{\rm SD} + V_{\rm VD}. \tag{10}$$

The $1/m^2$ spin-dependent $V_{\rm SD}$ and velocity-dependent $V_{\rm VD}$ potentials are derived from the semirelativistic reduction of the Bethe–Salpeter (BS) equation for the quark-antiquark connected amputated Green function or, equivalently at this level, from the semirelativistic reduction of the quark-antiquark scattering amplitude with an effective exchange equal to the BS kernel. Several ambiguities are involved in this procedure, due on one hand to the fact that we do not know the relevant confining Bethe–Salpeter kernel, on the other hand due to the fact that we have to get rid of the temporal (or energy Q_0, $Q = p_1 - p_1'$ being the momentum transfer) dependence of the kernel to recover a potential (instantaneous) description. It turns out that, at the level of the approximation involved, the spin-independent relativistic corrections at the order $1/m^2$ depend on the way in which Q_0 is fixed together with the gauge choice of the kernel. The Lorentz structure of the kernel is also not known.

On a phenomenological basis, the following ansatz for the kernel was intensively studied

$$I(Q^2) = (2\pi)^3 \left[\gamma_1^\mu \gamma_2^\nu P_{\mu\nu} J_v(Q) + J_s(Q)\right] \tag{11}$$

in the instantaneous approximation $Q_0 = 0$. Notice that the effective kernel above was taken with a pure dependence on the momentum transfer Q. But, of course, the dependence on the quark and antiquark momenta could have been more complicated. The vector kernel $\gamma_1^\mu \gamma_2^\nu P_{\mu\nu} J_v(Q)$ corresponds to the one gluon exchange, and e.g. in the Coulomb gauge $P_{\mu\nu}$ has the structure $P_{\mu\nu} = g_{\mu\nu} + \dfrac{J_v'(-\mathbf{Q}^2)}{J_v(-\mathbf{Q}^2)} Q_\mu Q_\nu - \dfrac{J_v'(-\mathbf{Q}^2)}{J_v(-\mathbf{Q}^2)} Q_0 (Q_\mu n_\nu + Q_\nu n_\mu)$ where the prime indicates the derivative and n_μ is the unit vector in the time direction. The scalar kernel $J_s(Q)$ accounts for the nonperturbative interaction. The semirelativistic reduction of the kernel (11) (in the instantaneous approximation, with the Coulomb gauge fixed for the vectorial part, in the centre of mass frame $\mathbf{p}_1 = -\mathbf{p}_2 = \mathbf{q}, \mathbf{p}_1' = -\mathbf{p}_2' = \mathbf{q}', \mathbf{Q} \equiv \mathbf{q} - \mathbf{q}'$ and in the equal mass case), gives

$$V_0 = \tilde{J}_s(r) + \tilde{J}_v(r) \tag{12}$$

$$\begin{aligned} V_{\mathrm{SD}} = {} & \frac{3}{m^2}\frac{1}{r}\left(\frac{d\tilde{J}_v}{dr} - \frac{1}{3}\frac{d\tilde{J}_s}{dr}\right) \mathbf{S}\cdot\mathbf{L} \\ & + \frac{1}{m^2}\left(\frac{1}{r}\frac{d\tilde{J}_v}{dr} - \frac{d^2\tilde{J}_v}{dr^2}\right) S_1^h \left(\frac{r^h r^k}{r^2} - \frac{\delta^{hk}}{3}\right) S_2^k + \frac{2}{3m^2}\Delta \tilde{J}_v(r)\mathbf{S}_1\cdot\mathbf{S}_2 \end{aligned} \tag{13}$$

$$V_{\mathrm{VD}} = \frac{1}{4m^2}\Delta(\tilde{J}_s + \tilde{J}_v) - \frac{1}{4m^2} q^h \tilde{J}_s q^h + \frac{1}{2m^2} q^h \left(\delta^{hk} - \frac{\partial_h \partial_k}{\partial^2}\tilde{J}_v\right) q^k \tag{14}$$

with $\tilde{J}_{v,s}(\mathbf{r}) \equiv \int d^3\mathbf{Q}\, e^{i\mathbf{Q}\cdot\mathbf{r}} J(\mathbf{Q})_{v,s}$. Taking $J_v = -\dfrac{1}{2\pi^2}\dfrac{4}{3}\dfrac{1}{\mathbf{Q}^2}$ and $J_s = -\dfrac{\sigma}{\pi^2}\dfrac{1}{\mathbf{Q}^4}$, V_0 reproduces the Cornell potential.

The confining part of the kernel is usually chosen to be a Lorentz scalar in order to match the data on the fine separation. Indeed, the ratio ρ_{FS} of the fine structure splitting

$$\rho_{FS} = \frac{M(^3P_2) - M(^3P_1)}{M(^3P_1) - M(^3P_0)} \tag{15}$$

is $\rho_{FS} = 0.8$ for a pure vector Coulomb exchange while the data give $\rho_{FS} \simeq 0.49, 0.66, 0.57$ for the $c\bar{c}(1P), b\bar{b}(1P)$ and $b\bar{b}(2P)$ respectively. Adding a scalar exchange gives a contribution to the spin orbit interaction that reduces the vector part and thus also the value of the ratio ρ_{FS} (Schnitzer (1978)). Moreover, since the two terms, vectorial short range and scalar long range exchange, contribute with opposite signs, we expect that at high orbital excitations, where the $Q\bar{Q}$ pair probes large average distances (see Fig. 6), the triplet multiplet inverts with reference to the ordering of the low excitation multiplets (i.e. $M(^3L_{L+1}) \geq M(^3L_L) \geq M(^3L_{L-1})$ at low L and the reverse at high L). The fine structure turns out to be a nice test for the form of the confining interaction (cf. e.g. Isgur (1998)).

We have presented a way to obtain phenomenologically the $1/m^2$ relativistic corrections. However, this is not really rewarding. In a confining interaction the average $\langle p^2 \rangle$ increases with the excited states (see Tab. 1) and so the pattern of the excited levels is likely to be considerably distorted if the nonperturbative relativistic corrections are not the appropriate ones. Indeed, fits made with only the static potential turn out to be better than fits made with the interaction (12)-(14) (Brambilla et al. (1990)). Moreover, there are characteristics of the spectrum, like the hyperfine separation as well as the leptonic decays, that are due to processes taking place at very short scale. In this case it is important to add higher order perturbative corrections to the one gluon exchange as well as the running of α_s. In the phenomenological potential framework it is somehow ambiguous how to take into account the running of $\alpha_s(\mu)$ as well as the scale μ.

In this section we realized that the description of the heavy meson spectrum turns out to be a priori a quite complicate problem with an interplay of different relevant scales as well as of perturbative and nonperturbative effects. The conclusion is that we badly need, on one hand a framework in which relativistic as well as perturbative corrections to the quark-antiquark interaction can be evaluated unambiguously and systematically and, on the other hand a clear, well founded and eventually computable approach to the long range quark-antiquark interaction which is essentially nonperturbative. We need this both at a concrete level, in order to make quantitative predictions in which the size of the neglected terms can be estimated, and both at a fundamental level, in order to use the spectrum to get some insight into the confinement mechanism.

In the next section we address the problem of how to study quark confinement beyond phenomenological models i.e. in a QCD based framework and we give a criterium that decides whether a gauge theory is confined or not.

3. Exercises

3.1 Consider the quark-antiquark scattering

$$q_i(p_1, \sigma_1) + \bar{q}_j(p_2, \sigma_2) \to q_k(q_1, \tau_1) + \bar{q}_l(q_2, \tau_2) \tag{16}$$

where $i, j, \cdots = 1, 2, 3$ label the colour indices. Remembering that the quarks in the meson are in a colour-singlet state and introducing the meson colour wavefunctions $\delta_{ij}/\sqrt{3}$, calculate the T-matrix element, extract the first contribution in the nonrelativistic limit and obtain, via Fourier transform, the perturbative one gluon exchange potential of Eq. (7). Show that the contribution of the annihilation graph vanishes.

3.2 Consider the average excitation energy in charmonium and bottomonium (e.g. $M'_\Upsilon - M_\Upsilon$). This should be of the order of the average kinetic energy $E \simeq mv^2$. Taking m to be roughly half of the ground state mass obtain the estimates (9) for the quark velocities.

3.3 Consider the same scattering of (16) with the exchange given in Eq. (11). Obtain the result (12)-(14) by computing the scattering matrix element of this process, expanding up to the $1/m^2$ order and Fourier transforming.

3.4 Consider the same scattering matrix of Exs. 3.1 and 3.3 but with a kernel of the type $I = \gamma_5^1 \gamma_5^2 V_p(Q)$. Show that there is no static potential in the nonrelativistic limit of the matrix element. Discuss the result in relation to the deuteron.

4 The Wilson Loop: Confinement and Flux Tube Formation

The most powerful technique in order to extract nonperturbative information from QCD is the lattice gauge theory approach. This has been undoubtedly successful and rewarding and has produced over the last years an impressive amount of results. Yet, in spite of almost two decades of intensive efforts the characteristics of QCD associated with colour confinement are still not understood. It is our belief that some insight in the mechanism of confinement cannot be obtained without developing, in strict connection with lattice QCD, also analytic methods. In this way information coming from the lattice can be inserted inside analytic models or vice versa lattice calculations can be used in order to interpret analytic models.

To this aim we need an unambiguous way of establishing a gauge invariant and systematic procedure to calculate the quark dynamics. In Sec. 5, we will show that this is feasible in the case of heavy quarks in which the whole dynamics can be reduced to few expectation values of chromoelectric and chromomagnetic fields that can be calculated analytically (once a model for the QCD vacuum is assumed) or numerically on the lattice. Then, the comparison between the two results supply us with hints about the mechanism of confinement.

The simplest manifestation of confinement in quenched QCD is the linear rising of the potential $V_0(r)$ between static colour sources in the fundamental representation. In this section we will show how this has been clearly proved and connected to the formation of a chromoelectric flux tube between the quarks. The question of the nature and the origin of these nonperturbative field configurations will be addressed in Sec. 6.

4.1 The QCD static potential and the Wilson loop

Let us consider a locally gauge invariant quark-antiquark singlet state[g]

$$|\phi^{lj}_{\alpha\beta}\rangle \equiv \frac{\delta_{lj}}{\sqrt{3}}\bar{\psi}^i_\alpha(x)U^{ik}(x,y,C)\psi^k_\beta(y)|0\rangle \tag{17}$$

where i, j, k, l are colour indices (that will be suppressed in the following), $|0\rangle$ denotes the ground state and the Schwinger string line has the form

$$U(x,y;C) = P\exp\left\{ig\int_y^x A_\mu(z)\,dz^\mu\right\}, \tag{18}$$

where A_μ is the gauge potential of Eq. (2), g the QCD coupling constant, and the integral is extended along the path C. The operator P denotes the path-ordering prescription[h] which is necessary due to the fact that A_μ are non-commuting matrices.

---

[g] The contribution of the string to the potential vanishes in the limit $T \to \infty$, Eichten et al. (1981) and Brambilla et al. (1999).

[h] Path ordering prescription means operatively that one has to decompose the path C connecting y with x into infinitesimal pieces, then take the exponential along the infinitesimal pieces, expand at the first order and order the factors according to their appearance along the path.

Remembering that under a $SU(3)$ gauge transformation $\mathcal{V}(\theta) = \exp(-i\theta) \simeq 1 - i\theta$, the gauge potential undergoes the transformation $A_\mu \to A_\mu + i[\theta, A_\mu] + g^{-1}\partial_\mu\theta$, we obtain the transformation law of the string

$$U'(x,y;C) = \exp\{i\theta(x)\}U(x,y;C)\exp\{-i\theta(y)\} \tag{19}$$

and then it is clear that (17) is a gauge-invariant state. Actually, it is a colour singlet and we are interested only in colour singlet being the only existing initial and final states.

The quark-antiquark potential can be extracted from the quark-antiquark Green function. A simple example clarifies in which way. Let us consider the following two-particle Green function

$$G(T) = \langle\phi(\mathbf{x},0)|\phi(\mathbf{y},T)\rangle = \langle\phi(\mathbf{x},0)|\exp(-iHT)|\phi(\mathbf{y},0)\rangle. \tag{20}$$

Inserting a complete set of energy eigenstates ψ_n with eigenvalues E_n and making a Wick rotation we find

$$\begin{aligned} G(-iT) &= \sum_n \langle\phi(\mathbf{x},0)|\psi_n\rangle\langle\psi_n|\phi(\mathbf{y},0)\rangle\exp(-E_nT) \\ &\to \langle\phi(\mathbf{x},0)|\psi_0\rangle\langle\psi_0|\phi(\mathbf{y},0)\rangle\exp(-E_0T) \quad \text{for} \quad T\to\infty \end{aligned} \tag{21}$$

which gives the Feynman–Kac formula for the ground state energy

$$E_0 = -\lim_{T\to\infty}\frac{\log G(-iT)}{T}. \tag{22}$$

The only condition for the validity of Eq. (22) is that the ϕ states have a non-vanishing component over the ground state. The same is still true for finite T if the overlap with the ground state is not too small. This is precisely the way in which hadron masses are computed on the lattice. Of course, many tricks are used in order to maximize the overlap with the ground state in consideration. If the ϕ state denotes a state of two exactly static particles interacting at a distance r, then the ground state energy is a function of the particle separation, $E_0 \equiv E_0(r)$, and gives the potential of the first adiabatic surface. With this in mind, we will perform in the remaining of this section an explicit evaluation of the quark-antiquark Green function for infinitely heavy quarks ($m_j \to \infty$) and for large temporal intervals ($T \to \infty$).

In the following we will be working in the Euclidean space taking advantage of the usual relation between Euclidean position x^E_μ, momentum k^E_μ, field A^E_μ, gamma matrices γ^E and the corresponding quantities in Minkowski space

$$t^E = it^M \qquad x^E_i = x^M_i$$

$$
\begin{aligned}
k_4^E &= -ik_0^M \qquad & k_i^E &= k_i \\
A_4^E &= -iA_0^M \qquad & A_i^E &= A_i^M \\
\gamma_4^E &= \gamma_0 \qquad & \gamma_i^E &= -i\gamma^i .
\end{aligned} \tag{23}
$$

Let us assume that at a time $t = 0$ a quark and an antiquark are created and that they interact while propagating for a time $t = T$ at which they are annihilated. Then $(x_j = (\mathbf{x}_j, T), y_j = (\mathbf{y}_j, 0))$

$$
\begin{aligned}
&G_{\beta_1\beta_2\alpha_1\alpha_2}(T) \\
&= \langle 0|\bar{\psi}_{\beta_2}(\mathbf{y}_2, 0)U(y_2, y_1)\psi_{\beta_1}(\mathbf{y}_1, 0)\bar{\psi}_{\alpha_1}(\mathbf{x}_1, T)U(x_1, x_2)\psi_{\alpha_2}(\mathbf{x}_2, T)|0\rangle \\
&= \frac{1}{Z}\int \mathcal{D}\psi\mathcal{D}\bar{\psi}\mathcal{D}A\, \bar{\psi}_{\beta_2}(\mathbf{y}_2, 0)U(y_2, y_1)\psi_{\beta_1}(\mathbf{y}_1, 0) \\
&\qquad \times\bar{\psi}_{\alpha_1}(\mathbf{x}_1, T)U(x_1, x_2)\psi_{\alpha_2}(\mathbf{x}_2, T)e^{-\int L\, d^4x}
\end{aligned} \tag{24}
$$

where L is the Euclidean version of Eq. (1), $L = L_{YM} + L_F = \frac{1}{4}F_{\mu\nu}F_{\mu\nu} + \bar{\psi}(\gamma_\mu D_\mu + m)\psi$. The indices α, β are spinor indices, while the detailed structure of the colour indices is not displayed. Since the action is quadratic in the quark fields, it is possible to perform the fermion integration

$$
\begin{aligned}
&G_{\beta_1\beta_2\alpha_1\alpha_2}(T) \\
&= \frac{1}{Z}\int \mathcal{D}A\, (\mathrm{Tr}\{S_{\alpha_2\beta_2}(x_2, y_2; A)U(y_2, y_1)S_{\beta_1\alpha_1}(y_1, x_1; A)U(x_1, x_2)\} \\
&\qquad -\mathrm{Tr}\{S_{\beta_1\beta_2}(y_1, y_2; A)U(y_2, y_1)\}\mathrm{Tr}\{S_{\alpha_2\alpha_1}(x_2, x_1; A)U(x_1, x_2)\}) \\
&\qquad \times \det K(A)e^{-\int L_{YM} d^4x}
\end{aligned} \tag{25}
$$

where the trace is over the colour indices and K is the fermionic determinant of the matrix $K_{\alpha x\beta y}(A) \equiv [\gamma_\mu D_\mu + m]_{\alpha\beta}\delta^4(x - y)$. In the following, we will assume the quenched approximation[i], $\det K = 1$. The second term in Eq. (25) describes quark-antiquark annihilation and hence appears only for quarks of the same flavour. Since this effect is dominated by the perturbative two or three gluons exchange in the s channel we will not consider this term any more

[i]In perturbation theory the logarithm of this determinant is given by the sum of Feynman diagrams consisting of fermion loops with an arbitrary number of fields A_μ attached to it. In the limit $m \to \infty$ this determinant approaches a constant (infinite but canceled by a factor in Z) and then $\det K = 1 +$ corrections of order $O(1/m^n)$. Therefore for heavy quarks the quenched approximation $\det K = 1$ makes sense. On the contrary the fermionic determinant associated to light quarks in principle cannot be neglected. This determinant will eventually be responsible for the breaking of the string between the quarks.

here. Then we obtain

$$G(T) \simeq \frac{1}{Z}\int \mathcal{D}A\,\mathrm{Tr}\{S(x_2,y_2;A)U(y_2,y_1)S(y_1,x_1;A)U(x_1,x_2)\}e^{-\int L_{YM}d^4x}. \tag{26}$$

In Eq. (26) $S(x,y;A)$ denotes the quark propagator in the presence of the gluon field A_μ. It obeys the equation

$$(\gamma_\mu D_\mu + m)S(x,y;A) = \delta^4(x-y). \tag{27}$$

This is in principle a system of coupled partial differential equations that cannot be solved in a closed form for an arbitrary A_μ. However, we are interested in the limit $m \to \infty$. In this approximation (Wilson (1974), Brown and Weisberger (1979)) we can replace S by the static solution S_0 obtained dropping the spatial part of the gauge-covariant derivative in (27) while maintaining the time component. This approximation maintains the manifest gauge invariance. Then we have

$$(\gamma_4 D_4 + m)S_0(x,y;A) = \delta^4(x-y) \tag{28}$$

which is an ordinary differential equation solvable in a closed form. Indeed, we can get rid of the A_4 in the equation making the ansatz

$$S_0(x,y;A) = \mathrm{P}\exp\left\{ig\int_{x_4}^{y_4} dt A_4(\mathbf{x},t)\right\}\hat{S}_0(x-y) \tag{29}$$

with $\hat{S}_0$ satisfying

$$(\gamma_4\partial_4 + m)\hat{S}(x-y) = \delta^4(x-y). \tag{30}$$

Therefore the solution has the form

$$S_0(x,y;A) = \delta^3(\mathbf{x}-\mathbf{y})\mathrm{P}e^{ig\int_{x_4}^{y_4} dt A_4(\mathbf{x},t)}\left\{\theta(x_4-y_4)\frac{1+\gamma_4}{2}e^{-m(x_4-y_4)}\right.$$
$$\left.+\theta(y_4-x_4)\frac{1-\gamma_4}{2}e^{-m(y_4-x_4)}\right\}. \tag{31}$$

This expression shows that the time evolution of a (infinitely) heavy quark field consists purely in the accumulation of phase determined by A_4 and the quark mass (cf. Ex. 4.1.1). The spatial delta function says that the infinitely heavy quark cannot propagate in space[j].

---

[j] For this solution the annihilation term in Eq. (24) does not give contribution since the $\mathbf{x} = \mathbf{y}$ condition is not satisfied.

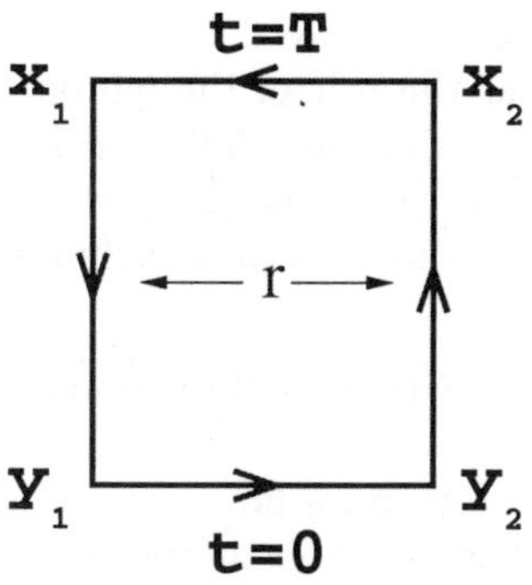

Figure 7: *Static Wilson loop with contour* Γ_0.

The quark-antiquark Green function is given by

$$G_{\beta_1\beta_2\alpha_1\alpha_2}(T) \overset{m_j\to\infty}{\longrightarrow} \delta^3(\mathbf{x}_1-\mathbf{y}_1)\delta^3(\mathbf{x}_2-\mathbf{y}_2)(P_+)_{\beta_1\alpha_1}(P_-)_{\alpha_2\beta_2} \times e^{-(m_1+m_2)T}\langle \mathrm{Tr}\,\mathrm{P}e^{ig\oint_{\Gamma_0} dz_\mu A_\mu(z)}\rangle \tag{32}$$

with $P_\pm \equiv (1\pm\gamma_4)/2$. The integral in Eq. (32) extends over the circuit Γ_0 which is a closed rectangular path with spatial and temporal extension $r = |\mathbf{x}_1 - \mathbf{x}_2|$ and T respectively, and has been formed by the combination of the path-ordered exponentials along the horizontal (=time fixed) lines, coming from the Schwinger strings, and those along the vertical lines coming from the static propagators (see Fig. 7). The brackets in (32) denote the pure gauge vacuum expectation value. In Euclidean space

$$\langle f[A]\rangle \equiv \frac{1}{Z}\int \mathcal{D}Af[A]e^{-\int d^4x L^E_{YM}}. \tag{33}$$

From Eq. (32) it is clear that the dynamics of the quark-antiquark interaction is contained in

$$W(\Gamma_0) = \mathrm{Tr}\,\mathrm{P}e^{ig\oint_{\Gamma_0} dz_\mu A_\mu(z)}. \tag{34}$$

This is the famous (*static*) Wegner–Wilson loop (Wegner (1971) and Wilson (1974)). In the limit of infinite quark mass considered, the kinetic energies of the quarks drop out of the theory, the quark Hamiltonian becomes identical with the potential (see Ex. 4.1.1) while the full Hamiltonian contains also all types of gluonic excitations. According to the Feynman–Kac formula the limit $T \to \infty$ projects out the lowest state i.e. the one with the "glue" in the ground state. This has the role of the quark-antiquark potential for pure mesonic states. Now, comparing Eq. (32) with Eq. (21) and considering that the exponential factor $\exp(-(m_1 + m_2)T)$ just accounts for the the fact that the energy of the quark-antiquark system includes the rest mass of the pair[k], we obtain

$$V_0(r) \equiv E(r) = -\lim_{T\to\infty} \frac{1}{T} \log\langle W(\Gamma_0)\rangle. \tag{35}$$

The quark degrees of freedom have now completely disappeared and the expectation value in (35) has to be evaluated in the pure Yang–Mills theory ((Wilson (1974), Brown and Weisberger (1979)). Notice that the potential is given purely in terms of a gauge invariant quantity (the Wilson loop precisely). In this way we have reduced the calculation of the static potential to a well posed problem in field theory: to obtain the actual form of V_0 we need to calculate the QCD expectation value of the static Wilson loop.

We conclude this section pointing out that Eq. (35) is rigorously true for static sources. For (realistic) heavy quarks with finite mass, the static potential, interpreted as the static limit of the potential appearing in the Schrödinger equation, could in principle not coincide exactly with Eq. (35). Actually it does not. The reason is that in QCD quarks in the static limit can still change colour by emission of gluons. This introduces a new dynamical scale in the evaluation of the potential from Eq. (35) of the order of the kinetic energy which is finite if the quarks have finite mass. In perturbative QCD this new scale is given by the difference between the singlet and the octet potential. Contributions of the same order of the kinetic energy are not of potential type and have to be explicitly subtracted out from Eq. (35). In the next section we will give the leading effect of this subtraction on the static Wilson loop. For an extended analysis we refer to Brambilla et al. (1999).

4.1 Exercises

4.1.1 Consider a particle of mass m moving in a potential $V(x)$ in one space dimension. The propagator is given by $K(x', t; x, 0) = \langle x'|\exp(-iHt)|x\rangle$ with $H = \dfrac{p^2}{2m} + V(x)$. Obtain the form of the propagator K in the static limit $m \to \infty$ and compare it with Eq. (31).

[k]Moreover $E(R)$ includes also self-energy effects which need to be subtracted when calculating the quark-antiquark potential.

4.1.2 Consider quenched QED. Show that the functional generator for the QED Lagrangian with a source term added of the form $J_\mu(x)A_\mu(x)$ with $J_\mu(x) = e\delta_{\mu 4}(\delta^3(\mathbf{x}-\mathbf{x}_1) - \delta^3(\mathbf{x}-\mathbf{x}_2))$, coincides with the vacuum expectation value of the static Wilson loop in the limit of infinite interaction time.

4.2 The Wilson loop in perturbative QCD

The Wegner–Wilson loop of contour Γ is defined in QCD as

$$W(\Gamma) \equiv \mathrm{Tr}\,\mathrm{P}\, e^{ig\oint_\Gamma dz_\mu A_\mu(z)} . \tag{36}$$

Due to the presence of the colour trace, it is a manifestly gauge invariant object. The field A_μ can be taken in any representation of $SU(3)$. When describing the quark-antiquark interaction, as in the present case, $A_\mu \equiv A_\mu^a \lambda^a/2$. The Wilson loop is called static when the integral is extended to a rectangular Γ_0 as in Fig. 7. Physically, in the static Wilson loop only the time component A_4 is relevant.

We know from the previous section that the vacuum expectation value of W on the QCD measure gives the static potential. However, we are in trouble when we step in to calculate it. Indeed, if we want to describe the long range quark-antiquark interaction, we should be able to calculate the Wilson loop in the region in which the running coupling constant is no longer small and we should be able to sum up all the relevant diagrams. Unfortunately, we have no methods at hand to sum up such contributions. Worse enough, also usual semiclassical approaches are not doomed to work in this case. We do not know of any dominant and confining configurations in the QCD measure and in the path integral that can make the work[l]! Hence, to obtain information on the behaviour of the Wilson loop in the nonperturbative region we have to resort either to strong coupling expansion and to lattice simulations, see Sec. 4.3 and Sec. 4.5, or to analytic models of the QCD vacuum, see Sec. 6.

However, in the weak coupling region the Wilson loop can be calculated perturbatively. Recently, the fully analytic calculation of the two-loop diagrams contributing to the static potential has been performed (Peter (1997), Schröder (1999)). We refer to the original papers for the details of the calculation. However, due to the high interest, we report here the final result (α_s is

[l]Instantons do not confine directly, i.e. give a zero string tension. Monopoles arise in the Abelian projection or after a dual transformation, see Sec.6.

in the $\overline{MS}$ scheme):

$$V_0(r) = -C_F \frac{\alpha_V(r)}{r} \tag{37}$$

$$\alpha_V(r) = \alpha_s(r) \left\{ 1 + \left(a_1 + \frac{\gamma_E \beta_0}{2} \right) \frac{\alpha_s(r)}{\pi} \right.$$

$$\left. + \left[\gamma_E \left(a_1 \beta_0 + \frac{\beta_1}{8} \right) + \left(\frac{\pi^2}{12} + \gamma_E^2 \right) \frac{\beta_0^2}{4} + b_1 \right] \frac{\alpha_s^2(r)}{\pi^2} \right\},$$

where γ_E is the Euler constant, $a_0 = 1$,

$$a_1 = \frac{31}{9} C_A - \frac{20}{9} T_F N_f$$

and

$$a_2 = \left(\frac{4343}{162} + 4\pi^2 - \frac{\pi^4}{4} + \frac{22}{3}\zeta_3 \right) C_A^2 - \left(\frac{1798}{81} + \frac{56}{3}\zeta_3 \right) C_A T_F N_f$$
$$- \left(\frac{55}{3} - 16\zeta_3 \right) C_F T_F N_f + \left(\frac{20}{9} T_F N_f \right)^2 .$$

$C_F = 4/3$ and $C_A = 3$ are the Casimir of the fundamental and of the adjoint representation respectively. Moreover in QCD we have $T_F = 1/2$.

From this result we learn that: the two-loop contribution is nearly as large as the one-loop term, both make the potential more attractive and eventually the perturbative potential seems to be reliable up to a distance $r\Lambda_{QCD} < 0.07$, which is considerably smaller than the average radius in quarkonia (see Fig. 6). For larger values a strong scale-dependence remains and the perturbation series breaks down above 0.1 fm. Hence, even the pure perturbative calculation indicates the need of a different long range approach[m].

We mention that the next perturbative correction to the static potential can be obtained only in an effective theory framework (pNRQCD, see Sec. 5). In that framework the leading log three-loop term has been very recently calculated in Brambilla et al. (1999). It amounts to a correction $C_A^3 \alpha_s^4(r)/12\pi \log r\mu'$ to α_V in Eq. (37), μ' being the scale of the matching.

[m] In Pineda and Yndurain (1998) the two-loop static potential and the one-loop relativistic perturbative corrections to the potential were used in order to calculate the ground state energies of bottomonium and charmonium and thus to obtain a value for the bottom and charm masses. Nonperturbative effects were encoded in the local gluon condensate. In the next section, we will show that nonperturbative contributions are actually carried by non-local quantities, Gromes (1982), like the Wilson loop, which can be approximated by local condensates only if the involved physical scales enable a local expansion. This is indeed the case of the bottomonium ground state.

Notice that the potential comes to depend on the infrared scale μ'. This signals the appearance at three-loop of a nonpotential type of contribution to the static Wilson loop which has been subtracted out explicitly at μ'.

In QED in the quenched approximation (i.e. neglecting light fermions) the (non-static) Wilson loop can be calculated analytically in a closed form. We sketch here the derivation. We have (in Euclidean space)

$$\langle W(\Gamma)\rangle = \frac{\int \mathcal{D}A\, e^{\frac{1}{2}\int d^4x A_\mu M_{\mu\nu} A_\nu + ie\oint dz_\mu A_\mu}}{\int \mathcal{D}A\, e^{\frac{1}{2}\int d^4x A_\mu M_{\mu\nu} A_\nu}} \tag{38}$$

where $M_{\mu\nu} \equiv \delta_{\mu\nu}\partial^2 - \partial_\mu\partial_\nu$ has been obtained by integrating by parts the original QED action. The integral over the A field in Eq. (38) is Gaussian and therefore can be performed provided that a gauge condition is imposed. However, since the Wilson loop is a gauge-invariant quantity, the choice of the gauge is immaterial. We obtain

$$\langle W(\Gamma)\rangle = \exp\left\{\frac{e^2}{2}\oint_\Gamma dx_\mu \oint_\Gamma dx'_\nu D_{\mu\nu}(x-x')\right\} \tag{39}$$

where $D_{\mu\nu}(x-x') \equiv \langle A_\mu(x)A_\nu(x')\rangle$ is the photon propagator. E.g. in Feynman gauge we have $D_{\mu\nu}(x) = -\frac{\delta_{\mu\nu}}{4\pi^2}\frac{1}{x^2}$. On a rectangular Wilson loop Eq. (39) becomes

$$\langle W(\Gamma_0)\rangle = \exp\left\{\frac{e^2}{4\pi r} T f(r,T)\right\} \tag{40}$$

with $f(T,r) = \frac{2}{\pi}\left[\arctan\frac{T}{r} - \frac{r}{2T}\log(1+\frac{T^2}{r^2})\right]$, $f \to 1$ for $T \to \infty$ getting back the Coulomb potential.

In the weak coupling region the vacuum expectation value of the (non-static) Wilson loop in QCD can be obtained by making a Gaussian approximation on the functional integral (i.e. by neglecting non-Abelian contributions). In this case, since the contribution coming from the Schwinger strings vanish in the limit $T \to \infty$ and taking advantage of the notation given in Fig. 8, we have

$$\begin{aligned}\langle W(\Gamma)\rangle &= e^{\frac{4}{3}g^2\oint_\Gamma dz^1_\mu \oint_\Gamma dz^2_\nu D_{\mu\nu}(z_1-z_2)} \\ &\xrightarrow{T\to\infty} e^{\frac{4}{3}g^2\int_{t_i}^{t_f} dt_1 \int_{t_i}^{t_f} dt_2 \dot{z}^1_\mu(t_1)\dot{z}^2_\nu(t_2) D_{\mu\nu}(z_1(t_1)-z_2(t_2))}\end{aligned} \quad , \tag{41}$$

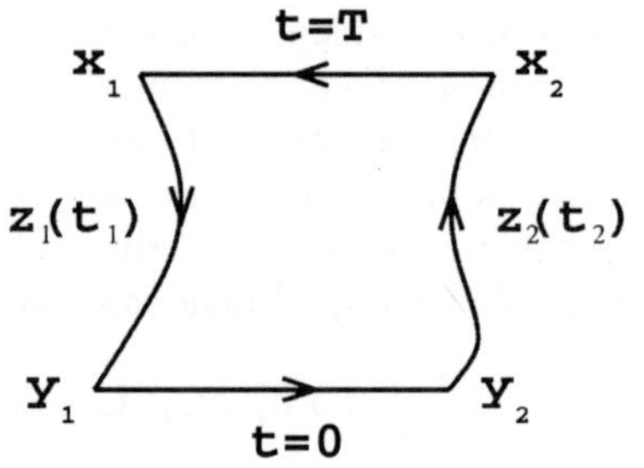

Figure 8: *Generalized Wilson loop with contour* Γ.

$D_{\mu\nu}$ being the gluon propagator.

4.2. Exercises

4.2.1 From Eqs. (35) and (41) obtain the QCD one gluon exchange contribution to the static QCD potential.

4.2.2 Calculate Eq. (40) from Eq. (39).

4.2.3 From the correspondent of Eq. (75) (Sec. 5) in Euclidean space, obtain the QCD V_0 and $V_{\rm VD}$ potentials using the weak coupling behaviour of the Wilson loop given in (41) with the gluon propagator first in the Coulomb gauge and then in the Feynman gauge. Demonstrate that, due to the gauge invariance of the Wilson loop, the two expression coincide. [Hint: perform the change of variables $t = (t_1 + t_2)/2; \tau = t_1 - t_2$ in the integrals in (41), expand $\mathbf{z}_j$ around t and integrate over $\tau^{+\infty}_{-\infty}$.]

4.3 Lattice formulation, strong coupling expansion and area law

Equation (35) is particularly useful on the lattice where the dynamical variables are unitary matrices associated with the links. We do not want to give here an introduction to lattice QCD (for this we refer the reader to Rothe (1992) and Montvay and Munster (1994)). However, in order to illustrate some interesting results, we recall the basic definitions and concepts.

Let us consider QCD in the pure gauge sector on a four-dimensional Euclidean discretized space-time, the "lattice" of step a. Lattice sites are denoted by n and lattice directions are denoted by μ, ν. We define the group element

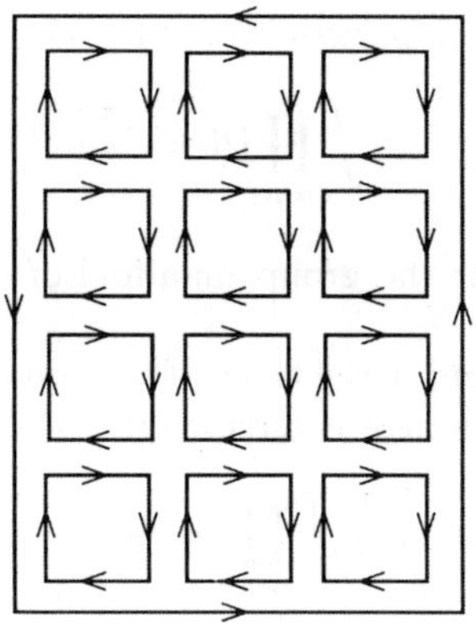

Figure 9: *Elementary plaquette on the lattice and tiling of the Wilson loop in strong coupling.*

associated with a link, from the lattice site n to $n+\hat{\mu}$ ($\hat{\mu}$ being a unit vector along the axis μ) as

$$U_\mu(n) \equiv U(n, n+\hat{\mu}) \simeq e^{igaA_\mu(n+\frac{\hat{\mu}}{2})} \qquad U_\mu^\dagger(n) \equiv U(n+\hat{\mu}, n). \tag{42}$$

These are the dynamical variables, the gluonic colour fields that relate the colour coordinate system at different space-time points. From (42) we see that they are represented as a 3×3 colour matrix that can be interpreted as the path-ordered exponential of the continuum colour fields $A_\mu(x)$. The gauge transformation on the lattice, $\mathcal{V}(x_n)$, acts directly on the link elements

$$U_\mu(n) \to \mathcal{V}^\dagger(n+\hat{\mu})U_\mu(n)\mathcal{V}(n). \tag{43}$$

The Yang–Mills action is

$$S = -\frac{\beta}{3}\sum_{\mu>\nu}\sum_{n} \mathrm{Re}\,\mathrm{Tr}\,U_{\mu\nu} \tag{44}$$

with $\beta = 6/g^2$. The elementary plaquette is the trace of the path-ordered product of links around a unit square (see Fig. 9)

$$U_{\mu\nu}(n) \equiv U_\mu(n)U_\nu(n+\hat{\mu})U_\mu^\dagger(n+\hat{\nu})U_\nu^\dagger(n); \qquad U_P \equiv \frac{1}{3}\mathrm{Re}\,\mathrm{Tr}\,U_{\mu\nu}. \tag{45}$$

In the continuum limit ($a \to 0$) expression (44) reduces to the usual Yang–Mills action[n]. The corresponding partition function is

$$Z = \int \prod_{n,\mu} \mathcal{D}U e^{-\beta \sum_P U_P} \tag{46}$$

where the integral is over the group manifold of colour $SU(3)$ for each link matrix U.

The Wilson loop is simply the trace of the product of the matrices $U(n,\mu)$ along the contour Γ which here is a rectangle. Then

$$\langle W(\Gamma)\rangle = \langle \mathrm{Tr} \prod_{l\in\Gamma} U(l,\mu_l)\rangle = \frac{1}{Z}\int \prod_{n,\mu} DU(n,n+\hat{\mu})\mathrm{Tr} \prod_{l\in\Gamma} U(l,\mu_l) e^{-\beta\sum_P U_P}. \tag{47}$$

Of particular interest is the strong coupling expansion on the lattice, which means to expand for large g (small β). This bears a relation to the string picture that, as we explained, characterizes the long range quark-antiquark interaction. In the strong coupling, we can expand the exponential of the action in (47)

$$\begin{aligned}\langle W(\Gamma)\rangle &= \frac{1}{Z}\int \prod_{n,\mu} DU(n,n+\hat{\mu})\mathrm{Tr} \prod_{l\in\Gamma} U(l,\mu_l) \\ &\times \left[1 - \beta\sum_P \mathrm{Tr}\, U_P + \frac{1}{2}\beta^2 \sum_P\sum_{P'} \mathrm{Tr}\, U_P \,\mathrm{Tr}\, U_{P'} + \cdots\right].\end{aligned} \tag{48}$$

Since each plaquette in the expansion costs a factor β, the leading contribution in the limit $\beta \to 0$ is obtained by paving the inside of the Wilson loop with the smallest number of elementary plaquettes yielding a non-vanishing value for the integral. Using the orthogonality relations supplied in Exercise 4.3.1, it is possible to show that the relevant configuration is the one presented in Fig. 9 and that

$$\langle W(\Gamma)\rangle \simeq \left(\frac{1}{g^2}\right)^{N_P}, \tag{49}$$

N_P being the minimal number of plaquettes required to cover the area enclosed by the path Γ (for more details see Creutz (1983)). This corresponds to the area law (Wilson (1974)) since the area enclosed by the path Γ is given by

[n] Of course there exists an infinite number of lattice actions that have the same naive continuum limit, in particular when one considers also the fermion contribution. To be sure that the given lattice action reproduces really QCD for $a \to 0$, the lattice theory should exhibit a critical region in parameter space where the correlation length diverges. See also Sec. 4.5.

$A(\Gamma) = a^2 N_P$. Furthermore, it is possible to demonstrate that the strong coupling expansion (48) has a finite radius of convergence. Hence, for g^2 large enough, the vacuum expectation value of the Wilson loop has the behaviour

$$\langle W(\Gamma)\rangle \simeq (g^2)^{-A(\Gamma)/a^2} = e^{-\frac{rT\log g^2}{a^2}} \tag{50}$$

where the last equality holds for a rectangular path $r \times T$. The behaviour of the Wilson loop given by Eq. (50) leads to a linear static potential with

$$\sigma \simeq \frac{\log g^2}{a^2}. \tag{51}$$

The layer of plaquettes giving this area contribution corresponds to a constant (chromo)electric field along the string connecting the quark and the antiquark. This suggests once again the relevance of a flux tube description of the non-perturbative interaction and is at the origin of the formulation of the flux tube model of Isgur et al. (1983).

However, we have to keep in mind that the continuum limit of lattice QCD is reached in the weak coupling limit, $\beta \to \infty$. Therefore it is impossible to extrapolate the strong coupling results directly to the continuum physics. Still, we can argue that a rather coarse lattice with large lattice spacing should already give some indicative results for a theory *without* phase transition. To such a lattice corresponds a large bare coupling g and hence the strong coupling result (50) should give the correct qualitative picture. Our philosophy is to take the behaviour (50) as a reliable suggestion to be used in the continuum physics.

The leading order strong coupling expansion of the Wilson loop in QED is identical to Eq. (50) and therefore produces a linear potential too. However, in the case of QED on the lattice a phase transition is clearly seen (Kogut et al. (1981)) when going from strong coupling to weak coupling. In QCD the analytic proof that there cannot be such a phase transition together with the finite radius of convergence of the strong coupling expansion would be equivalent to *a proof of confinement.* Such a proof does not exist up to now. However, the numerical lattice simulations present no hint of such a transition in the intermediate coupling region. On the contrary, the strong coupled behaviour $g^2(a) \sim e^{\sigma a^2}$ continuously goes into the weak coupling $g^2(a) \sim 1/\log a^{-1}$ as $a \to 0$. Moreover, within the coupling regions accessible to present day computers, there are already overlaps between lattice results and weak coupling expansion, the most impressive being the result of the alpha collaboration, see Capitani et al. (1998).

These results hint to the fact that QCD possesses both the property of asymptotic freedom and colour confinement. In other words the Wilson loop in QCD displays a perimeter law in weak coupling and an area law in strong coupling.

4.3 Exercises

4.3.1 The orthogonality properties of the group integral in $SU(3)$ are given by

$$\int dU(n,n+\hat{\mu})[U(n,n+\hat{\mu})]_{ij} = 0$$

$$\int dU(n,n+\hat{\mu})[U(n,n+\hat{\mu})]_{ij}[U^{\dagger}(n,n+\hat{\mu})]_{kl} = \frac{1}{3}\delta_{il}\delta_{jk}$$

$$\int dU(n,n+\hat{\mu})[U(n,n+\hat{\mu})]_{ij}[U(n,n+\hat{\mu})]_{kl} = 0$$

Using these relations justify the result (49).

4.4 Area law as a criterium for confinement, duality and the Wilson loop as an order parameter

To see whether QCD shows confinement, one can study the energy of a system composed of a quark and an antiquark along the lines exposed in Sec. 4.1. Then Eq. (35) tells us that it is the Wilson loop and its behaviour that determines the confinement property of the theory. In the previous section we have seen that in QCD in strong coupling expansion the Wilson loop in the fundamental representation obeys an area law behaviour and this in turn via Eq. (35) confirms the property of quark confinement.

For very large loops $\log\langle W(\Gamma)\rangle$ generally exhibits these two types of behaviour: it decreases either as the perimeter or as the area of Γ. In the first case, expansive loops are allowed and quark and antiquark can be far apart from each other. In the second case quark and antiquark propagate as a bound state. This is the Wilson criterion for confinement of electric charges (Wilson (1974))

$$\langle W(\Gamma)\rangle \sim e^{-KL(\Gamma)} \qquad \text{no confinement} \tag{52}$$

$$\langle W(\Gamma)\rangle \sim e^{-K'A(\Gamma)} \qquad \text{confinement} \tag{53}$$

with $L(\Gamma)$ the perimeter of Γ, $A(\Gamma)$ the minimal area enclosed by Γ, and K and K' dimensionful constants. These are statements about the response of the pure gauge vacuum to external perturbations.

This criterion inspires physical pictures of the QCD vacuum (see Sec. 6). In particular it was suggested by 't Hooft (see e.g. 't Hooft (1994)) that a pure non-Abelian gauge theory could display quite a complicate pattern of vacuum phases: 1) The Higgs mode. Only colour magnetic charges are confined. 2) The Coulomb mode, featuring ordinary massless "photons" and no superconductivity or confinement. 3) The confinement mode or "magnetic superconductor". Quarks (i.e. electric charges) are permanently confined. The properties of these phases are then expressed in terms of operators that create vortices of electric flux (Wilson loops $W_A(\Gamma)$, evaluated on the gauge fields A_μ) and operators that create vortices of magnetic flux ('t Hooft operators $W_C(\Gamma)$, evaluated on the dual potentials C_μ). Using the definition of $W_C(\Gamma)$ via its commutation relation with $W_A(\Gamma)$, 't Hooft (1979) showed that if $W_A(\Gamma)$ obeys an area law, then $W_C(\Gamma)$ necessarily satisfies a perimeter law. The phase in which the 't Hooft operator obeys an area law is the Higgs phase, while that one in which the Wilson loop obeys an area law is the confinement phase. 't Hooft result is a precise way of saying that the vacuum of a confining theory has the properties of a magnetic superconductor. The key ideas are the fact that a non-Abelian gauge theory can be viewed as an Abelian gauge theory enriched with Dirac magnetic monopoles and the concept of electric-magnetic duality. For QCD this implies that the vacuum behaves as a dual superconductor and confinement is explained through the monopole condensation, see Sec. 6. However, if in an Abelian theory it is possible to introduce dual field strengths and electric potentials C^μ dual to the ordinary (magnetic) potentials A^μ in a straightforward manner, in a non-Abelian theory it is not possible to express the dual potentials explicitly in terms of ordinary potentials. The explicit form of the exact Yang–Mills Lagrangian as a function of the C_μ^a fields is unknown. The construction of the long distance limit of the Lagrangian is based on the fact that the dual potentials are weakly coupled since the dual Wilson loop obeys a perimeter law. The quadratic part of the dual Lagrangian is thereby determined. The minimal extension can be constructed under requisite of dual gauge invariance. For a concrete construction of a QCD dual Lagrangian see Baker et al. (1985)-(1995); Maedan et al. (1989) and Sec. 6.

These ideas have originated a quite intensive research activity in the "nonperturbative" physics community in the last few years (in QCD mainly on the lattice while in supersymmetric field theories very promising analytic results are obtained just now, see e.g. Alvarez-Gaumé et al. (1997)). We will come back to this point in Sec. 6.2. Here we refer the reader to the historical papers of Nambu (1974), Mandelstam (1979) and 't Hooft (1982) and references therein.

4.5 Lattice simulations and lattice results

With the action of Eq. (44), using periodic boundary conditions in space and time and taking a lattice of finite volume and finite spacing, the system has a finite number of degrees of freedom: the gluon fields on the link and (if we add them) the quark fields at the lattice sites. The functional integral over the gauge fields is converted to a multiple integral with a positive definite integrand (in Euclidean space): Eq. (46). Unlike in continuum perturbation theory the calculation of averaged values of gauge invariant observables is done without any gauge-fixing.

For a lattice of L^4 sites with colour group $SU(3)$ the integral in (46) would be a $8 \times 4 \times L^4$ dimensional integral: the standard approach is to use a Monte Carlo approximation to the integrand. Precisely, a stochastic estimate of the integral is made from a finite number of samples (the "configurations") of equal weight. For more details see Rothe (1992). Here we are interested only in explaining that in this way we have at our disposal a set of samples of the vacuum, then, it is possible to evaluate the average of fields over these samples and obtain a nonperturbative evaluation of Green functions as well as of any field vacuum correlator[o].

In this section we present some lattice results on the calculation of the Wilson loop expectation value, the static potential, the flux tube configuration and the determination of the string tension σ. However, before, we comment briefly on the validity of the lattice results and give few concepts to enable the "reading" of those results.

The lattice is not the real world so that before extracting the physics we have to be sure that: 1) the Green functions have been extracted without contamination (the ground state mass should not be contaminated by pieces coming from the excited states); 2) the lattice size is big enough, we mean that the relevant physical distance should fit in! (finite size error); 3) the statistical errors of the calculation is under control (statistical error); 4) the lattice spacing is small enough, since we have to approach the continuum limit (discretization error). On the other hand the lattice spacing supplies an explicit ultraviolet cut-off.

The last condition is the most subtle one. Let us just sketch the idea. In order to extract the continuum limit from the lattice, it has to be shown that the results do not change if the lattice spacing is decreased further. However, the lattice spacing in physical units is not known directly. Actually it is mea-

[o] Analytic continuation from Euclidean to Minkowski time of Green functions that are only available on a finite set of points is not possible. Therefore, on the lattice one can only determine masses and on-shell matrix elements in a straightforward way. Real time processes like scattering, hadronic decays, are not directly accessible.

sured. The lattice simulations are performed at a fixed value of β (cf. Eq. (44)). We recall that $\beta = 6/g^2$ and therefore a large β corresponds to a small g^2. Ensembles at different values of a are obtained by using different bare coupling constants g in the action. However, the value of a for a given value of g is not known *a priori* but has to be obtained calculating a dimensionful parameter and comparing it to an experiment.

As an example let us consider the string tension σ. Measured in lattice units it is only a function of the bare coupling: $\hat{\sigma}(g)$, the $\hat{}$ denoting a dimensionless quantity. In physical units, however, it has the dimension of a (mass)^2, so that the physical string tension is given by

$$\sigma = \lim_{a\to 0} \frac{1}{a^2}\hat{\sigma}(g(a)). \tag{54}$$

It approaches a finite limit for $a \to 0$, if g is tuned with a in an appropriate way. By requiring that in this limit

$$a\frac{\partial}{\partial a}\sigma = 0 \tag{55}$$

we obtain

$$\begin{aligned} \sqrt{\sigma} &= c_\sigma \Lambda_L \\ \Lambda_L &= a^{-1} e^{\frac{-1}{2\beta_0 g^2}} (\beta_0 g^2)^{\frac{-\beta_1}{2\beta_0^2}} (1 + O(g^2)) \end{aligned} \tag{56}$$

where we have used the usual lattice convention for the beta function: $\beta(g) = -\beta_0 g^3 - \beta_1 g^5 + O(g^7)$ (which differs for a factor $1/(4\pi)^2$ from Eq. (3)) and Λ_L is independent of a and fixes the mass scale of the theory[p]. This equation tells us again that a perturbative calculation of σ is a priori impossible due to the non-analytic behaviour in g of Eq. (56). It is possible to show that all dimensionful physical quantities can be expressed in the form

$$\theta_{phys} = \lim_{a\to 0} \frac{1}{a^{d_\theta}}\hat{\theta}(g(a), a) = c_\theta (\Lambda_L)^{d_\theta} \tag{57}$$

d_θ being the naive dimension. Then, for small lattice spacing we can determine the universal function $g = g(a)$ by fixing the l.h.s. of (56) at the physical value of the string tension. This gives g as a function of $a\sqrt{\sigma}$

$$g^2(a) = -\frac{1}{2\beta_0 \log(a\Lambda_L)} \tag{58}$$

[p] **The appearance of a scale as Λ_L is well known from perturbative continuum QCD, where the necessity of renormalizing the theory also requires the introduction of a scale. Usually the relation between Λ_{QCD} in different regularization schemes and Λ_L is known in perturbation theory in the bare coupling. See however S. Capitani et al. (1998).**

(where Λ_L is now determined by the physical condition imposed) which ensures the finiteness of any observable and allows us to convert them to physical units[q].

The continuum limit has to be taken at a constant physical volume L^4. As a is decreased the number of lattice points $\hat{L} = L/a$ increases. Therefore, a finite computer limits the lattice spacing to $a \geq a_{\rm min} > 0$. In practice one is looking for scaling of the results within this window or at least checking that the results follow the expected leading order a dependence in order to safely extrapolate them to $a = 0$. A typical value for lattice simulations of QCD is $\beta \simeq 6$ and hence the bare coupling constant is $\alpha_s = g^2/(4\pi) \simeq 0.08$. In the early years of lattice QCD it was widely assumed that at least for $\beta \simeq 6$, two-loop perturbation theory can be applied to $a(g)$. Therefore, after having determined a at a coupling g, the Λ parameter was extracted and $a(g\prime)$ for $g\prime \neq g$ computed via perturbation theory. Nowadays, the lattice spacing is determined separately for each simulation point β by inputting one experimental value. In doing so, $a(g)$ is obtained as a function of the bare lattice g. One finds big deviations from perturbation theory (asymptotic scaling) that are related to the importance of the so called tadpole diagrams in lattice perturbation theory (see e.g. Michael (1997), Davies (1997) and Lepage et al. (1993) for these developments)[r].

The actual methods of simulating lattice QCD and extracting physical information have reached quite a high level of sophistication and we refer the interested reader to the reviews (e.g. Rothe (1992), Davies (1997), Montvay and Munster (1994)). The above discussion should be sufficient to make clear that: 1) lattice simulations are performed at a fixed value of β, 2) the value of β is connected with the lattice spacing a, 3) results relevant for continuum physics are effectively independent of a (scaling) 4) the extraction of the physical results requires to fix a and in the quenched approximation it may be dependent on the experimental quantity chosen to fix it. It is, therefore, important to know what quantity was chosen when looking at the lattice data.

Now let us come to some results. First, we discuss the lattice measure of σ. The lattice counterpart of Eq. (35) is

$$\hat{V}(\hat{r}) = -\lim_{\hat{T}\to\infty} \frac{1}{\hat{T}} \log W(\hat{r}, \hat{T}) \tag{59}$$

where $W(\hat{r}, \hat{T})$ denotes the expectation value of a Wilson loop with spatial and

[q] A corresponding statement is expected to hold if the action depends on several parameters, e.g. coupling constant and quark masses.

[r] One might ask if it was nonetheless possible to invert the $a(g)$ relation to obtain an $\alpha(q)$ and run the q to high momenta subsequently. For this purpose effective couplings other than g have been suggested and determined which show an improved asymptotic scaling behaviour.

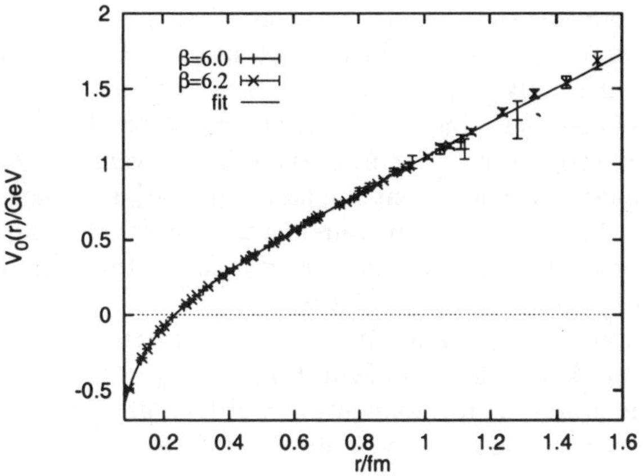

Figure 10: *Static potential at $\beta = 6.0$ and $\beta = 6.2$ in the quenched approximation. The solid line is a fit of the form of Eq. (61). Bali et al. (1997).*

temporal extension $\hat{r}$ and $\hat{T}$ respectively. Assuming that $W(\hat{r},\hat{T}) = \exp\{\hat{\sigma}\hat{r}\hat{T} - \hat{\delta}(\hat{r}+\hat{T}) + \hat{\gamma}\}$, the famous Creutz ratio

$$\chi(\hat{r},\hat{T}) = -\log\left(\frac{W(\hat{r},\hat{T})W(\hat{r}-1,\hat{T}-1)}{W(\hat{r},\hat{T}-1)W(\hat{r}-1,\hat{T})}\right) \tag{60}$$

coincides with the string tension, since the parameters $\hat{\delta}$ and $\hat{\gamma}$ drop out. The original calculation (Creutz et al. (1982)) was performed on a 6^4 lattice and the asymptotic scaling seemed to show up for values of β slightly below 6.0. This measurement of a string tension different from zero from the strong coupling region to the asymptotic scale region, without any indication of a phase transition in the intermediate region, constituted the first *evidence* of quark confinement in QCD.

In Fig. 10 we show the most recent lattice measurement of the quenched static potential from Eq. (59) on a hypercubic lattice $V = 16^4$ at $\beta = 6.0$ and $V = 32^4$ at $\beta = 6.2$. These values correspond to inverse lattice spacings $a^{-1} \simeq 2.1$ GeV and $a^{-1} \simeq 2.9$ GeV respectively. The scale is adjusted to optimally reproduce the bottomonium level splittings (Bali et al. (1997)). The fit curve corresponds to the parameterization

$$V_0(r) = -\frac{e}{r} + \sigma r + \frac{f}{r^2} \tag{61}$$

which clearly confirms the Cornell potential (8) (the $1/r^2$ correction, that accounts for the running of the coupling, is not meant to be physical but has been introduced to effectively parameterize the data within the given range of r). The parameters take the value: $e = 0.321$ and $\sigma = (468 \text{ MeV})^2$. The coefficient of the Coulomb term is quite far from the effective value for α_s coming from the potential models[s]. Notice that the lattice potential becomes clearly linear around 0.2 fm. Similar lattice measurements exist for the unquenched static potential (see e.g. Bali (1998)): the string is found to break down around a quark-antiquark distance of about 1.2 fm.

The static potential in Fig. 10 has been extracted as the ground state energy of the quark-antiquark configuration (cf. Eq. (22)), which in turn corresponds to the lowest energy configuration of the "glue" between the quarks. Yet, also the excited gluonic modes have been measured on the lattice and the corresponding potentials have been adiabatically extracted. This should corresponds to the potential of heavy hybrids. For further details of this confirmation of the excited structure of the flux tube we refer the reader to Michael (1997), Morningstar et al. (1998).

Lattice studies have been undertaken to probe the energy momentum tensor of the colour fields. The probe used is the gauge-invariant insertion of a plaquette in the presence of a static Wilson loop. Depending on the Lorentz orientation of the plaquette, this corresponds to the average of a (chromo)electric or a (chromo)magnetic field in the presence of quark sources. Lattice sum rules can be used to normalize these distributions and to relate them to the β function. For separations $r > 0.7$ fm, a string-like spatial distribution is found with a transverse rms width increasing very slowly with r and reaching a rather constant value between 1 and 2 fm. The physical value for this constant ranges between 0.5 and 0.75 fm. The averages of the squared components of the colour fields (which are gauge invariant quantities) are found to be roughly equal (i.e. $\langle E_i^2 \rangle \sim \langle B_j^2 \rangle$ in the presence of the Wilson loop, for $i, j = 1, 2, 3$). This implies that the energy density is much smaller than the action density (in Euclidean space). (See Sec. 6.3 for a discussion.) The result for the action density is presented in Fig. 11 and is quite impressive! The formation of the interquark flux tube is evident. The interpretation of this phenomenon is the following. When the distance between the quarks becomes larger than some critical value (connected to Λ_{QCD}) the branching of the gluons, due to the non-Abelian nature of QCD, becomes so intensive that it makes no sense to speak about individual gluons. A coherent effect develops with the subsequent formation of the flux tube. It is conjectured that a specific organization of the QCD vacuum makes this kind of configuration energetically favorable (see Sec. 6).

[s]Part of this discrepancy can be traced back to the quenched approximation.

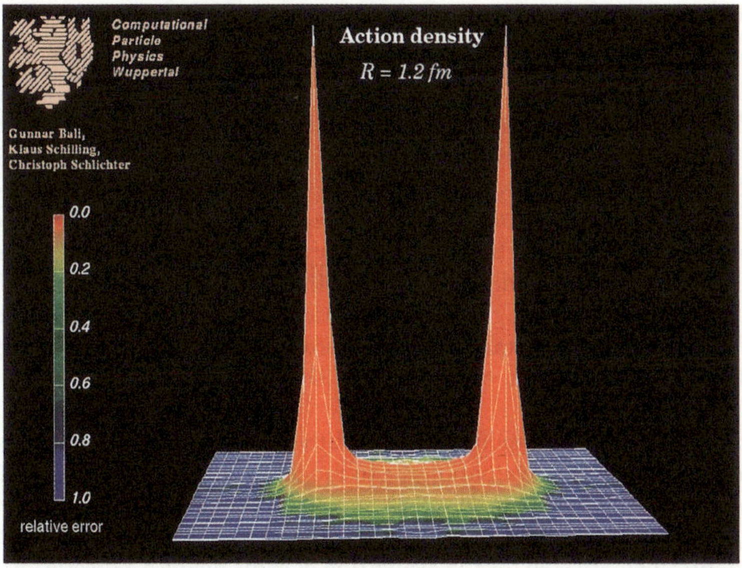

Figure 11: *Action density distribution between two static quarks in* $SU(2)$ *measured on a lattice* $V = 32^4$ *at* $\beta = 2.5$. *The physical quark-antiquark distance is 1.2 fm. Bali et al. (1995).*

However, lattice QCD seems more suitable to ask "what" and not "why". Here, we have shown some model independent results on the quark interaction that provide evidence of quark confinement but do not explain *why* quarks are confined. Eventually, to get some insight in the quark confinement mechanism, it is necessary to build up and use some analytic models. The role of the lattice measurements will then be to validate these models. The combination of analytic and lattice techniques will give us some information on the nature of the nonperturbative quark interaction.

Finally, we mention that there are a lot of measurements of hadron masses performed in lattice QCD as well as in lattice nonrelativistic QCD (NRQCD). However, since in these lectures we are more interested in the mechanism of confinement as well as in developing analytic approaches to it, we refer the reader to the literature (see e.g. Davies (1997)).

5 The Heavy Quark Interaction

In this part of the lectures we summarize some nonperturbative analytic results on the heavy quark interaction. With nonperturbative we mean the fact that they do not rely necessarily on a perturbative expansion in g. An analytic result of this type was already obtained for the static potential in Sec. 4 with Eq. (35). Here, we want to establish a systematic procedure to obtain relativistic corrections. In order to take full advantage from lattice calculations, our analytic approach will be manifestly gauge-invariant. In the next sections we present some model independent results: 1) the effective theories of QCD that enable the performing of a systematic expansion of the heavy quarks dynamics in some small parameters which are nonperturbative and 2) the exact and physically transparent expression for the quark-antiquark interaction at order $1/m^2$.

However, at the very moment we want to calculate the quark dynamics we have to resort to lattice evaluations or to models of the QCD vacuum (see Sec. 6). Nevertheless, the present approach allows us to gain something with respect to the standard lattice formulation. In fact, here the lattice simulations come in at an intermediate step for the evaluation of some definite expectation values of fields inserted in the static Wilson loop. These can be directly tested with the analytic (model dependent) results. From this comparison, in a process of model validation, we gain a deeper understanding of the confinement physics.

A remark: we have performed all the calculations of the last section in Euclidean space since lattice simulations (as all numerical evaluations) are done in Euclidean space. We no longer have this restriction in the next sections where we come back to the Minkowski space. We use the following notation for the vacuum expectation value (to be compared with Eq. (33))

$$\langle f[A]\rangle \equiv \frac{1}{Z}\int \mathcal{D}A f[A] e^{i\int d^4x L_{YM}}. \tag{62}$$

In most of the cases it will be straightforward to switch from Euclidean to Minkowski space by simply using the relations (23).

5.1 *Effective theories for heavy quarks*

We have seen that the physics of heavy quark bound states is complicated by the interplay of different characteristic scales like the heavy quark mass m, the momentum of the bound state mv, the energy of the bound state mv^2 (and in principle also Λ_{QCD}), being v the heavy quark velocity. These scales can be disentangled using effective theories of QCD. From the technical point of view, this simplifies considerably the calculation. From the conceptual point of view, this enables us to factorize the part of the interaction that we know (and we are able to calculate in perturbation theory) from the low energy part which is dominated by nonperturbative physics.

Nonrelativistic QCD (NRQCD) is an effective theory equivalent to QCD and constructed integrating out the high energy ($E > m$) degrees of freedom, thus making explicit the mass parameter. NRQCD has been extensively discussed at this school by Peter Lepage. We recall only few points useful to prepare the developments of the next sections. The interested reader is referred to Lepage (1996) and Thacker et al. (1991).

NRQCD was devised to be applied to lattice simulations, however, here we are interested in the definition of NRQCD in the continuum. In Sec. 4 we have considered the static limit of an infinitely massive quark. In that case the Dirac equation for the quark propagator in an external field is exactly solvable. Now, we want to calculate the subsequent corrections showing up when the quark mass is large but finite. In order to disentangle the dynamical scales it is convenient to use the heavy quark velocity, $v_Q \equiv v$, as an expansion parameter. Notice that, even if this seems to be analogous to what happens in QED where, e.g. in positronium, the expansion parameter is $v_e \simeq \alpha$, the quark velocity is also sensitive to the nonperturbative quark interaction and turns out to be a function of both α_s and Λ_{QCD} (or σ).

Let us consider Eq. (27) (now in Minkowski space) or equivalently the Lagrangian

$$L = \bar{\psi}(i\gamma^\mu D_\mu - m)\psi \,. \tag{63}$$

Applying a Foldy–Wouthuysen transformation and expanding in the inverse of the quark mass m, we obtain[t] up to order $1/m^2$

$$\begin{aligned} L = \psi^\dagger \Bigg(& m + iD_0 + \frac{\mathbf{D}^2}{2m} + c_1(m/\mu)\frac{(\mathbf{D}^2)^2}{8m^3} + c_2(m/\mu)\frac{g}{8m^2}(\mathbf{D}\cdot\mathbf{E} - \mathbf{E}\cdot\mathbf{D}) \\ & + ic_3(m/\mu)\frac{g}{8m^2}\boldsymbol{\sigma}\cdot(\mathbf{D}\times\mathbf{E} - \mathbf{E}\times\mathbf{D}) + c_4(m/\mu)\frac{g}{2m}\boldsymbol{\sigma}\cdot\mathbf{B} + \ldots \Bigg)\psi \\ & + \text{antiquark terms} \end{aligned} \tag{64}$$

In expanding the QCD Lagrangian in the inverse of the mass we have lost explicit renormalizability. The NRQCD Lagrangian (64) must be regularized. μ is an ultraviolet cut-off which restricts the momenta to the region $p \sim mv < \mu < m$. The effect of the excluded momenta, e.g. in gluon loops, is factorized in the (matching) coefficients which multiply the nonrelativistic operators. In Eq. (64) we call them $c_i(m,\mu)$. Such coefficients are straightforwardly calculated

[t] We neglect here operators involving more than 2 fermions. These are irrelevant to our discussion here, see Brambilla et al. (1998).

in perturbation theory by imposing that scattering amplitudes evaluated with (64) are equal to the same amplitudes evaluated in QCD order by order in α_s and in $1/m$ (see Manohar (1997)). This procedure is called *matching*. The coefficients c_i are normalized in such a way to be one (or zero) at tree-level.

The terms in the Lagrangian can be ordered in powers of the squared velocity of the heavy quark using the power counting rules for momentum and kinetic energy of Thacker et al. (1991): $\mathbf{D} \sim p \sim mv$, $K \sim mv^2$. From the lowest order field equation

$$\left(i\partial_0 - gA_0 - \frac{\mathbf{D}^2}{2m} \right) \psi = 0$$

we get

$$\begin{aligned} gA_0 \sim \partial_0 \sim K &= mv^2 \\ g\mathbf{E} = [D_0, \mathbf{D}] \sim pK &= m^2 v^3 \\ -ig\epsilon_{ijk} B^k = [D_i, D_j] \sim K^2 &= m^2 v^4. \end{aligned}$$

Following these power-counting rules the number of operators to be included in L can be truncated at a fixed order in v^2 depending on the precision we require on the calculation of the energy.

This power counting is not exact. It accounts only for the leading binding contributions. The reason is that two scales, the soft one $\sim mv$ and the ultrasoft $\sim mv^2$, are still mixed up in the NRQCD Lagrangian. An exact power counting can be achieved by integrating out from NRQCD the soft scale with the same procedure as the mass scale was integrated out from QCD in order to get NRQCD. In this way one obtains a further effective theory, called potential nonrelativistic QCD (pNRQCD) (Pineda and Soto (1998)) where only dynamical ultrasoft degrees of freedom are present. These are the heavy quark bound states (explicitly projected in colour singlet and octet states) and gluons propagating at the ultrasoft scale. Since nonrelativistic potentials get contributions only from the soft scale, in the pNRQCD Lagrangian potential and nonpotential contributions are explicitly disentangled. More precisely, the pNRQCD Lagrangian density with leading nonpotential corrections is given by (Brambilla et al. (1999)):

$$\begin{aligned} L = \mathrm{Tr} \left\{ S^\dagger \left(i\partial_0 - \frac{\mathbf{p}^2}{m} + \sum_n \frac{V_s^{(n)}(\mathbf{r}, \mathbf{p}, \boldsymbol{\sigma}; \mu')}{m^n} \right) S \right\} \\ + \mathrm{Tr} \left\{ O^\dagger \left(iD_0 - \frac{\mathbf{p}^2}{m} + \sum_n \frac{V_o^{(n)}(\mathbf{r}, \mathbf{p}, \boldsymbol{\sigma}; \mu')}{m^n} \right) O \right\} \end{aligned}$$

$$+gV_A(\mathbf{r};\mu')\mathrm{Tr}\left\{O^\dagger\mathbf{r}\cdot\mathbf{E}\,S+S^\dagger\mathbf{r}\cdot\mathbf{E}\,O\right\}$$
$$+g\frac{V_B(\mathbf{r};\mu')}{2}\mathrm{Tr}\left\{O^\dagger\mathbf{x}\cdot\mathbf{E}\,O+O^\dagger O\mathbf{r}\cdot\mathbf{E}\right\} \tag{65}$$

where $\mathbf{r} \equiv \mathbf{x}_1 - \mathbf{x}_2$ and $\mathbf{X} \equiv (\mathbf{x}_1 + \mathbf{x}_2)/2$, $S = S(\mathbf{r}, \mathbf{X}, t)$ and $O = O(\mathbf{r}, \mathbf{X}, t)$ are the singlet and octet wave function respectively. All the gauge fields in Eq. (65) are evaluated in $\mathbf{X}$ and t. In particular $\mathbf{E} \equiv \mathbf{E}(\mathbf{X}, t)$ and $iD_0O \equiv i\partial_0 O - g[A_0(\mathbf{X},t), O]$. The matching coefficients V_A and V_B are normalized in such a way that at the leading perturbative order they are one. μ' is the ultrasoft cut-off corresponding to the regularization of the Lagrangian (65): $mv > \mu' > mv^2$. At the leading order in the soft scale the equations of motion associated with the pNRQCD Lagrangian (65) are two decoupled nonrelativistic Schrödinger equations describing the propagation of a singlet and an octet bound state defined by the potentials $\sum V_s^n/m^n$ and $\sum V_o^n/m^n$ respectively. Next-to-leading nonpotential corrections involve the coupling of octet and singlet states via ultrasoft chromoelectric fields. We point out that: 1) The potentials have now the status of matching coefficients (in the matching from NRQCD to pNRQCD) and depend in general on the scale μ' (see Sec. 4.2). They can be calculated comparing Wilson loop functions in NRQCD and singlet/octet propagators in pNRQCD. The novel feature is that the matching takes place now in the low energy region and thus it can be even nonperturbative. In this case the potentials are given by expressions to be evaluated on the lattice. Notice that the potentials contain in their definition also the NRQCD coefficients $c_i(m/\mu)$. 2) the pNRQCD Lagrangian contains both potential and nonpotential contributions. In some kinematic regions they coincide with the Voloshin (1979) and Leutwyler (1981) corrections. 3) Only the ultrasoft scale is left. Hence each term in (65) has a definite power counting in v! In particular: $1/r \sim p \sim mv$, $V_{s,o}^{(n)} \sim mv^2$ and $F_{\mu\nu} \sim m^2v^4$.

In the lattice NRQCD approach the NRQCD Lagrangian of Eq. (64) is discretized on the lattice and the heavy hadron masses are obtained by performing lattice simulations. These are considerably less time consuming than the traditional QCD lattice simulations because the cut-off scale μ (smaller than m) makes possible to use coarse lattices. Since in the following we will not deal explicitly with nonpotential contributions we will not perform explicitly the matching from NRQCD to pNRQCD. As far as ultrasoft degrees of freedom are not considered, this corresponds only to a choice of language. Actually, getting the heavy quark potential from the NRQCD Lagrangian, in any way one is doing it, if done properly, is nothing else than performing the leading order matching with the pNRQCD Lagrangian. In particular we will show how to obtain the spin and velocity dependent singlet potentials up to order $1/m^2$. A possible method is to calculate these as corrections to the static propagator from the Lagrangian (64) (see Eichten et al. (1981), Tafelmayer (1986)). We will use a path integral formalism (Peskin (1983)). This approach has the advantage that the result, expressed in terms of deformed Wilson loops, is suitable to be used also in QCD vacuum models (see Sec. 6).

One remark at the end. The matching coefficients c_i contain all the fermionic loop contributions at the hard scale. If the matching between NRQCD and pNRQCD is nonperturbative, then the quark loop contributions at the low energy scale are contained in the functional measure of the object to be evaluated on the lattice (and it is indeed a hard task even for the lattice up to now). In the following section for simplicity we work in the quenched approximation.

5.2 The QCD spin-dependent and velocity-dependent potentials

Let us consider Eq. (64), taking the matching coefficients at tree level[u]. From Eq. (64) it follows that the heavy quark (of mass m_j) propagator K_j in external field (that now is reduced to a Pauli quark propagator, i.e. a 2×2 matrix in the spin indices) satisfies the Schrödinger equation

$$
\begin{aligned}
i\frac{\partial}{\partial x^0} K_j(x,y;A) &= H_{\rm FW} K_j(x,y;A) \\
&\equiv \Bigg[m_j + \frac{1}{2m_j}(\mathbf{p}_j - g\mathbf{A})^2 - \frac{1}{8m_j^3}(\mathbf{p}_j - g\mathbf{A})^4 - \frac{g}{m_j}\mathbf{S}_j \cdot \mathbf{B} + gA^0 \\
&\quad - \frac{g}{8m_j^2}(\partial_i E^i - ig[A^i, E^i]) + \frac{g}{4m_j^2}\varepsilon^{ihk} S_j^k \{(p_j - gA)^i, E^h\} \Bigg] K_j(x,y;A)
\end{aligned}
\tag{66}
$$

with the Cauchy condition

$$
K_j(x,y;A)|_{x^0=y^0} = \delta^3(\mathbf{x}-\mathbf{y}) \tag{67}
$$

where ε^{ihk} is the three-dimensional Ricci symbol. By standard techniques the solution of Eq. (66) (see Sakurai (1985) and Exercise 5.2.1) with the initial condition (67), can be expressed as a path integral in the phase space

$$
K_j(x,y;A) = \int_{\mathbf{z}_j(y^0)=\mathbf{y}}^{\mathbf{z}_j(x^0)=\mathbf{x}} \mathcal{D}[\mathbf{z}_j, \mathbf{p}_j] \, \mathrm{T}\exp\left\{ i \int_{y^0}^{x^0} dt \, [\mathbf{p}_j \cdot \dot{\mathbf{z}}_j - H_{\rm FW}] \right\}. \tag{68}
$$

[u] The matching coefficients can be simply included in the calculation (see Chen et al. (1995), Bali et al. (1997)). They are needed for example to obtain the one-loop perturbative behaviour of the potentials, see Brambilla et al. (1998)).

Here, the time-ordering prescription T acts both on spin and gauge matrices. The trajectory of the quark j in coordinate space is denoted by $\mathbf{z}_j = \mathbf{z}_j(t)$, the trajectory in momentum space by $\mathbf{p}_j = \mathbf{p}_j(t)$ and the spin by $\mathbf{S}_j$ (See Fig. 8). Standard path integral manipulations on Eq. (68) give

$$K_j(x,y;A)=\int_{\mathbf{z}_j(y^0)=\mathbf{y}}^{\mathbf{z}_j(x^0)=\mathbf{x}} \mathcal{D}[\mathbf{z}_j,\mathbf{p}_j]\,\mathrm{T}\exp\left\{i\int_{y^0}^{x^0} dt\,\left[\mathbf{p}_j\cdot\dot{\mathbf{z}}_j - m_j - \frac{\mathbf{p}_j^2}{2m_j} + \frac{\mathbf{p}_j^4}{8m_j^3}\right.\right.$$
$$-gA^0 + \frac{g}{m_j}\mathbf{S}_j\cdot\mathbf{B} + \frac{g}{2m_j^2}\mathbf{S}_j\cdot(\mathbf{p}_j\times\mathbf{E}) - \frac{g}{m_j}\mathbf{S}_j\cdot(\dot{\mathbf{z}}_j\times\mathbf{E})$$
$$\left.\left.+g\dot{\mathbf{z}}_j\cdot\mathbf{A} + \frac{g}{8m_j^2}(\partial_i E^i - ig[A^i,E^i])\right]\right\}. \tag{69}$$

Inserting Eq. (69) into expression (26) of the quark-antiquark Green function and taking $x_1^0 = x_2^0 = t_\mathrm{f}$, $y_1^0 = y_2^0 = t_\mathrm{i}$ with $T \equiv t_\mathrm{f} - t_\mathrm{i} > 0$, one obtains the two-particle Pauli-type propagator K in the form of a path integral on the world lines of the two quarks

$$K(\mathbf{x}_1,\mathbf{x}_2,\mathbf{y}_1,\mathbf{y}_2;T) = \int_{\mathbf{z}_1(t_\mathrm{i})=\mathbf{y}_1}^{\mathbf{z}_1(t_\mathrm{f})=\mathbf{x}_1} \mathcal{D}[\mathbf{z}_1,\mathbf{p}_1] \int_{\mathbf{z}_2(t_\mathrm{i})=\mathbf{y}_2}^{\mathbf{z}_2(t_\mathrm{f})=\mathbf{x}_2} \mathcal{D}[\mathbf{z}_2,\mathbf{p}_2]$$
$$\times\exp\left\{i\int_{t_\mathrm{i}}^{t_\mathrm{f}} dt \sum_{j=1}^{2}\left[\mathbf{p}_j\cdot\dot{\mathbf{z}}_j - m_j - \frac{\mathbf{p}_j^2}{2m_j} + \frac{\mathbf{p}_j^4}{8m_j^3}\right]\right\}$$
$$\times\left\langle \frac{1}{3}\mathrm{Tr}\,\mathrm{T_s}\,\mathrm{P}\exp\left\{ig\oint_\Gamma dz^\mu\,A_\mu(z) + \sum_{j=1}^{2}\frac{ig}{m_j}\int_{\Gamma_j} dz^\mu\right.\right.$$
$$\left.\left.\times\left(S_j^l\hat{F}_{l\mu}(z) - \frac{1}{2m_j}S_j^l\varepsilon^{lkr}p_j^k F_{\mu r}(z) - \frac{1}{8m_j}D^\nu F_{\nu\mu}(z)\right)\right\}\right\rangle, \tag{70}$$

where the dual tensor field is defined to be $\hat{F}^{\mu\nu} \equiv \varepsilon^{\mu\nu\rho\sigma}F_{\rho\sigma}/2$ Here $\mathrm{T_s}$ is the time-ordering prescription for spin matrices, P is the path-ordering prescription for gauge matrices along the loop Γ, Γ_1 denotes the path going from $(t_\mathrm{i},\mathbf{y}_1)$ to $(t_\mathrm{f},\mathbf{x}_1)$ along the quark trajectory $(t,\mathbf{z}_1(t))$, Γ_2 the path going from $(t_\mathrm{f},\mathbf{x}_2)$ to $(t_\mathrm{i},\mathbf{y}_2)$ along the antiquark trajectory $(t,\mathbf{z}_2(t))$ and Γ is the path made by Γ_1 and Γ_2 closed by the two straight lines joining $(t_\mathrm{i},\mathbf{y}_2)$ with $(t_\mathrm{i},\mathbf{y}_1)$ and $(t_\mathrm{f},\mathbf{x}_1)$ with $(t_\mathrm{f},\mathbf{x}_2)$ (see Fig. 8). Finally Tr denotes the trace over the gauge matrices. Note that the right-hand side of (70) is manifestly gauge invariant.

Defining the angular bracket term in Eq. (70) as (more rigorously this should correspond to the leading matching between NRQCD and pNRQCD)

$$\left\langle \frac{1}{3}\mathrm{Tr}\,\mathrm{T_s}\,\mathrm{P}\exp\ldots \right\rangle = \mathrm{T_s}\exp\left[-i\int_{t_\mathrm{i}}^{t_\mathrm{f}} dt\, V_{\mathrm{Q\bar{Q}}}(\mathbf{z}_1,\mathbf{z}_2,\mathbf{p}_1,\mathbf{p}_2,\mathbf{S}_1,\mathbf{S}_2)\right]; \qquad (71)$$

we obtain that

$$i\frac{\partial}{\partial T}K = \left[\sum_{j=1}^{2}\left(m_j + \frac{\mathbf{p}_j^2}{2m_j} - \frac{\mathbf{p}_j^4}{8m_j^3}\right) + V_{\mathrm{Q\bar{Q}}}\right]K, \qquad (72)$$

where $V_{\mathrm{Q\bar{Q}}}$ is *the complete QCD (quenched) quark-antiquark potential at the order* v^4. Expanding the logarithm of the left-hand side of (71) up to $1/m^2$, we find[v]

$$\begin{aligned}\int_{t_\mathrm{i}}^{t_\mathrm{f}} dt\, V_{\mathrm{Q\bar{Q}}} = i\log\langle W(\Gamma)\rangle - \sum_{j=1}^{2}\frac{g}{m_j}\int_{\Gamma_j} dz^\mu \Big(S_j^l \,\langle\!\langle \hat{F}_{l\mu}(z)\rangle\!\rangle \\ -\frac{1}{2m_j}S_j^l\varepsilon^{lkr}p_j^k\,\langle\!\langle F_{\mu r}(z)\rangle\!\rangle - \frac{1}{8m_j}\,\langle\!\langle D^\nu F_{\nu\mu}(z)\rangle\!\rangle\Big) \\ -\frac{1}{2}\sum_{j,j'=1}^{2}\frac{ig^2}{m_j m_{j'}}\mathrm{T_s}\int_{\Gamma_j} dx^\mu \int_{\Gamma_{j'}} dx'^\sigma\, S_j^l\, S_{j'}^k \\ \times\Big(\langle\!\langle \hat{F}_{l\mu}(z)\hat{F}_{k\sigma}(z')\rangle\!\rangle - \langle\!\langle \hat{F}_{l\mu}(z)\rangle\!\rangle\,\langle\!\langle \hat{F}_{k\sigma}(z')\rangle\!\rangle\Big), \qquad (73)\end{aligned}$$

where we recall that

$$\langle f(A)\rangle \equiv \frac{1}{3}\mathrm{Tr}\,\mathrm{P}\,\frac{\displaystyle\int \mathcal{D}A\, e^{iS_{\mathrm{YM}}(A)} f(A)}{\displaystyle\int \mathcal{D}A\, e^{iS_{\mathrm{YM}}(A)}}$$

and we have introduced the vacuum expectation value in presence of quarks (i.e. of the Wilson loop)

$$\langle\!\langle f(A)\rangle\!\rangle \equiv \frac{\displaystyle\int \mathcal{D}A\, e^{iS_{\mathrm{YM}}(A)}\mathrm{Tr}\,\mathrm{P}\, f(A)\exp\left[ig\oint_\Gamma dz^\mu A_\mu(z)\right]}{\displaystyle\int \mathcal{D}A\, e^{iS_{\mathrm{YM}}(A)}\mathrm{Tr}\,\mathrm{P}\exp\left[ig\oint_\Gamma dz^\mu A_\mu(z)\right]}.$$

[v] Notice that $\int_{\Gamma_j} dz^\mu f_\mu(z) = (-1)^{j+1}\int_{t_\mathrm{i}}^{t_\mathrm{f}} dt(f_0(z_j) - \dot{\mathbf{z}}_j\cdot\mathbf{f}(z_j))$, where $z_j = (t,\mathbf{z}_j(t))$. The factor $(-1)^{j+1}$ accounts for the fact that world line Γ_2 runs from t_f to t_i. We also use the notation $z'_j = (t',\mathbf{z}_j(t'))$.

In this way one obtains from QCD the static, spin-dependent and velocity dependent terms that control the quarkonium spectrum and that were introduced on a pure phenomenological basis in Sec. 3, cf Eqs. (12)–(14):

$$V_{Q\bar{Q}} = V_0 + V_{VD} + V_{SD} \,. \tag{74}$$

These terms have a physical direct interpretation.

The spin independent part of the potential, $V_0 + V_{VD}$, is obtained in (73) from the expansion of $\log\langle W(\Gamma)\rangle$ for small velocities $\dot{\mathbf{z}}_1(t) = \mathbf{p}_1/m_1$ and $\dot{\mathbf{z}}_2(t) = \mathbf{p}_2/m_2$:

$$i\log\langle W(\Gamma)\rangle = \int_{t_i}^{t_f} dt\; (V_0(r(t)) + V_{VD}(\mathbf{r}(t)))\,, \tag{75}$$

where V_0 is the static part and $\mathbf{r}(t) \equiv \mathbf{z}_1(t) - \mathbf{z}_2(t)$.

The spin-dependent part, V_{SD}, contains for each quark terms analogous to those one would obtained by making a Foldy–Wouthuysen transformation of a Dirac equation in an external field $\langle\!\langle F_{\mu\nu}\rangle\!\rangle$, along with an additional term V_{SS} having the structure of a spin-spin interaction. Therefore we can write

$$V_{SD} = V_{LS}^{MAG} + V_{\rm Thomas} + V_{\rm Darwin} + V_{SS} \tag{76}$$

using a notation which indicates the physical significance of the individual terms (MAG denotes Magnetic). The correspondence between (76) and (73) is given by

$$\int_{t_i}^{t_f} dt V_{LS}^{MAG} = -\sum_{j=1}^{2} \frac{g}{m_j} \int_{\Gamma_j} dz^\mu S_j^l \,\langle\!\langle \hat{F}_{l\mu}(z)\rangle\!\rangle\,, \tag{77}$$

$$\int_{t_i}^{t_f} dt V_{\rm Thomas} = \sum_{j=1}^{2} \frac{g}{2m_j^2} \int_{\Gamma_j} dz^\mu S_j^l \varepsilon^{lkr} p_j^k \,\langle\!\langle F_{\mu r}(z)\rangle\!\rangle\,, \tag{78}$$

$$\int_{t_i}^{t_f} dt V_{\rm Darwin} = \sum_{j=1}^{2} \frac{g}{8m_j^2} \int_{\Gamma_j} dz^\mu \langle\!\langle D^\nu F_{\nu\mu}(z)\rangle\!\rangle\,, \tag{79}$$

$$\int_{t_i}^{t_f} dt V_{SS} = -\frac{1}{2}\sum_{j,j'} \frac{ig^2}{m_j m_{j'}} \mathrm{T_s} \int_{\Gamma_j} dz^\mu \int_{\Gamma_{j'}} dz'^\sigma\, S_j^l\, S_{j'}^k \Big(\langle\!\langle \hat{F}_{l\mu}(z)\hat{F}_{k\sigma}(z')\rangle\!\rangle - \langle\!\langle \hat{F}_{l\mu}(z)\rangle\!\rangle\,\langle\!\langle \hat{F}_{k\sigma}(z')\rangle\!\rangle\Big)\,. \tag{80}$$

If we had worked in QED, we would have obtained the same formal result. The point is that in QED one can calculate perturbatively the field strength expectation values in the presence of the Wilson loop, here one can rely on a perturbative calculation only for very short interquark distances, shorter than the typical radius of the bound system. Therefore, we have to obtain a nonperturbative evaluation of $\langle\langle F\rangle\rangle$ and $\langle\langle FF\rangle\rangle$ that, together with the Wilson loop contain, all the relevant information on the heavy quark dynamics. Notice that the expression for the potential contains only manifestly gauge-invariant quantities.

We have at our disposal two ways of obtaining the nonperturbative quark interaction. First, we can exactly translate all our results in terms of field strength expectation values in presence of a *static* Wilson loop. These are plaquette insertions in the static Wilson loop and have been evaluated on the lattice (Bali et al. (1997)). The interesting fact is that we can also perform an analytic evaluation making only an assumption on the nonperturbative behaviour of the Wilson loop. This is due to the fact that all the expectation values of Eq. (77)-(80) can be obtained as functional derivatives of $\log\langle W(\Gamma)\rangle$ with respect to the path, i.e. with respect to the quark trajectories $\mathbf{z}_1(t)$ or $\mathbf{z}_2(t)$. In fact let us consider the change in $\langle W(\Gamma)\rangle$ induced by letting $z_j^\mu(t) \to z_j^\mu(t) + \delta z_j^\mu(t)$ where $\delta z_j^\mu(t_{\rm i}) = \delta z_j^\mu(t_{\rm f}) = 0$, then we have

$$g\langle\langle F_{\mu\nu}(z_j)\rangle\rangle = (-1)^{j+1}\frac{\delta i \log\langle W(\Gamma)\rangle}{\delta S^{\mu\nu}(z_j)}, \tag{81}$$

$$\delta S^{\mu\nu}(z_j) = dz_j^\mu \delta z_j^\nu - dz_j^\nu \delta z_j^\mu,$$

and varying again the path

$$g^2\left(\langle\langle F_{\mu\nu}(z_1)F_{\lambda\rho}(z_2)\rangle\rangle - \langle\langle F_{\mu\nu}(z_1)\rangle\rangle\langle\langle F_{\lambda\rho}(z_2)\rangle\rangle\right) = -ig\frac{\delta}{\delta S^{\lambda\rho}(z_2)}\langle\langle F_{\mu\nu}(z_1)\rangle\rangle. \tag{82}$$

Therefore, to obtain the whole quark-antiquark potential no other assumptions are needed than the behaviour of $\langle W(\Gamma)\rangle$. *In particular all contributions to the spin dependent part of the potential can be expressed as first and second variational derivatives of* $\log\langle W(\Gamma)\rangle$. *The obtained expressions are correct up to order* v^4. *Higher order corrections can in principle be included systematically in the same way.*

We can use this result as a laboratory to understand confinement. In fact any assumption on the QCD vacuum, i.e. on the nonperturbative behaviour of the Wilson loop, is put in direct connection, on one hand with the lattice evaluation and on the other hand with the phenomenological data. We address this issue in Sec. 6.

We conclude this section presenting one of the most common representations of the $1/m^2$ potentials (Eichten et al. (1981) and Barchielli et al. (1988))

which we will use in Sec. 6:

$$
\begin{aligned}
V_{\mathrm{SD}} = & \frac{1}{8}\left(\frac{1}{m_1^2}+\frac{1}{m_2^2}\right)\Delta\left[V_0(r)+V_{\mathrm{a}}(r)\right] \\
& + \left(\frac{1}{2m_1^2}\mathbf{L}_1\cdot\mathbf{S}_1 - \frac{1}{2m_2^2}\mathbf{L}_2\cdot\mathbf{S}_2\right)\frac{1}{r}\frac{d}{dr}\left[V_0(r)+2V_1(r)\right] \\
& + \frac{1}{m_1 m_2}\left(\mathbf{L}_1\cdot\mathbf{S}_2 - \mathbf{L}_2\cdot\mathbf{S}_1\right)\frac{1}{r}\frac{d}{dr}V_2(r) \\
& + \frac{1}{m_1 m_2}\left(\frac{\mathbf{S}_1\cdot\mathbf{r}\,\mathbf{S}_2\cdot\mathbf{r}}{r^2} - \frac{\mathbf{S}_1\cdot\mathbf{S}_2}{3}\right)V_3(r) + \frac{1}{3m_1 m_2}\mathbf{S}_1\cdot\mathbf{S}_2\,V_4(r) \qquad (83)
\end{aligned}
$$

with $\mathbf{L}_j = \mathbf{r}\times\mathbf{p}_j$ and

$$
\begin{aligned}
V_{\mathrm{VD}} = & \sum_{j=1}^{2}\frac{V_{\prime}(r)}{m_j} \\
& + \frac{1}{m_1 m_2}\left\{V_{\prime\prime}(r) + \mathbf{p}_1\cdot\mathbf{p}_2 V_{\mathrm{b}}(r) + \left(\frac{\mathbf{p}_1\cdot\mathbf{p}_2}{3} - \frac{\mathbf{p}_1\cdot\mathbf{r}\;\mathbf{p}_2\cdot\mathbf{r}}{r^2}\right)V_{\mathrm{c}}(r)\right\} \\
& + \sum_{j=1}^{2}\frac{1}{m_j^2}\left\{V_{\prime\prime\prime}(r) + p_j^2 V_{\mathrm{d}}(r) + \left(\frac{p_j^2}{3} - \frac{\mathbf{p}_j\cdot\mathbf{r}\;\mathbf{p}_j\cdot\mathbf{r}}{r^2}\right)V_{\mathrm{e}}(r)\right\}. \qquad (84)
\end{aligned}
$$

The brackets $\{\cdots\}$ mean here an ordering prescription between position and momentum operators. The functions $V_i(r)$ contain all the dynamics and are given by expectation values of electric and magnetic field insertion in the static Wilson loop. Explicit expressions can be found in Eichten et al. (1981), Gromes (1984), Barchielli et al. (1988), (1990), Brambilla et al. (1990), (1993), (1997a).

Exercises

5.2.1 Consider a free particle in one dimension. Using Eq. (68) calculate the free particle propagator. [Hint: use the discretized version of the path integral. See e.g. Sakurai (1985) for details].

5.2.2 Demonstrate Eqs. (81) and (82) first in QED and then in QCD.

6 Modelling the QCD vacuum

The limitations of a perturbative approach are appreciated when considering the vacuum. In perturbation theory the vacuum is approximated as an empty state with rare quark or gluon loop fluctuations. This in turn means that quarks and gluons are allowed to propagate freely. We have seen that experimentally this is not the case. The true, nonperturbative vacuum could be better imagined as a disordered medium with whirlpools of colour on different scales, thus densely populated by fluctuating fields whose amplitude is so large that they cannot be described by perturbation theory. Such a vacuum would be responsible for the fact that quark and gluons are confined. Such a vacuum would be responsible for the area law behaviour of the Wilson loop.

It is established that the QCD vacuum is (phenomenologically) characterized by various nonperturbative condensates, for a recent review see Shifman (1998). (An introduction to the topic has been given at this school by Anatoly Radyushkin.) Half-dozen of them are known: the gluon condensate $F_2 \equiv \langle \frac{\alpha_s}{\pi} F^a_{\mu\nu}(0) F^a_{\mu\nu}(0)\rangle$, the quark condensate $\langle \bar{q}q\rangle$, the mixed condensate $\langle \bar{q}\sigma_{\mu\nu}F_{\mu\nu}q\rangle$ and so on. Physically, the gluon condensate measures the vacuum energy density ε_{vac}. Indeed, due to the scale anomaly of QCD, the trace of the energy-momentum tensor is given by

$$\theta^\mu_\mu = \frac{\beta(\alpha_s)}{2\alpha_s} F^a_{\mu\nu}(0) F^a_{\mu\nu}(0). \tag{85}$$

Then, in the lowest order expansion of the beta function, we get

$$\varepsilon_{vac} \equiv \frac{1}{4}\langle \theta^\mu_\mu \rangle \simeq -\frac{\beta_0}{32}\langle \frac{\alpha_s}{\pi} F^a_{\mu\nu}(0) F^a_{\mu\nu}(0)\rangle. \tag{86}$$

Therefore, the nonperturbative gluon condensate shifts the vacuum energy downwards, making it advantageous. We remark that the negative sign in (86) is due to the asymptotic freedom! The vacuum fields fluctuate and these fluctuations contribute to the vacuum energy density. High energy modes of the fluctuating fields are in the weak coupling regime and can be dealt with perturbation theory as usual. The low frequency modes are responsible for the peculiar properties of the vacuum medium.

In the sum rule approach of Shifman, Vainshtein and Zakharov (1979) the effects caused by the vacuum fields are parameterized into few local vacuum condensates. This approach describes quite successfully the low-lying hadrons where the QCD string is not so relevant. The underlying assumption is that the characteristic frequencies of the valence quarks in the bound state are larger than the characteristic scale parameter of the vacuum medium. In other words, the valence quark pair injected in the vacuum is assumed to perturb it only slightly. However, sum rules cannot tell us anything about confinement. This is due to the fact that there is no local order parameter for confinement. But, as discussed in particular in Sec. 4.4, confinement manifests itself in the area law of the Wilson loop which is a nonlocal quantity[w]. In general ***supplying an analytic form for the nonperturbative Wilson loop average amounts to defining a model of the QCD vacuum.*** In the previous section we have seen that the static and the semirelativistic quark-antiquark interaction can be expressed in terms of the Wilson loop only. Therefore, it is tempting to explore the confinement mechanism using the simplest case of the heavy quark interaction and building analytic models of the QCD vacuum to be tested directly on the lattice and on the phenomenology.

In Sec. 6.1 we present a pedagogical model, the so-called Minimal Area Law model, which takes seriously the lattice results and assumes that the logarithm of the Wilson loop is simply given by a constant (the string tension) times the minimal area enclosed by the Wilson loop. This model will turn out to be very instructive for three reasons: 1) it gives the feeling of how a real calculation is done in practice, 2) it gives at the end a concrete and non-trivial form for the QCD potential, 3) it shows how the flux tube degrees of freedom emerge. In Sec. 6.2 we briefly summarize the main points underlying the issue of the dual Meissner effect as the mechanism of confinement. In order to give a concrete idea of how this mechanism could lead to quark confinement, we present the ordinary Abelian Higgs model of superconductivity. We also briefly explain how a mechanism similar to the Abelian Higgs model arises in QCD using the 't Hooft Abelian projection idea. We present some lattice results obtained in Abelian projection on the interquark flux tube distribution and on the static potential. In Sec. 6.3 we discuss the flux tube structure at the light of some general low energy theorems. Finally in Sec. 6.4 we only summarize the main ingredients of some other analytic models of the QCD vacuum, in particular Dual QCD and the Stochastic Vacuum Model.

6.1 Minimal Area Law Model (MAL)

In this model (Brambilla et al. (1993)) $\langle W(\Gamma)\rangle$ is approximated by the sum of a short range part given at the leading order by the perturbative gluon propagator $D_{\mu\nu}$ and a long-range part given by the value of the minimal area enclosed by the deformed Wilson loop of fixed contour Γ (see Fig. 8) plus a perimeter contribution $\mathcal{P}$:

$$\begin{aligned} i\log\langle W(\Gamma)\rangle &= i\log\langle W(\Gamma)\rangle^{\rm SR} + i\log\langle W(\Gamma)\rangle^{\rm LR} \\ &= -\frac{4}{3}g^2 \oint_\Gamma dx_1^\mu \oint_\Gamma dx_2^\nu \, iD_{\mu\nu}(x_1-x_2) + \sigma S_{\rm min} + \frac{C}{2}\mathcal{P}. \quad (87) \end{aligned}$$

[w] We remark that instead the local quark condensate is the order parameter of chiral symmetry breaking.

Denoting by $u^\mu = u^\mu(s,t)$ the equation of a typical surface of contour Γ ($s \in [0,1]$, $t \in [t_{\rm i}, t_{\rm f}]$, $u^0(s,t) = t$, $\mathbf{u}(1,t) = \mathbf{z}_1(t)$, $\mathbf{u}(0,t) = \mathbf{z}_2(t)$) and defining $\mathbf{u}_{\rm T} \equiv \mathbf{u} - (\mathbf{u}\cdot\mathbf{n})\,\mathbf{n}$ with $\mathbf{n} = (\partial\mathbf{u}/\partial s)|\partial\mathbf{u}/\partial s|^{-1}$, we can write:

$$S_{\rm min} = \min \int_{t_{\rm i}}^{t_{\rm f}} dt \int_0^1 ds \left[- \left(\frac{\partial u^\mu}{\partial t}\frac{\partial u_\mu}{\partial t}\right)\left(\frac{\partial u^\mu}{\partial s}\frac{\partial u_\mu}{\partial s}\right) + \left(\frac{\partial u^\mu}{\partial t}\frac{\partial u_\mu}{\partial s}\right)^2\right]^{\frac{1}{2}}$$

$$= \min \int_{t_{\rm i}}^{t_{\rm f}} dt \int_0^1 ds \left|\frac{\partial \mathbf{u}}{\partial s}\right| \left\{ 1 - \left[\left(\frac{\partial \mathbf{u}}{\partial t}\right)_{\rm T}\right]^2\right\}^{\frac{1}{2}}, \tag{88}$$

which coincides with the Nambu–Goto action. Up to the relative order $1/m^2$ (v^2 in the velocity) the minimal surface can be identified exactly with the surface spanned by the straight-line joining $(t, \mathbf{z}_1(t))$ to $(t, \mathbf{z}_2(t))$ with $t_{\rm i} \le t \le t_{\rm f}$. The generic point of this surface is

$$u^0_{\rm min} = t \qquad \mathbf{u}_{\rm min} = s\ \mathbf{z}_1(t) + (1-s)\ \mathbf{z}_2(t)\,, \tag{89}$$

with $0 \le s \le 1$ and $\mathbf{z}_1(t)$ and $\mathbf{z}_2(t)$ being the positions of the quark and the antiquark at the time t. Then, the expression for the minimal area at order $1/m^2$ in the MAL turns out to be

$$S_{\rm min} = \int_{t_{\rm i}}^{t_{\rm f}} dt\, r \int_0^1 ds\, [1 - (s\ \dot{\mathbf{z}}_{1\rm T} + (1-s)\ \dot{\mathbf{z}}_{2\rm T})^2]^{\frac{1}{2}}$$

$$= \int_{t_{\rm i}}^{t_{\rm f}} dt\, r \left[1 - \frac{1}{6}\left(\dot{\mathbf{z}}_{1\rm T}^2 + \dot{\mathbf{z}}_{2\rm T}^2 + \dot{\mathbf{z}}_{1\rm T}\cdot\dot{\mathbf{z}}_{2\rm T}\right) + \cdots \right], \tag{90}$$

where $r = |\mathbf{z}_1 - \mathbf{z}_2|$. The perimeter term is given by

$$\mathcal{P} = |\mathbf{x}_1 - \mathbf{x}_2| + |\mathbf{y}_1 - \mathbf{y}_2| + \sum_{j=1}^2 \int_{t_{\rm i}}^{t_{\rm f}} dt \sqrt{\dot{z}_j^\mu \dot{z}_{j\mu}}\,. \tag{91}$$

In the limit of large time interval $t_{\rm f} - t_{\rm i}$ the first two terms in the right-hand side of Eq. (91) can be neglected. By expanding also Eq. (91) up to $1/m^2$, we obtain

$$i \log\langle W(\Gamma)\rangle^{\rm LR} = \int_{t_{\rm i}}^{t_{\rm f}} dt\, \sigma r \left[1 - \frac{1}{6}\left(\dot{\mathbf{z}}_{1\rm T}^2 + \dot{\mathbf{z}}_{2\rm T}^2 + \dot{\mathbf{z}}_{1\rm T}\cdot\dot{\mathbf{z}}_{2\rm T}\right)\right]$$

$$+ \frac{C}{2}\sum_{j=1}^2 \int_{t_{\rm i}}^{t_{\rm f}} dt \left(1 - \frac{1}{2}\dot{\mathbf{z}}_j\cdot\dot{\mathbf{z}}_j\right). \tag{92}$$

For what concerns the perturbative part in the limit of large $t_f - t_i$ the only non-vanishing contribution to the Wilson loop is given by

$$i \log\langle W(\Gamma)\rangle^{\rm SR} = -\frac{4}{3} g^2 \int_{t_i}^{t_f} dt_1 \int_{t_i}^{t_f} dt_2 \; \dot{z}_1^{\mu}(t_1) \; \dot{z}_2^{\nu}(t_2) \; i D_{\mu\nu}(z_1 - z_2). \tag{93}$$

In the infinite time limit this expression is still gauge invariant. Expanding $z_2(t_2)$ around t_1 it is possible to evaluate explicitly from Eq. (93) the short-range potential up to a given order in the inverse of the mass (this was the task of Exercise 4.2.3). Self-energy terms are neglected.

Eventually, in the MAL model the following static and velocity dependent potentials are obtained:

$$V_0 = -\frac{4}{3}\frac{\alpha_s}{r} + \sigma r + C\,, \tag{94}$$

$$V_b(r) = \frac{8}{9}\frac{\alpha_s}{r} - \frac{1}{9}\sigma r\,, \qquad V_c(r) = -\frac{2}{3}\frac{\alpha_s}{r} - \frac{1}{6}\sigma r\,,$$
$$V_d(r) = -\frac{1}{9}\sigma r - \frac{1}{4}C\,, \qquad V_e(r) = -\frac{1}{6}\sigma r\,, \tag{95}$$

which fulfill the relations (Barchielli et al. (1990))

$$V_d(r) + \frac{1}{2}V_b(r) + \frac{1}{4}V_0(r) - \frac{r}{12}\frac{dV_0(r)}{dr} = 0, \tag{96}$$

$$V_e(r) + \frac{1}{2}V_c(r) + \frac{r}{4}\frac{dV_0(r)}{dr} = 0. \tag{97}$$

By evaluating the functional derivatives of the Wilson loop, as given by Eqs. (81)-(82), we obtain also the spin-dependent potentials

$$\Delta V_a(r) = 0, \quad \frac{d}{dr}V_1(r) = -\sigma, \quad \frac{d}{dr}V_2(r) = \frac{4}{3}\frac{\alpha_s}{r^2},$$
$$V_3(r) = 4\frac{\alpha_s}{r^3}, \quad V_4(r) = \frac{32}{3}\pi\alpha_s\delta^3(\mathbf{r}). \tag{98}$$

These potentials reproduce the Eichten–Feinberg–Gromes results (Eichten and Feinberg (1981), Gromes (1984)) and fulfill the Gromes relation

$$\frac{d}{dr}\left[V_0(r) + V_1(r) - V_2(r)\right] = 0. \tag{99}$$

Notice that, as a consequence of the vanishing of the long-range behaviour of the spin-spin potential V_{SS} and of the spin-orbit magnetic potential $V_{LS}^{\rm MAG}$ in this model, there is no long-range contribution to V_2, V_3 and V_4. Furthermore,

V_1 has only a nonperturbative long-range contribution, which comes from the Thomas precession potential (78).

The MAL model strictly corresponds to the Buchmüller picture (see Fig. 4) (Buchmüller (1982)) where the magnetic field in the comoving system is taken to be equal to zero. Let us first notice that the perimeter contributions at the $1/m^2$ order can be simply absorbed in a redefinition of the quark masses $m_j \to m_j + C/2$. Then, let us consider the moving quark and antiquark connected by a chromoelectric flux tube and let us describe the flux tube as a string with transverse velocity $\mathbf{v}_{\rm T}$. At the classical relativistic level the system is described by the flux tube Lagrangian (Olsson et al. (1993))

$$L = -\sum_{j=1}^{2} m_j \sqrt{1-\mathbf{v}_j^2} - \sigma \int_0^r dr' \sqrt{1-\mathbf{v}_{\rm T}'^2}\,, \tag{100}$$

with $\mathbf{v}'_{\rm T} = \mathbf{v}_{1\rm T}\, r'/r + \mathbf{v}_{2\rm T}(1 - r'/r), \quad 0 < r' < r$. The semirelativistic limit of this Lagrangian gives back the nonperturbative part of the V_0 and $V_{\rm VD}$ potentials in the MAL model (94) (95) (notice that the minimal area law in the straight-line approximation is the configuration given by a straight flux tube). The remarkable characteristics of the obtained $V_{\rm VD}$ potential is the fact that it is proportional to the square of the angular momentum and so takes into account the energy and the angular momentum of the string

$$V_{\rm VD}^{\rm LR} = -\frac{1}{12 m_1 m_2}\frac{\sigma}{r}(\mathbf{L}_1\cdot\mathbf{L}_2 + \mathbf{L}_2\cdot\mathbf{L}_1) - \sum_{j=1}^{2}\frac{1}{6m_j^2}\frac{\sigma}{r}\mathbf{L}_j^2\,. \tag{101}$$

Finally, the nonperturbative spin-dependent part of the potential in this intuitive flux tube picture simply comes from the Buchmüller ansatz that the chromomagnetic field is zero in the comoving framework of the flux tube.

We notice that even if V_1 seems to arise from an effective Bethe–Salpeter kernel which is a scalar and depends only on the momentum transfer, a simple convolution kernel (see Eq. (11)) cannot reproduce the correct velocity dependent potential (101) or equivalently (95). Nevertheless the behaviour (101) is important to reproduce the spectrum[x].

The static, spin-dependent and velocity-dependent potentials have been recently evaluated on the lattice (Bali et al. (1997)) and, at the present level of accuracy, confirm the prediction of this simple model.

---

[x]The extension of the Wilson loop approach to the relativistic treatment of heavy-light bound states supports the fact that the nonperturbative kernel *is not* a scalar convolution kernel, cf. Brambilla (1998) and Brambilla et al. (1997b).

6.2 Dual Meissner effect: a simple example

In the previous sections we have seen that the main characteristic of the quark nonperturbative interaction is the chromoelectric flux tube formation (see Fig. 11). The formation of a flux tube is reminiscent of the Meissner effect in (ordinary) superconductivity (for a review see Weinberg (1986)). As it is well-known, the superconducting media do not tolerate the magnetic field. If one imposes a certain flux of magnetic field through such a medium, the magnetic field will be squeezed into a thin tube carrying all the magnetic flux. The superconducting phase is destroyed inside the tube. A well-known example of the string-like solution of the classical equations of motion is given by the Abrikosov string (Abrikosov (1957)). Superconductivity is caused by condensation of the Cooper pairs (pairs of electric charges). If there were monopoles and antimonopoles in the superconducting medium, then a string would be formed and confine them. Therefore, to explain the confinement of electric charges (the quarks), we need a condensate of (chromo)magnetic monopoles (see Fig. 12). This is the simple qualitative idea suggested by Nambu (1974) and 't Hooft (1976) and Mandelstam (1976). Then, the vacuum of QCD behaves like a dual superconductor, where the word "dual" here means that the role of electric and magnetic quantities is interchanged with respect to an ordinary superconductor (see Fig. 12). In the next section we present a concrete model for a superconductor.

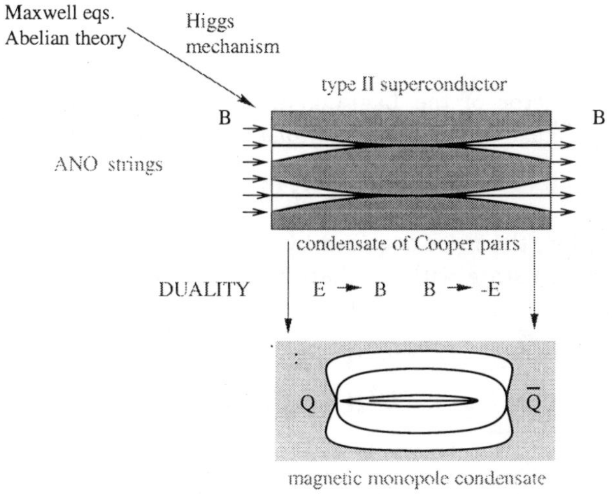

Figure 12: *The QCD vacuum as a "dual" superconductor .*

Abelian Higgs Model

A relativistic version of the Ginzburg–Landau model for superconductivity is the Abelian Higgs model:

$$L = -\frac{1}{4}F^{\mu\nu}F_{\mu\nu} + \frac{1}{2}(D^{\mu}\Phi)^{\dagger}(D_{\mu}\Phi) - U(\Phi) \tag{102}$$

where Φ is the Higgs field, in this case a complex (charged) scalar field describing Cooper pairs (i.e. condensate of electrons pairs in a lattice of positive ions, a superconducting solid) of charge $q = 2e$, e is the electron charge. $D_{\mu}\Phi = (\partial_{\mu} - iqA_{\mu})\Phi$ is the covariant derivative, and $U(\Phi) = \frac{\lambda'}{4}(\Phi^{\dagger}\Phi - \mu^2)^2$ is the Higgs potential. In Nielsen and Olesen (1973) it was shown that the Lagrangian (102) allows for vortex-line solutions. These vortex-lines solutions were approximately identified with the Nambu string.

The Lagrangian L is invariant under local $U(1)$ transformations. However, if $\mu^2 > 0$, there is a condensation of Cooper pairs signalled by the vacuum expectation value $\langle\Phi\rangle \neq 0$. This corresponds to the spontaneous breaking of the U(1) electric symmetry[y]. As a consequence the photon becomes massive inside a superconducting body and an external magnetic field can penetrate the body only to a finite depth λ equal to the inverse of the photon mass (Meissner effect). In fact putting $\Phi = \rho e^{i\theta}$, $\rho > 0$, and $\tilde{A}_{\mu} = (A_{\mu} - \partial_{\mu}\theta)$, L can be rewritten in the gauge-invariant form

$$L = -\frac{1}{4}F^{\mu\nu}F_{\mu\nu} + \frac{M^2}{2}\tilde{A}^{\mu}\tilde{A}_{\mu} + \tilde{L}[\rho] \tag{103}$$

where $\tilde{L}$ is the sector of the Lagrangian describing the propagation of the massive field ρ (mass M_{ϕ}). The equations of motion for the electromagnetic field read

$$\partial^{\mu}F_{\mu\nu} + M^2\tilde{A}_{\nu} = 0 \tag{104}$$

with $M^2 = q^2\langle\Phi\rangle^2$, the mass acquired by the photon.

In a stationary state with no charges $A_0 = 0$, $\partial_0\mathbf{A} = 0$, Eq. (104) gives $(\mathbf{H} = \nabla \wedge \mathbf{A})$

$$\nabla \wedge \mathbf{H} + M^2\tilde{\mathbf{A}} = 0 \tag{105}$$

$$\nabla^2\mathbf{H} + M^2\mathbf{H} = 0 \tag{106}$$

[y]This is often referred as spontaneous breaking of the local gauge symmetry. However, this statement is somewhat inaccurate. In this kind of theories the vacuum *does not* break local gauge invariance. Any state in the Hilbert space that fails to be invariant under local gauge transformation is an unphysical state. The vacuum is entirely gauge invariant at variance with what happens in theories with a global symmetry. See 't Hooft (1994).

Eq. (105) means that a permanent current (London current) $\mathbf{j} = M^2\tilde{\mathbf{A}}$ exists, and since $\mathbf{E} = 0$ and $\mathbf{E} = \sigma_c \mathbf{j}$ (σ_c is the conductivity), $\sigma_c = 0$.

The key parameter is the order parameter $\langle\Phi\rangle$ which signals the Higgs phenomenon. There are two characteristic lengths in the system related to $\langle\Phi\rangle$: the correlation length of the Φ field (or the inverse Higgs mass) $\Lambda = 1/M_\phi$, $M_\phi = \lambda'\langle\Phi\rangle$, and the penetration depth of the photon, $\lambda = 1/M$. If $\lambda > \Lambda$ the superconductor is called of type II, and the formation of Abrikosov flux tubes is favored in the process of penetrating the material with a magnetic field. If the opposite inequality holds, $\lambda < \Lambda$, it happens that, when the magnetic field is increased, there is an abrupt penetration of it at some value and superconductivity is destroyed. The superconductor is of type I.

Summarizing, in this model the condensation of electric charges leads to the formation of a quantized flux tubes (see Exercise 6.2.1.1) whose radius and shape are controlled by Λ and λ. If a magnetic monopole and antimonopole were introduced into such a superconducting medium they would be connected by a flux tube of finite energy per unit length (finite string tension). Thus magnetic monopoles would be confined due to a linear potential.

Exercises

6.2.1.1 Show that a side consequence of the Meissner effect is the flux quantization. [Hint: Calculate the integral $\oint \tilde{\mathbf{A}} \cdot d\mathbf{x}$ around a large circle centered on the flux tube.]

6.2.1.2 Consider the Maxwell equations in a relativistic medium without sources. Instead of expressing as usual the $\mathbf{E}$ and $\mathbf{B}$ fields in terms of the magnetic potential A_μ, express the $\mathbf{D} = \varepsilon\mathbf{E}$ and $\mathbf{H} = \mathbf{B}/\mu$ fields in terms of a (dual) electric potential C_μ. Write down the Maxwell equations in a covariant form in terms of the field strength tensor of C_μ. How do you have to modify the definition of this field strength tensor in the presence of sources?

The QCD Vacuum

We have seen in the previous section how in ordinary superconductivity the condensation of electric charges gives origin to monopole confinement. On the other hand, in all the theories allowing for an analytic proof of electric charge confinement, like compact electrodynamics, the Georgi–Glashow model

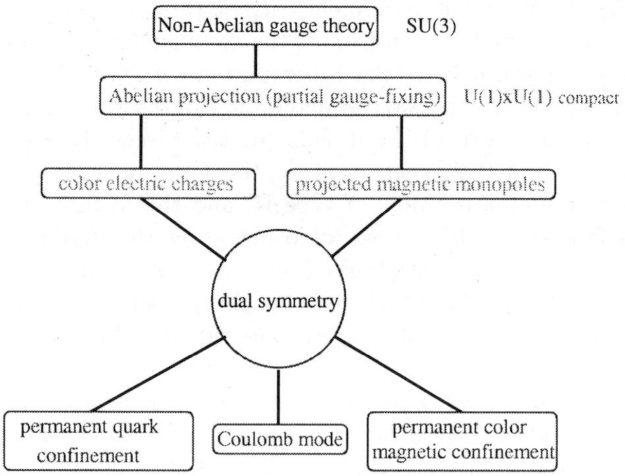

Figure 13: *Schematic diagram illustrating how a Non-Abelian gauge theory reduces to an Abelian theory with electric charges and monopoles and the pattern of vacuum phases (see Sec. 4.4).*

or some supersymmetric Yang–Mills theories (see Alvarez-Gaumé (1997)), this is due to the condensation of monopoles. However, if we want to apply this idea to QCD we have first to understand in which way we can obtain Abelian degrees of freedom and monopoles in QCD, since *in the QCD Lagrangian there are no Higgs fields!*. Subsequently we have to prove that the monopoles actually do condense. The application of this idea to non-Abelian gauge theories is based on the so-called Abelian projection, see 't Hooft (1981). The Abelian projection is a partial gauge fixing (of the off-diagonal components of the gauge field) under which the Abelian degrees of freedom remain unfixed. In QCD the Abelian projection reduces the $SU(3)$ gauge symmetry to a $U(1)^2$ gauge symmetry. The Abelian projection monopoles appear as topological quantities in the residual Abelian channel. QCD is then reduced to an Abelian theory with electric charges and monopoles, see Fig. 13. Precisely, it can be regarded as an Abelian gauge theory with magnetic monopoles and charged matter fields (quarks and off-diagonal gluons). Then, the dual superconductor picture is realized if these Abelian monopoles condense: this causes confinement of the particles that are electrically charged with respect to the above "photons". In this scenario large distance (low momentum) properties of QCD are carried by the Abelian degrees of freedom (Abelian dominance) and specifically by the monopole configurations (monopole dominance), see Suzuki (1993).

We do not have enough space to give further details. The reader is referred to Ex. 6.2.2.1 and to the reviews e.g. of Bali (1998), Chernodub et al. (1997), Di Giacomo (1998), Haymaker (1998). In the last years intensive work has been done to collect information on the quark confinement mechanism via lattice measurements. Depending on the picture, the excitations giving rise to confinement are thought to be magnetic monopoles, instantons, dyons, centre vortices (see Faber (1998)), etc. The above ideas are not completely disjoint and do not necessarily exclude each other. The above mentioned topological excitations indeed, are found to be correlated with each other in the lattice studies (e.g. correlations between instantons and monopoles). Many questions are still not completely settled. Here, we would like to summarize briefly the present understanding and to show some of the lattice measurements made in Abelian projection.

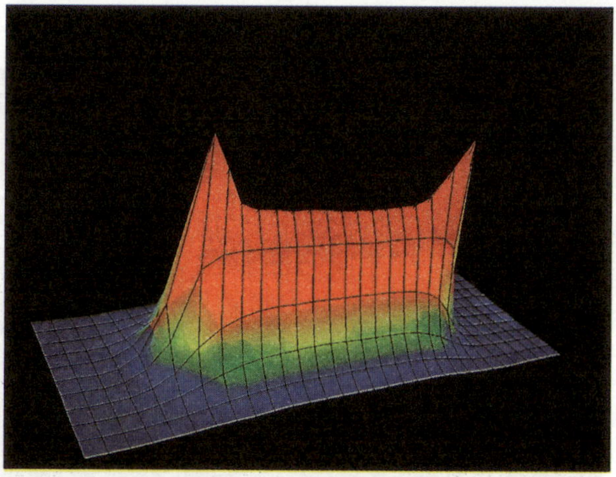

Figure 14: *Action density distribution in Maximal Abelian Projection of* $SU(2)$ *measured on a lattice* $V = 32^4$ *at* $\beta = 2.5115$. *The physical quark-antiquark distance is* 1.2 *fm. Figure provided by G. Bali.*

First, we mention some general problems one encounters in this approach. 1) The identification of "photon fields" and monopoles is a gauge invariant process. However, the choice of the operator that defines the Abelian projection is somehow ambiguous. There are many different ways of making the Abelian projection. The physics, e.g. the monopole condensation, should be independent of the gauge fixing. 2) The actual composition of the condensate should be known. 3) The origin of the monopole potential that yields to a non-vanishing vacuum expectation value of the magnetic condensate should be understood.

On the other hand it is clear that: 1) monopoles are condensed in the confinement phase and this independently of the choice of the Abelian projection. 2) Monopoles are responsible for the main part of the nonperturbative dynamics. In particular Abelian and monopoles dominance seem to hold (in the Maximal Abelian Projection (MAP)): expectation values of physical quantities in the non-Abelian theory coincide with (or are very close to) the expectation values of the corresponding Abelian operators in the Abelian theory obtained via Abelian projection. In other words, disregarding the off-diagonal gluons, the long range features of a $SU(N)$ gauge theory are reproduced while short range features may be altered. Monopole dominance means that the same result can be approximately calculated in terms only of the monopole currents extracted from the Abelian fields.

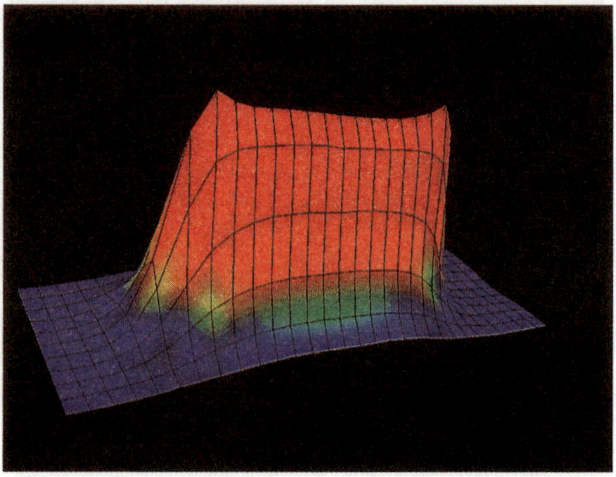

Figure 15: *Monopole contribution to the action density distribution in Maximal Abelian Projection of* $SU(2)$*. Parameters as in Fig. 14.*

As a general conclusion one can say that *the lattice data in MAP show that the lattice gluodynamics is in a sense equivalent at large distances to the Dual Abelian Higgs model at the border between superconductor of type I and superconductor of type II.* More precisely the Abelian monopole action has been extracted from $SU(2)$ lattice data in MAP and mapped into an Abelian Higgs model. From this the effective string theory has even been reconstructed and it occurs that the classical string tension of the string model is close to the quantum string tension of $SU(2)$ lattice gluodynamics, cf. Chernodub et al. (1999). The field distributions in MAP are found to satisfy the dual Ginzburg Landau equations and the coherence and the penetration lengths have been measured, cf. Bali (1998).

In Fig. 14 we present the action density distribution between two static quarks in $SU(2)$ in MAP to be compared on one hand with the action density in Fig. 11 and on the other hand with the monopole contribution in Fig. 15. We see that still the "photon" (neutral gluon) contributes to the self-energy of charges: the monopole part is free of self-energy while the "photon" part is free of string tension. Electric flux tubes (cf. Fig. 16) are found to be significantly thinner after Abelian projection.

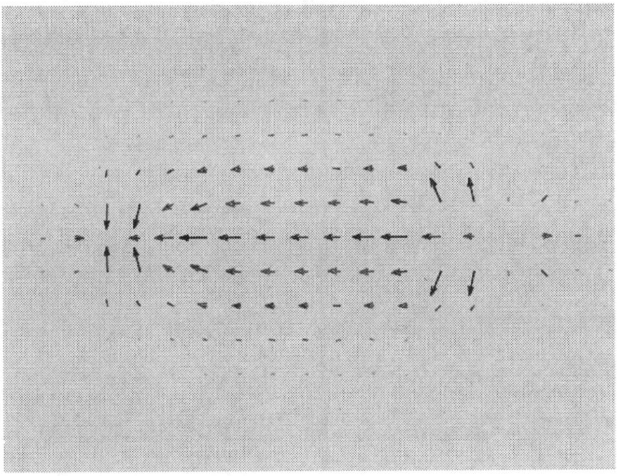

Figure 16: *Electric field at 12 lattice units between two static quarks in* $SU(2)$ *measured on a lattice* $V = 32^4$ *at* $\beta = 2.5115$. *The lattice spacing is fixed on* $\sigma = (440\ MeV)^2$. *The physical quark-antiquark distance is 1 fm. Figure provided by G. Bali.*

At the end of this section we present two plots of the static potential. Fig. 17 is a plot of the static potential calculated by means of either non-Abelian or Abelian projected Wilson loops. If one takes into account that in the Abelian projected case all the "charged components" of A_μ are neglected, one finds an impressive agreement between the two curves. Fig. 18 shows the static potential calculated in the Abelian projection. The photon contribution and the monopole contribution have been calculated separately. This plot confirms the action density result discussed above. Moreover, it is possible to quantify $\sigma_{U(1)} \simeq 92\% \, \sigma_{SU(2)}$ and $\sigma_{mon} \simeq 95\% \, \sigma_{U(1)}$. Notice that in the Abelian projection the coefficient of the Coulombic part comes out to be smaller by more than a factor of two.

The original phenomenological Cornell potential that was confirmed by the lattice simulations on the static Wilson loop in Sec. 4.5, is now understood in terms of monopole contributions while the monopole currents are the origin of the chromoelectric flux tube.

It is also possible to construct infrared effective "analytic" models based upon the assumptions and on the results presented above and to use them to explain the QCD low energy physics. As already discussed, in the Wilson loop formalism only an assumption on the Wilson loop behaviour is necessary. This is outlined in the next sections.

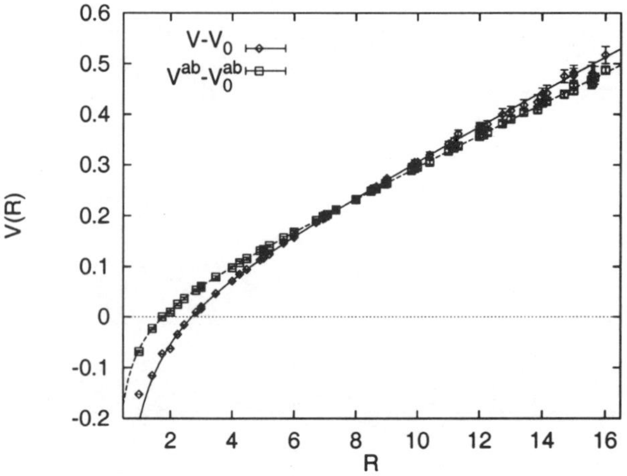

Figure 17: *Static potential in SU(2) (diamonds) and in the Abelian projection of SU(2) (squares). (in lattice units, a $\simeq$ 0.081 fm). Bali et al. (1996).*

Exercises

6.2.2.1 This is a simple example of Abelian projection for SU(2). Consider the following condition $F_{12}(x)$ = diagonal matrix. Show that F_{12} is still

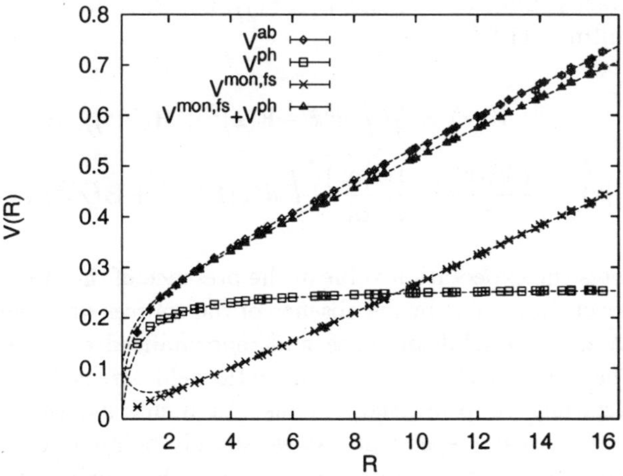

Figure 18: *The Abelian projected $SU(2)$ potential (diamonds) in comparison with the "photon" contribution (squares), the monopole contribution (crosses) and the sum of the two parts (triangles) (in lattice units, $a = 0.081$ fm). No self-energy constant have been subtracted. Bali et al. (1996).*

invariant under the U(1) gauge transformation $\Omega(x)_{ij}$ defined as

$$\Omega_{12} = \Omega_{21} = 0 \qquad \Omega_{11} = \Omega_{22}^* = e^{i\alpha(x)} \qquad \alpha \in [0, 2\pi).$$

Define $A_\mu^\pm \equiv A_\mu^1 \pm A_\mu^2$, $1, 2, 3 =$ colour indices. Show that A_μ^3 transforms as a photon under Ω while $A_\mu^\pm$ transform as charged fields. Define an Abelian field strength tensor in terms of A_μ^3 and show that, if the matrix of the gauge transformation Ω contains singularities, also the Abelian field strength tensor may contain singularities (monopoles).

6.3 Low energy theorems and flux tube

We have seen that one of the relevant features of confinement is the interquark flux tube formation, see Figs. 11, 16 in $SU(2)$ and Figs. 14, 15 in the MAP of $SU(2)$. Typical quantities of interests are the transverse extent of the flux tube and the nature of the colour fields (i.e. electric or magnetic), see Bali et al. (1995) and Green et al. (1997). There are two low energy theorems relating the potential of a static quark-antiquark pair with the total energy and the action stored in the flux tube between the sources (in lattice QCD these are know as Michael's sum rules (Michael (1987)), Dosch et al. (1995), Novikov et al. (1981), Shifman (1998),

$$V_0(r) = \frac{1}{2}\langle \int d^3x(-\mathbf{E}(x)^2 + \mathbf{B}(x)^2)\rangle_{r\times T}$$
$$V_0(r) + r\frac{\partial V_0(r)}{\partial r} = \frac{1}{2}\frac{\beta(\alpha_s)}{\alpha_s}\langle \int d^3x(\mathbf{E}(x)^2 + \mathbf{B}(x)^2)\rangle_{r\times T}. \tag{107}$$

$\langle \cdots \rangle_{r\times T}$ denotes the expectation value in the presence of the static Wilson loop where the expectation value in the absence of the sources has been subtracted. The formulas are in Euclidean space and renormalized composite operators are used. The squared electric and magnetic field strengths are separately not renormalization group invariant. Thus statements like $\langle g^2\mathbf{B}^2\rangle_{r\times T} \simeq 0$ or $\langle g^2\mathbf{E}_\perp^2\rangle_{r\times T} \simeq 0$ are scale dependent. As we already pointed out, on the lattice the averages of the squared components of the colour fields are found to be roughly equal (i.e. $\langle E_i^2\rangle_{r\times T} \simeq \langle B_j^2\rangle_{r\times T}$, $i,j = 1,2,3$), while in the greater part of the models the flux tube is mainly made of the longitudinal electric field. Eqs. (107) fixes the scale at which each of these situations can be fulfilled. In particular for models, typically describing the long range behaviour with $V_0 \simeq \sigma r$, one has $\beta/\alpha_s \simeq -2$. Taking the three-loop beta function, this gives $\alpha_s \simeq 0.6$.

6.4 Models of the QCD Vacuum

We briefly mention some analytic models of the QCD vacuum. We refer the reader to the papers quoted below for further details.

Dual QCD (DQCD)

Impressive progresses have been done recently towards an understanding of confinement via duality and monopoles condensation in supersymmetric theories, see e.g. Alvarez-Gaumé et al. (1997). However, it is also possible to construct a dual effective theory of long distance Yang–Mills theory, Baker et al. (1986-1996), see also Maedan et al. (1989). We call this theory Dual QCD (DQCD).

DQCD is a concrete realization of the Mandelstam and 't Hooft dual superconductor mechanism of confinement. It describes the QCD vacuum as a dual superconductor on the border between type I and II. The Wilson loop approach supplies a simple method to connect averaged local quantities in QCD

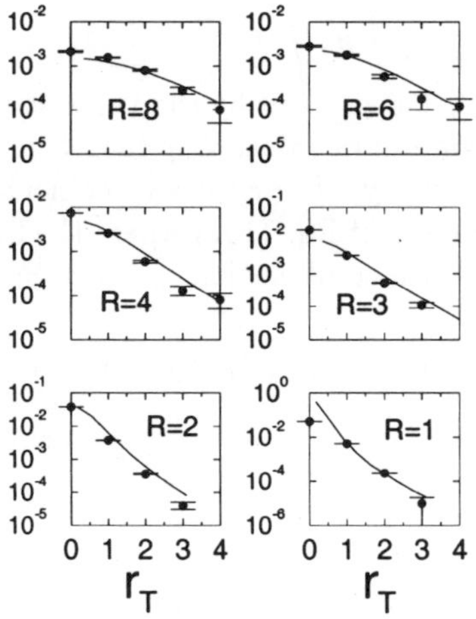

Figure 19: *On a semi-log scale, a comparison of the total energy profile for dual QCD (solid line) and in the lattice predictions (solid circles) for R (=$q\bar{q}$ distance) ranging from 8 down to 1 lattice units as a function of the transverse coordinate r_T (with respect to the interquark string). All profiles are in lattice units with a $\simeq 0.6$ Gev^{-1}. From Green et al. (1997).*

and in dual QCD. For large loops we assume

$$\langle W(\Gamma)\rangle \simeq \frac{\int \mathcal{D}C\,\mathcal{D}B\,\exp\left[i\int dx L(G^{\rm S}_{\mu\nu})\right]}{\int \mathcal{D}C\,\mathcal{D}B\,\exp\left[i\int dx L(0)\right]} \tag{108}$$

where L is the effective dual Lagrangian. The fundamental variables are an octet of dual potential C_μ coupled minimally to three octets of scalar Higgs field B_i carrying magnetic colour charge, the dual coupling constant being $e = 2\pi/g$. Notice that the dual transformation exchange the strong coupling limit of QCD with the weak coupling limit of DQCD. The monopole fields B_i develop nonvanishing expectation values B_{0i} giving rise to massive C_μ and to a dual Meissner effect. Dual potentials couple to electric colour charges like ordinary potentials couple to monopoles, i.e. C_μ couple to the quark-antiquark pair via a Dirac string connecting the pair. It turns out that, for the description of a quark-antiquark state, the relevant subset of the theory is a (dual) Abelian Higgs model. In this case indeed, the effective Lagrangian is explicitly given by $L(G^{\rm S}_{\mu\nu}) = 2\ {\rm Tr}\{-\frac{1}{4}G^{\mu\nu}G_{\mu\nu} + \frac{1}{2}(D_\mu B_i)^2\} - U(B_i)$, where $G_{\mu\nu} = (\partial_\mu C_\nu - \partial_\nu C_\mu + G^S_{\mu\nu})$. The dual field is directed along the hypercharge matrix Y in the colour space, $G^{\rm S}_{\mu\nu}(x) = g\,\epsilon_{\mu\nu\alpha\beta}\int ds \int d\tau\, \frac{\partial y^\alpha}{\partial s}\frac{\partial y^\beta}{\partial \tau}\delta(x - y(s,\tau))\,Y$, ($y(s,\tau)$ is a world sheet with boundary Γ swept out by the Dirac string) and $U(B_i)$ is the Higgs potential with the minimum values B_{0i} chosen in order to completely break the dual $SU(3)$ symmetry. It essentially coincides with the Abelian Higgs Lagrangian (plus sources) of Eq. (102) where the fields A_μ play the role of the dual potentials C_μ and a combination of fields B_i plays the role of the Higgs field Φ.

From (108) it follows that (see Baker et al. (1996))

$$\langle\langle F_{\mu\nu}(z_j)\rangle\rangle = \frac{2}{3}\,\varepsilon_{\mu\nu\rho\sigma}\langle\langle G^{\rho\sigma}(z_j)\rangle\rangle_{Dual}\,. \tag{109}$$

Therefore, it is possible to relate averaged values of local quantities in QCD and in the dual theory. The nonperturbative parameters are the v.e.v. of the Higgs field and the coupling constant of the Higgs potential (from these the penetration length and the correlation length can be constructed). Flux tube configurations with finite $r_{MS} \sim 1/M$, M = mass of the dual gluon, arise (Baker et al. (1996)) from the numerical solution of the classical dual Ginzburg–Landau type of equations obtained from L. The flux tube profile agrees well with lattice data (Green et al. (1996)), see Fig. 19. It is the presence of the Higgs field which confines the transverse energy distribution in a flux tube. The string tension σ comes from the integral of the exponentially decreasing energy distribution.

Stochastic Vacuum Model

The Stochastic Vacuum Model (Dosch (1987), Dosch et al. (1988)) is based on the idea that low frequency contributions in the functional integral can be taken into account by a simple stochastic process with a converging cluster expansion. Under this assumption the Wilson loop manifests an area law behaviour and therefore linear confinement. The model does not give rise to confinement for an Abelian gauge theory. In order to make quantitative predictions it is convenient to make a more radical assumption, namely that all higher cumulants can be neglected as compared to the two point function. This assumption appears to be in agreement with a recent lattice analysis (Bali et al. (1998)). Then, the stochastic process is Gaussian and all fields correlators are reduced to products of two point functions by factorization. This means

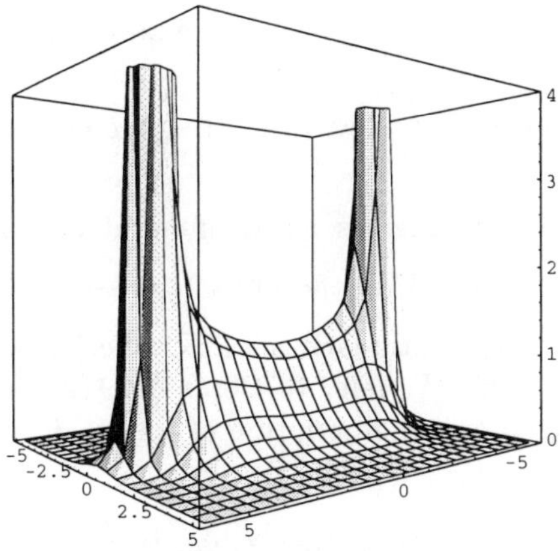

Figure 20: *The energy density distribution in* GeV/fm^3 *caused by the nonperturbative correlator* D *and the colour Coulomb contribution* D_1 *with* $\alpha_s = 0.57$. *In the figure* $a = T_g = 0.35 fm$, *the* $q\bar{q}$ *distance is 9* T_g. *From Rüter et al. (1995).*

that the Wilson loop can be approximated simply by (in Euclidean space)

$$\langle W(\Gamma)\rangle \simeq \exp\left[-\frac{g^2}{2}\int_{S(\Gamma)} dS_{\mu\nu}(x)\int_{S(\Gamma)} dS_{\lambda\rho}(y)\langle F_{\mu\nu}(x)U(x,y)F_{\lambda\rho}(y)U(y,x)\rangle\right] \tag{110}$$

where $S(\Gamma)$ is a surface with contour Γ, typically the minimal area surface. The nonperturbative dynamics is given in terms of one unknown function only: the non-local gluon condensate

$$g^2\langle U(0,x)F_{\mu\nu}(x)U(x,0)F_{\lambda\rho}(0)\rangle = \Big\{(\delta_{\mu\lambda}\delta_{\nu\rho} - \delta_{\mu\rho}\delta_{\nu\lambda})(D(x^2) + D_1(x^2))$$

$$+(x_\mu x_\lambda \delta_{\nu\rho} - x_\mu x_\rho \delta_{\nu\lambda} + x_\nu x_\rho \delta_{\mu\lambda} - x_\nu x_\lambda \delta_{\mu\rho})\frac{d}{dx^2}D_1(x^2)\Big\}. \tag{111}$$

Eq. (111) is the most general Lorentz decomposition of the two-point correlator. The dynamics is contained in the form factors D and D_1. The function D is responsible for the area law and confinement.

Lattice simulations (for a sum-rule analysis see Dosch, Eidemüller and Jamin (1998)) show that the D and D_1 functions exhibit a long-range exponential fall off $\simeq F_2 \exp\{-|x|/T_g\}$ where the correlation length T_g is about 0.2 $\div$ 0.3 fm (D'Elia et al. (1997)). The model has thus two nonperturbative parameters F_2 and T_g. The static potential is given by

$$V_0(r) \simeq F_2 \int_0^\infty d\tau \int_0^r d\lambda\,(r-\lambda)D(\tau^2+\lambda^2) \tag{112}$$

and the string tension σ emerges as an integral on the function D, $\sigma \simeq F_2 \int_0^\infty d\tau \int_0^\infty d\lambda\ D(\tau^2+\lambda^2)$, in the limit $T_g/r \to 0$. The field distribution between the quark and the antiquark is a flux tube (Rüter et al. (1995)) with $r_{MS} \simeq 1.8\,T_g$, see Fig. 20. We refer to Dosch (1994), Nachtmann (1996) and Simonov (1996) for a complete description of the model and its applications.

Finally, we mention in this neither exhaustive nor complete list, the flux tube model of Isgur et al. (1983). This model is extracted from the strong coupling limit of the QCD lattice Hamiltonian. A N-body discrete string-like Hamiltonian describes the gluonic degrees of freedom. The limit $N \to \infty$ corresponds to a localized string with an infinite number of degrees of freedom.

All the mentioned models allow to obtain via Eqs. (73), (81), (82), the complete semirelativistic quark interaction. For a discussion and a comparison of the result see Baker et al. (1997), Brambilla et al. (1997a) and Brambilla (1998). A relation between DQCD and SVM has been established in Baker et al. (1998).

Here, we only stress that we need two parameters like T_g and F_2 to control the structure of the flux tube. Had we only one parameter, like the string tension σ, we could only encode the information of a constant energy density in the flux tube. However, the whole structure is important and also the information about the width of the flux tube (typically proportional to T_g) has to be considered.

Exercises

6.4.1 Show that in the limit $T_g \to 0$ the Wilson loop in Eq. (110) is given by $\exp\{-\sigma S\}$ with σ as defined above and S the surface enclosed by the loop Γ.

6.4.2 Show that using the lattice parameterization of D given above, with $T_g \simeq 0.3$ fm and $F_2 \simeq 0.048\ \text{GeV}^4$, the Stochastic Vacuum Model gives a string tension σ compatible with the phenomenological value of Eq. (5).

6.5 *String description of QCD*

An effective string theory of strong interaction has been proposed first by Nambu and Goto in 1960 to explain the fact that the hadrons lie on Regge trajectories (see Sec. 2). Nambu was the first who explicitly constructed an effective theory of the Abelian Higgs model working in the limit of infinite Higgs mass which turned out to be coincident with the Nambu–Goto string Lagrangian. Recent works have shown that properly summing over all the string positions to account for the fluctuation of the string generates an effective string in four dimension free of the conformal anomaly. Results exist both in the limit of infinite Higgs mass (Akhmedov et al. (1996)) and without taking that singular limit (Baker et al. (1999)). A phenomenological description of hybrids has been obtained using a Nambu–Goto action in Allen et al. (1998).

7 Summary

These lectures have been devoted to a summary of our understanding of the heavy hadron spectrum in QCD. We started from some phenomenological guess of the interquark static potential inspired from a naive flux tube picture. Then, we managed to give a well founded definition of such a potential in field theory and to "derive" the quark confinement property in strong coupling expansion. To obtain the form of the potential we resorted to the lattice formulation and to numerical techniques that eventually confirmed the linear rising of the phenomenological static potential. The relativistic corrections were added in the framework of effective field theories establishing a systematic way of estimating the neglected terms. This allowed us to obtain a gauge-invariant field theory based definition of the semirelativistic quark interaction devised for lattice simulations as well as for analytic calculations. Eventually this procedure enabled us to connect directly the QCD vacuum with the spectrum. At the end we examined the Wilson loop in the Abelian projection suggested by 't Hooft, i.e. under the assumption that the QCD vacuum is a dual superconductor, and we found that our initial Cornell potential seems to be the output of dominant monopoles configurations.

The aim of these lectures has been to show how heavy quark bound states offer an ideal situation where exact results from QCD (effective field theories like NRQCD and pNRQCD) can be used to explore the QCD vacuum either with lattice numerical tools or by means of analytic models. Therefore many theoretical ideas and techniques can be and have been used in order to explain that kind of systems. All of them cooperate to enlarge our predictability and to open new perspectives. It is not surprising that, in spite of several still open problems, some remarkable progress have been achieved.

Acknowledgments

We gratefully acknowledge interesting discussions with Marshall Baker, Manfried Faber, Dieter Gromes, Khin Maung, Martin Zach and the participants at the HUGS'98 School. We thank Gunnar Bali for making available to us his lattice data, for many enlightening discussions and for making many useful comments and suggestions especially on the lattice part of these lectures. We acknowledge M. Baker and D. Gromes for reading the manuscript and making useful comments. It is a pleasure for N. B. to thank José Goity for the invitation to give these lectures and for the perfect organization of the school. N. B. acknowledges the support of the European Community, Marie Curie fellowship, TMR contract No. ERBFMBICT961714; A. V. acknowledges the FWF contract No. 9013.

References

1. A. Abrikosov, *JETP* **32**, 1442 (1957).
2. T. Allen, M. G. Olsson and S. Veseli, Phys. Lett. B **434**, 100 (1998); `hep-ph/9810363`.
3. L. Alvarez-Gaumé, F. Zamora, *(Lectures)* `hep-th/9709180`.
4. E. T. Akhmedov, M. N. Chernodub, M. I. Polikarpov and M. A. Zubkov, Phys. Rev. D **53**, 2087 (1996).
5. G. S. Bali, C. Schlichter and K. Schilling, *Phys. Rev.* D **51**, 5165 (1995).
6. G. S. Bali, V. G. Bornyakov, M. Müller–Preussker and K. Schilling, *Phys. Rev.* D **54**, 2863 (1996).
7. G. S. Bali, A. Wachter and K. Schilling, *Phys. Rev.* D **56**, 2566 (1997); G. Bali and Boyle, `hep-lat/9809180`.
8. G. S. Bali, N. Brambilla and A. Vairo, *Phys. Lett.* B **421**, 265 (1998).
9. G. S. Bali, `hep-ph/9809351`.
10. M. Baker, J. S. Ball and F. Zachariasen, *Phys. Lett.* B **152**, 351 (1985).
11. M. Baker, J. Ball and F. Zachariasen, *Phys. Rev.* D **51**, 1968 (1995); *Phys. Rev.* D **44**, 3328 (1991).
12. M. Baker, J. S. Ball and F. Zachariasen, *Int. Jour. Mod. Phys.* **A11**, 343 (1996a).
13. M. Baker, J. S. Ball, N. Brambilla, G.M. Prosperi and F. Zachariasen, *Phys. Rev.* D **54**, 2829 (1996b); M. Baker, J. S. Ball, N. Brambilla, A. Vairo, *Phys. Lett.* B **389**, 577 (1996b).
14. M. Baker, J. S. Ball and F. Zachariasen, *Phys. Rev.* D **56**, 4400 (1997).
15. M. Baker, N. Brambilla, H. G. Dosch and A. Vairo, *Phys. Rev.* D **58**, 034010 (1998).
16. M. Baker and R. Steinke, in preparation (1999).

17. A. Barchielli, E. Montaldi and G. M. Prosperi, *Nucl. Phys.* B **296**, 625 (1988); Erratum ibid. B **303**, 752 (1988).
18. A. Barchielli, N. Brambilla and G. M. Prosperi, *Nuovo Cimento* **103** A, 59 (1990).
19. I. Bigi, Y. Dokshitzer, V. Khoze, J. Kuhn and P. Zerwas, *Phys. Lett.* B **181**, 157 (1986).
20. N. Brambilla and G. M. Prosperi, *Phys. Lett.* B **236**, 69 (1990).
21. N. Brambilla and G. M. Prosperi, *Phys. Rev.* D **48**, 2360 (1993); *Phys. Rev.* D **46**, 1096 (1992).
22. N. Brambilla, P. Consoli, G. M. Prosperi, *Phys. Rev.* D **50**, 5878 (1994).
23. N. Brambilla and A. Vairo, *Phys. Rev.* D **55**, 3974 (1997a).
24. N. Brambilla and A. Vairo, *Phys. Lett.* B **407**, 167 (1997b).
25. N. Brambilla and A. Vairo, `hep-ph/9809230`; A. Vairo, `hep-ph/9809229`.
26. N. Brambilla, `hep-ph/9809263`.
27. N. Brambilla, A. Pineda, J. Soto, A. Vairo, `hep-ph/9903355` and in preparation.
28. L. S. Brown and Weisberger, *Phys. Rev.* D **20**, 3239 (1979).
29. W. Buchmüller and S. H. Tye, *Phys. Rev.* D **24**, 132 (1981).
30. W. Buchmüller, *Phys. Lett.* B **112**, 479 (1982).
31. S. Capitani, M. Lüscher, R. Sommer and H. Wittig, `hep-lat/9810063` and refs. therein.
32. Yu-Qi Chen, Yu-Ping Kuang and R. J. Oakes, *Phys. Rev.* D **52**, 264 (1995) 264; Yu-Qi Chen and Yu-Ping Kuang, *Z. Phys.* C **67**, 627 (1995). Some errors related to the reparameterization invariance where corrected in M. Finkemeier, H. Georgi and M. McIrvin, *Phys. Rev.* D **55** 6933 (1997).
33. M. Chernodub and M. Polikarpov, *(Lectures)* `hep-th/9710205`.
34. M. Chernodub, S. Kato, N. Nakamura and M. Polikarpov, `hep-lat/9902013`.
35. M. Creutz and K. Moriarty, *Phys. Rev.* D **26**, 2166 (1982).
36. M. Creutz, *Quarks, gluons and lattices*, (Cambridge University Press, Cambridge, 1983).
37. C. Davies, *Lectures given at the 1997 Schladming School*, Schladming, Austria, (Springer, Berlin, 1997).
38. A. Di Giacomo, `hep-lat/9802008`.
39. H. G. Dosch, *Phys. Lett.* B **190**, 177 (1987).
40. H. G. Dosch and Yu. A. Simonov, *Phys. Lett.* B **205**, 339 (1988).
41. H. G. Dosch, *Prog. Part. Nucl. Phys.* **33**, 121 (1994); H. G. Dosch, E. Ferreira and A. Krämer, *Phys. Rev.* D **50**, 1992 (1994).
42. H. G. Dosch, O. Nachtmann and M. Rüter, `hep-ph/9503386`.

43. H. G. Dosch and S. Narison, *Phys. Lett.* B **417**, 173 (1998).
44. H. G. Dosch, M. Eidemüller and M. Jamin, `hep-ph/9812417`.
45. M. D'Elia, A. Di Giacomo and E. Meggiolaro, *Phys. Lett.* B **408**, 315 (1997) and refs. therein.
46. E. Eichten, K. Gottfried, T. Kinoshita, K. D. Lane and T. M. Yan, *Phys. Rev.* D **17**, 3090 (1978); D **21**, 203 (1980).
47. E. Eichten and F. Feinberg, *Phys. Rev.* D **23**, 2724 (1981).
48. M. Faber, J. Greensite and S. Olejnik, `hep-lat/9810008` and refs. therein.
49. S. Godfrey and N. Isgur, *Phys. Rev.* D **32** 189 (1985).
50. S. Godfrey and J. Napolitano, `hep-ph/9811410`.
51. A. M. Green, C. Michael and P. S. Spencer, *Phys. Rev.* D **55**, 1216 (1997).
52. D. Gromes, *Phys. Lett.* B **115**, 482 (1982).
53. D. Gromes, *Z. Phys.* C **26**, 401 (1984); **22** 265 (1984).
54. D. J. Gross and F. Wilczek, *Phys. Rev. Lett.* **30**, 1343 (1973).
55. R. W. Haymaker, to be published in *Phys. Rep.*, `hep-lat/9809094`.
56. N. Isgur, *Phys. Rev.* D **57**, 4041 (1998).
57. N. Isgur and J. Paton, *Phys. Lett.* B **124**, 247 (1983); N. Isgur and J. Paton, *Phys. Rev.* D **31**, 2910 (1985).
58. M. Jamin and A. Pich, `hep-ph/9810259`.
59. R. D. Kenway, `hep-lat/9810054`.
60. J. Kogut, R. Pearson and J. Shigemitsu, *Phys. Lett.* B **98**, 63 (1981).
61. G. P. Lepage and P. Mackenzie, *Phys. Rev.* D **48**, 2250 (1993).
62. G. P. Lepage, *Lectures given at the 1996 Schladming School*, Schladming, Austria, (Springer, Berlin, 1996).
63. H. Leutwyler, *Phys. Lett* B **98**, 447 (1981).
64. H. Leutwyler, *Phys. Lett.* B **378**, 313 (1996).
65. W. Lucha, F. F. Schöberl and D. Gromes, *Phys. Rep.* **200**, 127 (1991).
66. S. Maedan and T. Suzuki, *Prog. Theor. Phys.* **81**, 229 (1989).
67. S. Mandelstam, *Phys. Rep.* C **23**, 245 (1976).
68. S. Mandelstam, *Phys. Rev.* D **19**, 2391 (1979).
69. A. Manohar, *Phys. Rev.* D **56**, 230 (1997).
70. A. Martin, *Phys. Lett.* B **93**, 338 (1980).
71. R. McClary and N. Byers, *Phys. Rev.* D **28**, 1692 (1983).
72. C. Michael, *Nucl. Phys.* B **280**, 13 (1987).
73. C. Michael, `hep-ph/9710429`; `hep-ph/9809211`.
74. C. Morningstar, K. J. Juge and J. Kuti, `hep-ph/9809015`.
75. I. Montvay and G. Munster, *Quantum fields on a lattice*, (Cambridge Univ. Press, Cambridge, 1994).

76. O. Nachtmann, *Lectures given at the 1996 Schladming School*, Schladming, Austria, (Springer, Berlin, 1996).
77. Y. Nambu, *Phys. Rev.* D **10**, 4262 (1974).
78. H. B. Nielsen and P. Olesen, *Nucl. Phys.* B **61**, 45 (1973).
79. V. A. Novikov, M. A. Shifman, A. I. Vainshtein, V. I. Zakharov, *Nucl. Phys.* B **191**, 301 (1981).
80. M. G. Olsson and K. Williams, *Phys. Rev.* D **48**, 417 (1993); M. G. Olsson and S. Veseli, *Phys. Rev.* D **51**, 3578 (1995); M. G. Olsson, *Int. J. Mod. Phys.* A **12**, 4099 (1997).
81. PDG (1998), Particle data Group. See review of particle Physics C. Caso et al., *The European Physical Journal* C **3** 1 (1998) or http://pdg.lbl.gov/.
82. M. A. Peskin, in *Proc. 11th SLAC Summer Institute*, PUB-3273, (SLAC, 1983).
83. M. Peter, *Nucl. Phys.* B **501**, 471 (1997).
84. A. Pineda and F. J. Yndurain, `hep-ph/9812371` and refs. therein.
85. A. Pineda and J. Soto, *Nucl. Phys.* **B** (Proc. Suppl.) **64**, 428 (1998); *Phys. Rev.* D **59**, 016005 (1999).
86. H. D. Politzer, *Phys. Rev. Lett.* **30**, 1346 (1973).
87. C. Quigg, *Physics Today* **50**, 20 (1997).
88. C. Quigg and J. L. Rosner, *Phys. Lett.* B **71**, 153 (1977).
89. J. Richardson, *Phys. Lett.* B **82**, 272 (1979).
90. H. Rothe, *Lattice gauge theories. An Introduction*, (World Scientific, Singapore, 1992).
91. M. Rüter and H. G. Dosch, *Z. Phys.* C **66**, 245 (1995).
92. J. J. Sakurai, *Modern Quantum Mechanics*, (The Benjamin/Cummings Pub. Comp., 1985).
93. Y. Schröder, *Phys. Lett.* B **447**, 321 (1999) where an error in Peter (1997) is corrected.
94. H. J. Schitzner, *Phys. Lett.* B **76**, 461 (1978).
95. M. Shifman, A. Vainshtein and V. Zakharov, *Nucl. Phys.* B **147**, 385 (1979); ib. 448.
96. M. Shifman, *Prog. Theor. Phys. Suppl.* **131**, 1 (1998).
97. Yu. Simonov, *Phys. Usp.* **39**, 313 (1996).
98. T. Suzuki, *Nucl. Phys.* B (Proc. Suppl.) **30**, 176 (1993).
99. R. Tafelmayer, *Diploma Thesis*, Heidelberg (1986).
100. G. 't Hooft, in *High Energy Physics*, ed. Zichichi (Editrice Compositori, Bologna, 1976).
101. G. 't Hooft, *Nucl. Phys.* B **153**, 141 (1979).
102. G. 't Hooft, *Nucl. Phys.* B **190**[FS3], 455 (1981).

103. G. 't Hooft, *Physica Scripta* **25**, 133 (1982).
104. G. 't Hooft, *Under the spell of the gauge principle*, (World Scientific, Singapore, 1994).
105. B. Thacker and G. Lepage, *Phys. Rev.* D **43**, 196 (1991); G. P. Lepage, L. Magnea, C. Nakhleh, U. Magnea and K. Hornbostel, *Phys, Rev.* D **46**, 4052 (1992).
106. M. B. Voloshin, *Nucl. Phys.* B **154**, 365 (1979).
107. S. Weinberg, *Prog. of Theor. Phys. Suppl.* **86**, 43 (1986).
108. F. J. Wegner, *J. Math. Phys.* **12**, 2259 (1971).
109. K. Wilson, *Phys. Rev.* D **10**, 2445 (1974).

ELECTRO- AND PHOTOPRODUCTION OF MESON AND BARYON RESONANCES

JIM NAPOLITANO
Department of Physics, Applied Physics, and Astronomy
Rensselaer Polytechnic Institute
Troy, NY 12180-3590, USA
E-mail: napolj@rpi.edu

I discuss the physics of quantum mechanical resonances, and the phenomenology of their excitation and observation in the absorption of photons on nucleons. I begin with a discussion of resonance phenomena in classical physics, in quantum mechanics, and finally incorporating relativistic effects. As an example, I consider in detail $J^{PC} = 1^{--}$ states produced in the reaction $e^+e^- \to \gamma^\star \to \pi^+\pi^-$. Next, the phenomenology of meson and baryon resonances are discussed in the context of the quark model. Finally, I discuss baryon resonances and their partial wave analysis in $\pi^\pm p \to \pi N$ reactions, and then in the more complicated, but still straightforward, case of $\gamma p \to \pi N$. I conclude with a topical discussion of current research areas, including electroproduction of baryon resonances, and of quasi-twobody photoproduction processes, and the isobar model partial wave analysis, in the reaction $\gamma p \to p\pi^+\pi^-$.

1 Introduction

These lectures are *not* about current, state-of-the-art results in nuclear physics. Instead, I will try to provide the necessary groundwork so that you can read about these results in the literature, and have some familiarity with the phenomenological tools and language used by those authors.

José Goity asked me originally to talk about "Electroproduction of Baryon Resonances", but I lengthened the title in order to talk about less. That sounds contradictory, I realize, but the point is that before I could talk about "electroproduction", I felt I first needed to talk about "photoproduction". Also, the principles of overlapping, relativistic Breit-Wigner resonances are so easily discussed in terms of e^+e^- annihilation (in which only one partial wave, namely $J^{PC} = 1^{--}$, is present), it seemed clear to me that I should talk about mesons before discussing baryons.

The result was three lectures. The first, which actually used up the better part of two hours, introduces the concept of resonance from the ground up. Interference between overlapping resonances is described using the reaction $e^+e^- \to \pi^+\pi^-$, and an exercise is left to the student to carry out some of these fits. We were also able to touch on the concept of ρ/ω interference, and the idea of isospin violation in the strong interactions.

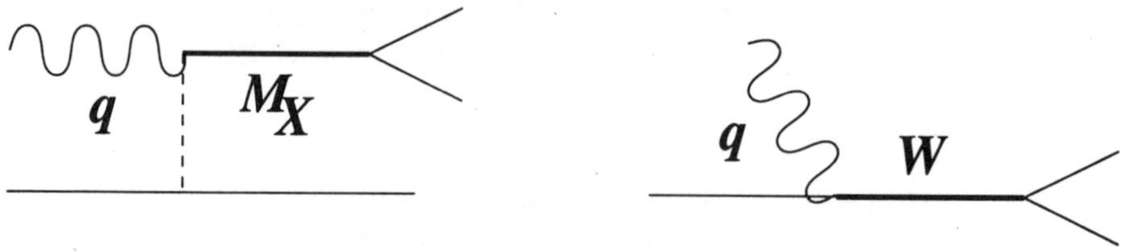

Figure 1: Production of hadronic resonances in γp reactions. On the left, a meson resonance with mass M_X is created, while on the right, a baryon resonance with mass W is formed. In both cases, the four-momentum of the photon is q, with $q^2 = 0$.

The second lecture, one hour, concentrated on the quark model and how we use it to describe the excited states of hadrons. This is, after all, the most common language used to discuss the resonance phenomena we're after in the first place.

Finally, the third lecture used a bit more than an hour to talk about the quantum mechanics of baryon resonances and introduce the idea of partial wave analysis. This begins with a discussion of baryon resonance production in πp scattering, but then moves on to discuss photoproduction reactions. I then finish up with a brief discussion about some of the most exciting[a] current pieces of research going on in the field.

2 Preliminaries

The first lecture actually started out with a brief survey of the field, but in this paper I will separate that part out into this section.

Hadronic resonance photoproduction is shown in Fig. 1. A meson resonance with mass M_X is produced when the incident photon exchanges energy and momentum with the target proton. On the other hand, if the photon is absorbed by the proton, an excited baryon resonance is formed. In either case, meson or baryon, the excited state represented by the resonance decays strongly to two or more particles. For real photons, as pictured here, $q^2=0$ and the photon only has two transverse degrees of freedom.

The electroproduction process, Fig. 2, is similar. In this case, there are two important differences compared to Fig. 1, both due to the fact that the photon is virtual. First, we now have[b] $q^2 < 0$, and its value can be adjusted experimentally independent of M_X or W. Second, there is an additional degree of freedom for the virtual photon, namely its longitudinal polarization. These

[a]Most exciting to me, of course!

[b]I use the metric in which the square of the four momentum of an on shell particle is positive.

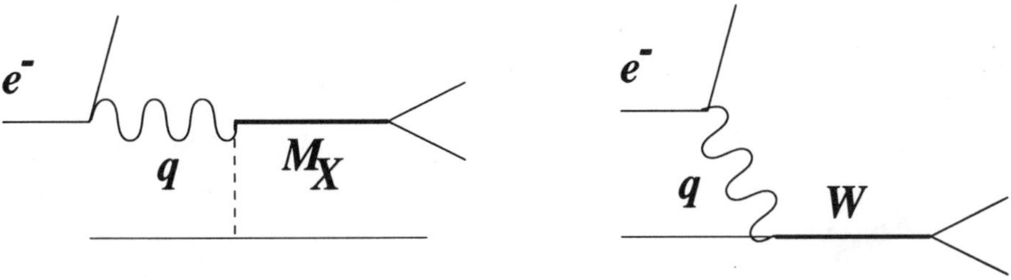

Figure 2: Production of hadronic resonances in $p(e,e')$ reactions. On the left, a meson resonance with mass M_X is created, while on the right, a baryon resonance with mass W is formed. In both cases, the four-momentum of the photon is q, with $q^2 < 0$.

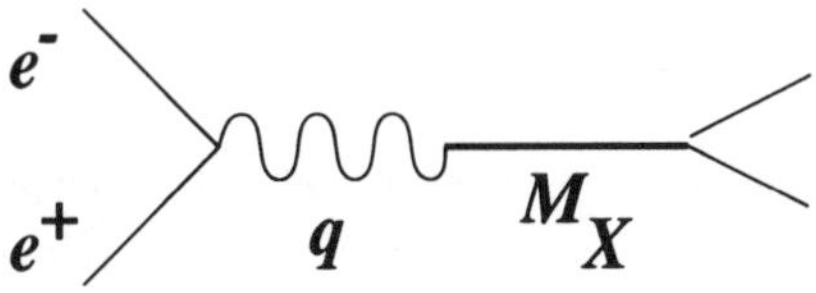

Figure 3: Production of a meson resonance in e^+e^- annihilation. In this case, $q^2 = M_X^2 > 0$.

two features extend the capability of electroproduction over photoproduction, but also lead to some complications in the analysis.

Finally, the production of meson resonances in e^+e^- annihilation is shown in Fig. 3. The kinematics here are unique, as shown in the caption. I'm including this only because we'll be using it as an example of overlapping resonances in the first lecture.

Of course, the literature is full of data on these sorts of reactions, although the more complex ones are pretty much starved for statistics. (That's a big reason we built CEBAF, after all.) Perhaps the first, most instructive thing to look at is just the total cross section for all γp reactions, pictured in Fig. 4. The bumps in the cross section at lower energy are generally referred to as the "first", "second", and "third" resonance regions. (These are baryon resonances; I won't be spending much time on meson resonances in photoproduction reactions on protons.) However, I emphasize that *bump hunting is not enough!* As I will discuss, one needs to carry out a partial wave analysis in exclusive reactions in order to see what quantum mechanical resonances are in fact under those bumps. It turns out that although the first resonance region is dominated by one state, the $\Delta(1232)$, the second is mainly two strongly overlapping states, while the third is quite complex.

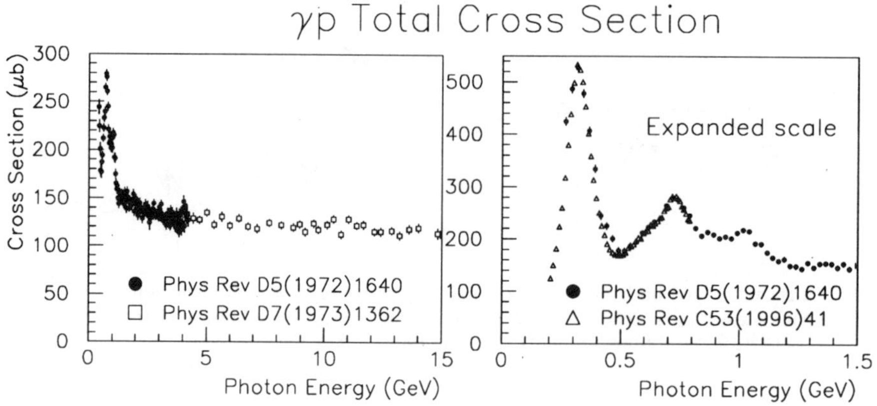

Figure 4: The total cross section for $\gamma p \to$ "anything". The "bumps" at lower photon energy are interpreted as baryon resonances with mass $W = \sqrt{s} = \left[2m_p E_\gamma + m_p^2\right]^{1/2}$. At higher photon energies, the cross section is dominated by meson resonance production, who's mass will depend on other kinematic variables besides the incident photon energy E_γ.

Here are some places to look for more representative data. The total cross sections for γp and γd reactions (where the latter essentially provides information on γn) can be found in the Particle Data Group compilation[1]. Detailed measurements at lower energy, including some complex nuclei, have been measured at Mainz[2]. The total *virtual* photoproduction cross section has been studied by Stoler[3]. The total cross section for the reaction $\gamma p \to n\pi^+$ was measured by the Cambridge Bubble Chamber group[4], who also measured the reaction[4,5,6] $\gamma p \to p\pi^+\pi^-$. Multipion photoproduction was also studied in the old days at DESY[7]. Finally, the reaction $e^+e^- \to \pi^+\pi^-$, in the region relevant to the overlapping resonances I'll talk about, was measured in Novosibirsk[8] and at Orsay[9].

3 Lecture One: Resonances

3.1 Review of mechanical resonance

Everyone thinks they know what a "resonance" is, but when you're dealing with many wide, overlapping and interfering states, it's important to be careful. Let's start with the basics and take a quick look at mechanical resonance. That will make most of the important points very clear.

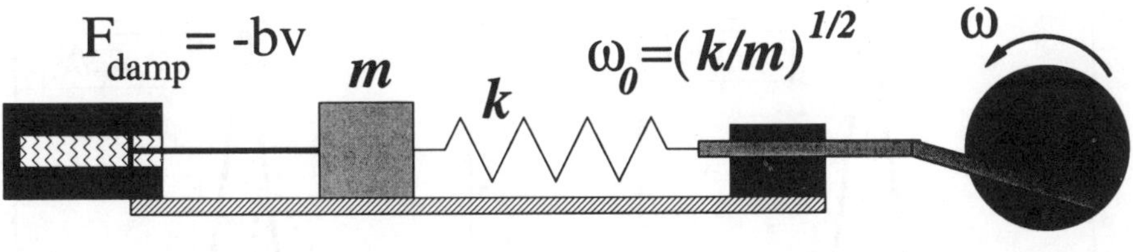

Figure 5: A classical mass and spring oscillator. The natural frequency is $\omega_0 = (k/m)^{1/2}$. The oscillator is driven by a cam attached to a wheel rotating at an adjustable angular velocity ω. The effect of the cam is to shift the equilibrium point of the oscillator. A dashpot imparts a viscous friction force $-bv = -b\dot{x}$.

Figure 5 shows a simple "forced" mass and spring system. The horizontal position of the mass m is specified by x. The forces on the mass include the damping force $-bv = -b\dot{x}$ and the spring force $-k(x - x_{eq})$. The spring equilibrium position x_{eq} moves as $x_{eq} = x_0 \cos(\omega t) = \mathrm{Re}\left(e^{-i\omega t}\right)$, where the frequency ω is adjustable. Newton's second law for the mass then gives

$$F_{spring} + F_{damp} = -k\left(x - x_0 e^{i\omega t}\right) - b\dot{x} = m\ddot{x}$$

This is rearranged to give

$$\ddot{x} + \gamma\dot{x} + \omega_0^2 x = \omega_0^2 x_0 e^{-i\omega t} \tag{1}$$

where $\omega_0^2 = k/m$ and $\gamma = b/m$, and we understand that only the real part of the forcing term is important.

This is most easily solved (ignoring transient solutions) using the ansatz $x = Ae^{-i\omega t}$ where the complex amplitude $A = |A|e^{i\phi}$ is a function of the driving frequency ω. Inserting this into Eq. 1 gives

$$A = \frac{\omega_0^2 x_0}{(\omega_0^2 - \omega^2) - i\gamma\omega} \tag{2}$$

The complex amplitude A is plotted in Fig. 6. In one case, we plot the magnitude of the amplitude and also the phase ϕ. We also plot the real and imaginary parts of A. Both forms are commonly seen in the literature.

There are three important points to glean from this plot:

1. The amplitude has a "peak" very near the natural frequency ω_0. That is, the magnitude of the amplitude, as well as its imaginary part, go through a maximum here, and the real part passes through zero.

2. The "width" of the peak is determined essentially by γ, which closely approximates the "full width at half maximum" of the amplitude.

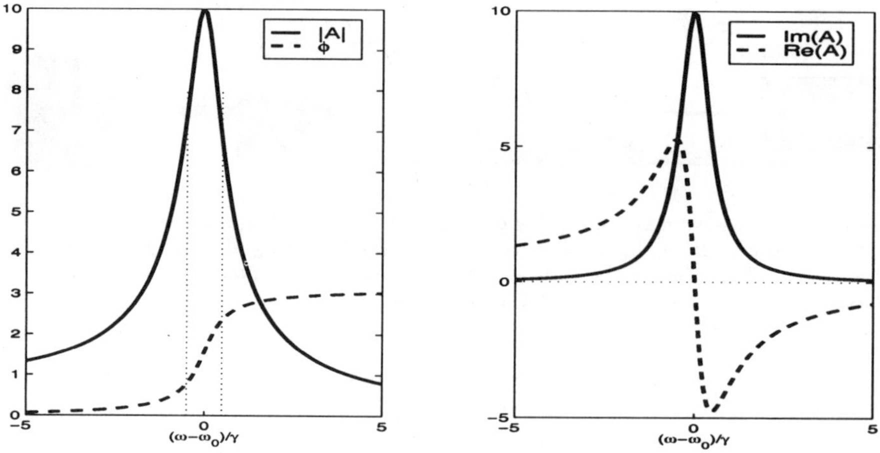

Figure 6: Response of a forced mechanical oscillator. The amplitude is plotted in units of the driving amplitude x_0. The driving frequency is plotted relative to the "natural" frequency $\omega_0 = \sqrt{k/m}$, and is scaled by the "width" $\gamma = b/m$. The complex phase ϕ is plotted in radians. The vertical dotted lines on the left are drawn at $\pm 1/2$, that is, $\omega = \omega_0 \pm \gamma/2$.

3. The phase increases through 180°, passing through 90° at the peak of the resonance.

These are three necessary characteristics of a resonance, mechanical or otherwise. Generalizing to hadron spectroscopy, it is not merely enough to identify a local maximum in the cross section (i.e. a peak in the intensity, or $|\text{amplitude}|^2$). The phase motion, consistent with the width of the resonance, is also a necessary condition.

3.2 Quantum Mechanical Resonance: The Nonrelativistic and Relativistic Breit-Wigner Formulae

In elementary quantum mechanics, we write Schrödinger's equation as

$$i\hbar \frac{d\Psi}{dt} = H\Psi = E\Psi$$

where H is the (Hermitian) Hamiltonian operator and the energy E is the eigenvalue of the Hamiltonian. This leads to so-called "stationary states" $\Psi(t)$ with the form

$$\Psi(t) = \Psi(0)e^{-i\omega t}$$

with $\omega \equiv E/\hbar$. For such states, the probability $|\Psi(t)|^2 = |\Psi(0)|^2$ does not change with time. Hence the name, "stationary" states.

Resonance phenomena appear naturally in quantum mechanics as soon as we allow for states which give probabilities that decay away in time, that is, states with finite lifetime. For a state with mean life τ, we expect

$$|\Psi(t)|^2 = |\Psi(0)|^2 e^{-t/\tau} = |\Psi(0)|^2 e^{-\Gamma t}$$

where $\Gamma \equiv 1/\tau$ and so the state with eigenvalue[c] $E_R = \omega_R$ is then

$$\Psi(t) = \Psi(0)e^{-i\omega_R t}e^{-(\Gamma/2)t} = \Psi(0)e^{-(iE_R+\Gamma/2)t}$$

In order to study this process in terms of energy $E = \omega$ instead of time t, we Fourier transform the wave function and have

$$\chi(E) = \int_0^\infty \Psi(t)e^{i\omega t}dt = \frac{\chi_0}{(E - E_R) + i(\Gamma/2)}$$

This function has qualitiatively the same behavior as we saw previously for mechanical resonance. In particular, the parameter Γ is called the "width" of the resonance and the functional form is called the "Breit-Wigner" resonance shape.

Soon we will be talking about the spectroscopy of very short lived particles which decay to other particles, with the energy release comparable to the masses of the particles involved. Clearly, simple nonrelativistic quantum mechanics will not do the job. There are many ways to approach the problem relativistically. (I rather like the treatment in the relatively recent quantum field theory textbook by Peskin and Schroeder[10].) Here I will just give a plausibility argument for the correct form for a Relativistic Breit-Wigner.

We want to interpret a resonance as an unstable particle of mass M and four-momentum $p = (p^0, \vec{p})$. Our goal is a relativistically covariant form for the resonance. Consider the term $p^2 - M^2 + iM\Gamma$ with

$$\begin{aligned}
p^2 - M^2 &= {p^0}^2 - \left(|\vec{p}|^2 + M^2\right) \\
&= {p^0}^2 - E_{\vec{p}}^2 \\
&= \left(p^0 + E_{\vec{p}}\right)\left(p^0 - E_{\vec{p}}\right) \\
&\approx (2E_0)(E - E_0)
\end{aligned}$$

[c] As is customary and prudent, we will henceforth take $\hbar = c = 1$. If that bothers you, I'm sorry, but you'll just have to get used to it in this business.

for a resonance in *energy* near $E_0 \equiv E_{\vec{p}}$, the energy of the decaying particle which we are studying. The form suggested above then becomes

$$p^2 - M^2 + iM\Gamma \approx 2E_0 \left[E - E_0 + i\frac{(M/E_0)\Gamma}{2}\right]$$

This is just the denominator of the nonrelativistic Breit-Wigner form, accounting for a broadening of the width Γ, based on the time dilation factor M/E_0. This gives us the form we will use.

To be explicit, the "Relativitic Breit-Wigner Amplitude Formula" is[10,11,12]

$$A = \frac{g_A e^{i\phi_A}}{s - M^2 + iM\Gamma(s)} \tag{3}$$

where $s = p^2$ is the square of the four-momentum of the resonance "particle". The particle "mass" M is interpreted as the position of the pole in the complex amplitude. The decay rate will depend of the invariant mass of the particle (i.e. s) and so we expect[12] to have some energy dependence to the width. We generally write

$$\Gamma(\omega) = \Gamma_0 \left(\frac{q}{q_0}\right)^{2\ell+1} \left[\frac{\rho(\omega)}{\rho(\omega_0)}\right]$$

where q is the center-of-mass three-momentum in the decay, ℓ is the orbital angular momentum in the decay, and $\omega = \sqrt{s}$. Also, $q = q_0$ and $\omega = \omega_0$ when $s = M^2$, and $\rho(\omega)$ is a slowly varying function that is determined either empirically, or phenomenologically for some specific resonance or decay. In many cases, however, we ignore this energy dependence altogether and simply use a single value for the width Γ. You can usually get away with this because, for relatively narrow resonances, there is not much variation over the active range.

3.3 Example and Exercise: The Timelike Pion Form Factor

We can now apply this formalism to some real data. The main point I want to make here is that the "interference" between complex resonant amplitudes can bring in some rather dramatic effects.

Soon we will discuss baryon resonances. An important point there, as well as in the case of meson resonance production in general, is that broad states with different quantum numbers will overlap in the reaction in question. Before the reaction can be analyzed in terms of resonances, the amplitudes corresponding to different quantum numbers have to be separated. This procedure is generally called "Partial Wave Analysis". In the example we discuss right now, however, we do not need to carry out such an analysis, so I think it is a good place to start working with quantum mechanical resonance with short lived particles.

The reaction we will study here is a specific case of Fig. 3, namely $e^+e^- \to \pi^+\pi^-$. The intermediate states X are simple, and have only one set of quantum numbers, for a few reasons. First, these states couple directly to a single (virtual) photon, and therefore have the same quantum numbers of the photon, namely $J^{PC} = 1^{--}$. By selecting mesons that decay to $\pi^+\pi^-$, we can only make states with relative $\pi\pi$ orbital angular momentum $L = 1$ and isospin $I = 1$. We call such states "ρ meson resonances". The best known such resonance is just called the ρ. It has a mass of $M = 770$ MeV and a width $\Gamma = 150$ MeV.

This data is generally analyzed in terms of the "pion form factor", where the hadronic part of the reaction in Fig 3 is separated away from the (calculable) electromagnetic part and some kinematic effects. That is, the complex amplitude we will study is $|F_\pi|$ where the reaction cross section is written

$$\begin{aligned}\sigma(e^+e^- \to \pi^+\pi^-) &= \frac{\pi}{3}\frac{\alpha^2}{s}\beta_\pi^3 \, |F_\pi|^2 \\ &= \sigma(e^+e^- \to \mu^+\mu^-)\frac{\beta_\pi^3}{4} \, |F_\pi|^2\end{aligned}$$

Seeing this written in terms of the $\mu^+\mu^-$ cross section makes it clear how the experimentalists can determine $|F_\pi(s)|^2$ without directly measuring the cross section. At one particular value of the center of mass energy $\sqrt{s}$, both $\pi^+\pi^-$ and $\mu^+\mu^-$ pairs are identified, and the ratio of the production rates, after some corrections for efficiency and kinematics, gives the the square of the form factor at that $\sqrt{s}$.

The available data on $|F_\pi|^2$ is shown in Fig. 7, including the fit (described below) to three distinct, interfering relativistic Breit-Wigner resonances. The points up to $M \approx 1500$ MeV are from two experiments at the VEPP-2 facility at Novosibirsk[8], and the higher energy data are from the DM2 experiment at Orsay[9]. These measurements get pretty hard when the cross sections are small, and not all the points at higher energies are in agreement with each other. We won't use all of them when we fit the data for resonances.

We will focus on two aspects of these data. One is the general body of data over all energies, and the number of ρ resonances needed to describe it. Secondly, we will examine the region close to the peak of the dominant $\rho(770)$ resonance, and the effects of isospin violation.

Both of these analyses are done in the same way, namely by writing the form factor in terms of a sum over n interfering complex amplitudes, where n

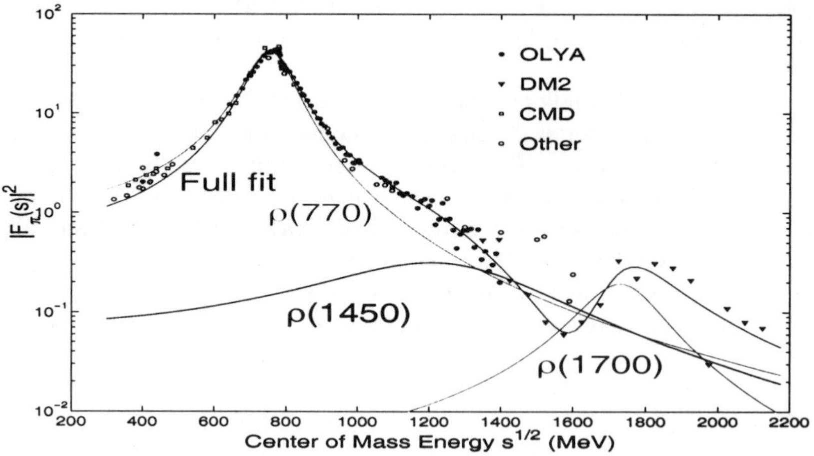

Figure 7: Data for the reaction $e^+e^- \to \pi^+\pi^-$.

is the number of resonances needed to describe the data. That is,

$$F_\pi(s) = \sum_{k=1}^{n} A_k$$

with the amplitude A_k given by Eqn. 3, in particular

$$A_k = \frac{g_k e^{i\phi_k}}{s - M_k^2 + iM_k\Gamma_k(s)}$$

For the mass dependent width, we use

$$\Gamma_k(s) \approx \Gamma_{0_k} \left(\frac{s - 4m_\pi^2}{M_k^2 - 4m_\pi^2}\right)^3 \left(\frac{M_k}{\sqrt{s}}\right)$$

where we take the function $\rho(\omega)$ as determined by Jackson[12] for a $J^P = 1^-$ resonance. This all leaves us with a total of $4n - 1$ free parameters, namely M_k, Γ_{0_k}, g_k, ϕ_k. Note that one of the phases is necessarily arbitrary so we simply set $\phi_1 = 0$.

Carrying out these fits were exercises we did at the end of lecture, and immediately following it. We used the program PAW for the fitting. (The source code is available from me if anyone would like it.) The first fit was just that which yielded the results in Fig. 7. That is, we used $n = 3$, and carried

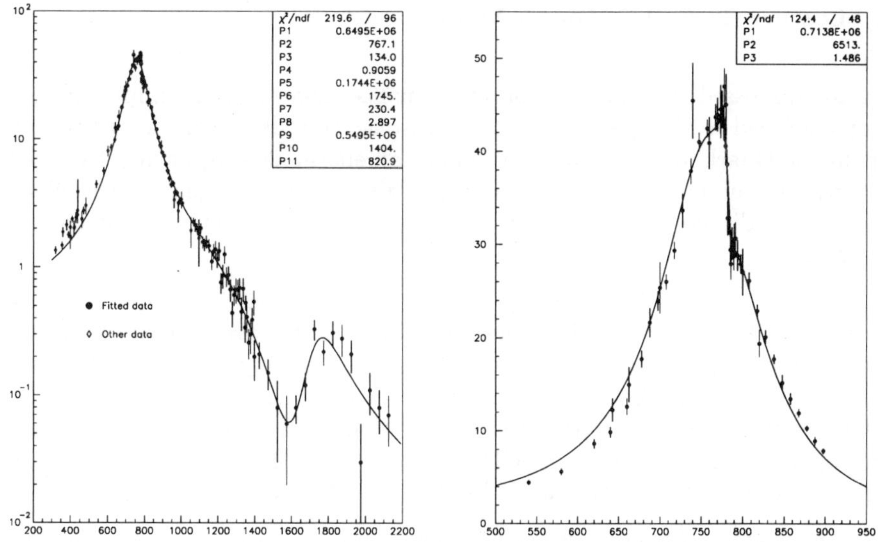

Figure 8: Fits to the pion form factor data. On the left is a free fit to three interfering resonances. The parameters $P1 \ldots P3$, $P5 \ldots P7$, and $P9 \ldots P11$, correspond to the normalization, mass, and width for each of the three resonances, while $P4$ and $P8$ are the relative phase. On the right, the fit demonstrates the phenomenon of ρ-ω interference, fitting two resonances with the mass and width of the ρ and ω, but with arbitrary normalizations ($P1$ and $P2$) and relative phase ($P3$).

out a fit with 11 free parameters, corresponding to the three ρ resonances. The direct output of PAW is shown on the left in Fig. 8.

It was left as an exercise to the student, to examine the region near the peak of the dominant $\rho(770)$. I was very pleased that just about everyone got very interested in this problem. The shape of the peak is clearly distorted, as see on the right side of Fig. 8. This is because the ω meson actually interferes with the ρ through isospin violation in the strong interactions. The effect is small (isospin is a very good, if not perfect, symmetry for the strong interactions) but the ω is very narrow, so the effect is rather striking. For more details of the behavior of the pion form factor in this region, I suggest looking at the analyses by Bernicha, *et al.*[13] and by Gardner and O'Connell[14].

4 Lecture Two: The Quark Model

Having discussed the phenomenon of resonance, and seeing an important example applied to the spectroscopy of ρ mesons, we took a little detour to talk about the Quark Model. This is, of course, an enormous subject, but I wanted to bring it out for two reasons. First, it gives us the language with which we tend to discuss the phenomenological implications of meson and baryon resonances.

Secondly, however, the Quark Model provides us with an important motivation for looking in detail at these resonances in the first place. As we will see, the Quark Model works very well, but that is largely a surprise. The *real* theory of the strong interactions is Quantum Chromodynamics (QCD), and QCD looks very different from the Quark Model, even though "quarks" are an important element of both QCD and the Quark Model. The point is, the "quarks" of the two theories are quite different objects.

The degrees of freedom we call "quarks" in the Quark Model are massive, weakly interacting things we call "constituent quarks". On the other hand, the "quark" degrees of freedom in QCD are light, strongly interacting fields, called "current quarks". (Some people use the notation Q to represent constituent quarks and q for current quarks.) They are similar, however, in that both are pointlike, spin-$\frac{1}{2}$ objects with an additional quantum number called "color". It is not understood in any detail, how QCD turns itself into the Quark Model, although some good plausibility arguments are beginning to emerge[15].

A key point is that QCD has degrees of freedom that are *not* present in the Quark Model. These are the "gluonic" degrees of freedom associated with the gluon fields in the theory. It should also be possible for multiple quark, or multiple quark and gluon, fields to combine and create "molecular" and "hybrid" states. The question is, therefore, why does the Quark Model work so well while it ignores some degrees of freedom that ought to be there? There are a number of possible answers to these questions, but most are rather model dependent.

Experimentally, we attack this problem by trying to find states that, in one way or another, do not correspond to the Quark Model. This always involves excitations, since it seems clear that the ground states of the theory only require consituent quark degrees of freedom. These excitations are broad, strongly decaying states, and therefore are studied as resonances. Some examples of specific areas of study include searches for "exotic mesons" and "glueballs"[15] as well as the "missing baryons"[16,17].

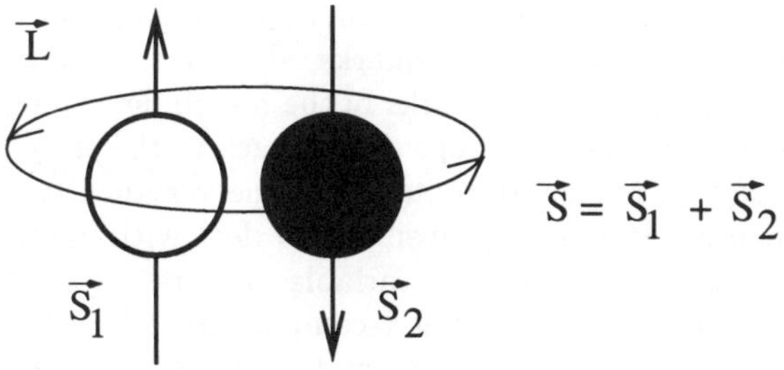

Figure 9: The angular momenta contributing to the quantum numbers of mesons, as $q\bar{q}$ pairs in the quark model.

4.1 Mesons in the Quark Model

Mesons are strongly interacting subatomic particles with integer spins. They are constructed in the Quark Model out of color singlet pairs of quarks and antiquarks. A full description of the phenomenology of meson spectroscopy in the quark model, including relativstic effects, was carried out Godfrey and Isgur[18]. (For a recent review of how well this phenomenology compares to experiment, see a paper[15] I recently wrote with Steve Godfrey.) We'll just cover the basics here.

The quantum numbers of mesons in the quark model are based on the simple diagram shown in Fig. 9. The total spin of the meson is $\vec{J} = \vec{L} + \vec{S}$ where the orbital angular momentum is $\vec{L}$ and the spin of total spin vector is the sum of the spins of the $q\bar{q}$ pair, that is $\vec{S} = \vec{S}_1 + \vec{S}_2$. The intrinsic parity of this fermion-antifermion pair is $P = (-1)^{L+1}$. Consequently, ground state $(L = 0)$ mesons have negative parity, regardless of their spin structure, and this is well borne out by experiment.

The two types of quarks making up the pair, say $q_1\bar{q}_2$, may or may not be the same. If they are the same, that is q_2 is just the antiparticle of q_1, then the meson is "self conjugate" because it is an eigenstate of the charge conjugation operator. (Such mesons are, by definition, neutral in all additive quantum numbers like charge and strangeness.) In this case, we can define a charge conjugation quantum number, namely the eigenvalue of the the operator, and we find $C = (-1)^{L+S}$. (For derivations of these quantum number formulae on the basis of first principles, see any one of a number of upper level textbooks. I would recommend the books by Wong[19] and Halzen and Martin[20].)

Examples include the π^+ (the $u\bar{d}$ combination with $L = S = 0$ so that $J^{PC} = 0^{-+}$), the ρ^- ($\bar{u}d$ with $L = 0$, $S = 1$, and $J^{PC} = 1^{--}$), the K_0 ($s\bar{d}$ with $L = S = 0$, and $J^{PC} = 0^{-+}$), and the $b_1^+(1235)$ ($u\bar{d}$ with $L = 1$,

$S = 0$, and $J^{PC} = 1^{+-}$). We generally talk about meson "nonets" for the nine possible combinations of u, d, and s quarks, all with the same L, S, and J.

The combination of u and d quarks in the meson determines the "isospin" of the state. Isospin is an old concept dating back to the early days of nuclear physics, but today we realize its roots are in the very nearly equal masses of the u and d quarks. It is still convenient to deal with isospin as if it were indeed a "spin" angular momentum variable in a special space, where the u and d quarks each have isospin $\frac{1}{2}$ but z-components $+\frac{1}{2}$ and $-\frac{1}{2}$ respectively. The $\bar{u}$ and $\bar{d}$ have the opposite values, and all the other quarks have isospin zero.

So, for example, the pion (π) and rho (ρ) are each considered a single particle with three different isospin states, all with isospin one, and with quark makeup $u\bar{d}$ (for π^+ and ρ^+), $\bar{u}d$ (π^- and ρ^-), and $(u\bar{d} - \bar{u}d)/\sqrt{2}$ (π^0 and ρ^0). The isospin zero combination would be[d] $(u\bar{d} + \bar{u}d)/\sqrt{2}$ and this does indeed seem to be the case for the ω meson, the isospin "partner" of the $S = 1$ ρ. However, it is also clear that for the $S = 0$ pions, there is a lot of mixing with the $s\bar{s}$ combination (which also, of course, has isospin zero) and the two physical states that result are the η and the η'. For $S = 1$, the $s\bar{s}$ state is the ϕ meson, which seems to have very little mixing with the ω.

A quantum number called G-parity was invented to describe the charge conjugation of particles which are indeed not self-conjugate. We define $G = (-1)^{L+S+I} = C(-1)^I$, where I is the isospin, for $I = 0$ or $I = 1$ mesons (i.e., no kaons).

This is a good time to point out that isospin is *not* and exact symmetry of the strong interactions, because the u quark is indeed slightly lighter than the d quark. (Isospin is also broken by the electromagnetic interaction because the u and d quarks have different charges, but this is a separate matter.) One very good example experimentally is the mixing between the ρ^0 and the ω which we saw in the first lecture.

Finally, we can make a connection to the pion form factor analysis we did in the first lecture. First, because the final state was produced through an intermediate photon, the mesons we create have to have the same quantum numbers as that photon, that is, $J^{PC} = 1^{--}$. Isospin, however, is irrelevant for photons, but by selecting $\pi^+\pi^-$ (which must have relative $\ell = 1$ to give the right J^{PC}) we force the intermediate meson to have isospin one. (Pions are bosons, so the wave function must be symmetric under their interchange. This means that if ℓ is odd, then so must be the isospin.) Therefore, the intermediate meson states we produce must have $I^G J^{PC} = 1^+1^{--}$. These quantum numbers specify ρ mesons.

---

[d]For a discussion of the relative phases, see Wong[19].

So, then, why might we expect *three* ρ states, as indicated by the data in Fig. 7, on the basis of the Quark Model? Certainly, there should be the ground state ρ, with $L = 0$ and $S = 1$, or, in spectroscopic notation, the 1^3S_1 state. We might also expect a radial excitation of this state, namely the 2^3S_1. Finally, it is also possible to get these quantum numbers with $L = 2$ and $S = 1$, i.e. 1^3D_1. (This state should also have $J^{PC} = 2^{--}$ and 3^{--} partners, according to the Quark Model.) Consequently, we have an explanation for the three states! We are presumably observing the 1^3S_1 ($\rho(770)$), 2^3S_1 ($\rho(1450)$), and 1^3D_1 ($\rho(1700)$), isospin one $q\bar{q}$ states.

Of course, there are important questions. How do we make these assignments for the observed states, and are they possibly mixtures? Why do we expect there to be no more (or no less) radial or orbital excitations at these energies? The answers lie in the dynamics of the Quark Model (and whether or not it is indeed a complete description of QCD!), which we will discuss after taking a quick look at baryons.

4.2 Baryons in the Quark Model

Baryons are strongly interacting subatomic particles with *half-integer* spins. They are more complicated objects than mesons. It will help to have the background in mesons, however. In fact, a relativized Quark Model for baryons was developed by Capstick and Isgur[21], based on the success of the model for mesons.

One basic tenet of the Quark Model is that physical states must be the result of quarks combined in "color singlets". That is, the color degree of freedom (the "charge" in QCD) must be hidden. This is easy for mesons, where the color of the q is cancelled out by the color of the $\bar{q}$. In order to have half-integer spins, though, we need to have an odd number of quarks (or quarks and antiquarks). Baryons in the Quark Model are made of three quarks, and the three colors of QCD make it possible to combine them in a color singlet fashion. That's all we'll say about that; for our purposes we will just take it that baryons are $q_1q_2q_3$ combinations in the Quark Model.

There are two regular patterns of quantum numbers that are easy to understand for baryons in the Quark Model. The first regards isospin. If all three quarks are u or d, then we can have either $I = \frac{1}{2}$ or $I = \frac{3}{2}$. For example, the uud combination (in the ground state) can either be the $J = \frac{1}{2}$ proton ($I = \frac{1}{2}$) or the $J = \frac{3}{2}$ Δ^+ ($I = \frac{3}{2}$). I like to refer to the excitations of these as N^* or Δ^* respectively, but a lot of people use N^* for any baryon excitation. Also note that for baryons with only two u or d quarks, such as uds, there can be $I = 0$ (ground state Λ) and $I = 1$ (Σ^0) combinations, but we won't be discussing strange baryons in these lectures.

The second pattern is not on as profound a footing, but does point towards the dynamics. As turned out to be so fruitful in nuclear physics, we classify the excitations of qqq baryons in terms of a "shell model"[22] based on the energy levels of the three dimensional harmonic oscillator. (For a nice discussion of the 3D harmonic oscillator in the context of nuclear and particle physics, see the textbook by Frauenfelder and Henley.[23]) The ground state of this system, with principle quantum number $N = 0$, has all three quarks in a relative $L = 0$ state and positive parity. These correspond to the nucleon (i.e. p or n) and $\Delta(1232)$ for the two isospin combinations. The first excitation level, $N = 1$, is reached by including one unit of orbital angular momentum between the quarks, and has negative parity and lies at an energy $\hbar\omega$ above the ground state. These correspond to a set of negative parity $N^\star$ states with masses between 1500 and 1700 MeV and spins $\frac{1}{2}$, $\frac{3}{2}$, and $\frac{5}{2}$, and two negative parity $\Delta^\star$ states with spins $\frac{1}{2}$ and $\frac{3}{2}$ at about the same mass. The $N = 2$, or $2\hbar\omega$, states are also seen, *but many of the predicted states have not yet been observed.* This is the so-called "missing baryons" problem, which we will discuss in a bit more detail in the third lecture.

Finally, a word about notation. So much of the work in baryon spectroscopy has been done with in the $\pi +$ nucleon (πN) system, that some people tend to refer $N^\star$ and $\Delta^\star$ states in terms of their πN quantum numbers. The notation is $(\ell_{\pi N})(2J)(2I)$, where ℓ is the relative orbital angular momentum between the pion and nucleon. So, for example, the proton is a "$P11$" state, while the $\Delta(1232)$ is "$P33$". On the other hand, one of the first $N^\star$ states is called the $S11(1535)$. One should be careful not to confuse the πN spectroscopic notation with that for the quark model states.

4.3 Formulations of the Quark Model

Of course, to really use the Quark Model for testable predictions, one needs to come up with a dynamical model. That is, some Hamiltonian that gives masses of the physical mesons and baryons, as well as some decay model that predicts their branching ratios to various final states. As mentioned above, this has been done in detail for mesons[18] and for baryons[21], but here I just want to give you the general idea. I will use mesons (i.e. $q\bar{q}$ systems) for most of the illustrations.

One reason the quark model is so impressive is that very simple things work very, very well. For example, suppose we come up with an "Idiot Quark Model" (IQM) that has almost no physics in it, but also a bare minimum of

Table 1: Some predictions of the Idiot Quark Model.

Hadron	Quarks	Predicted Mass (MeV)	Measured Mass (MeV)
ϕ	$s\bar{s}$	1014	1020
$D_S^\star$	$c\bar{s}$	2132	2112
$\Sigma^\star$	uds	1354	1385
Ω^-	sss	1598	1672
$\Lambda_c^\star$	udc	2472	2600
J/ψ	$c\bar{c}$	3250	3095

free parameters. For example,

$$M(q_1\bar{q}_2) = m_1 + m_2 - B \tag{4}$$
$$M(q_1q_2q_3) = m_1 + m_2 + m_3 - B \tag{5}$$

which predicts the masses of mesons and baryons, based only on the masses of the quarks, and on the binding energy B assumed the same for mesons and baryons. If we determine the parameters using the $\rho(770)$, $\Delta(1232)$, $K^\star(892)$, and $D^\star(2007)$ (all hadrons with aligned quark spins) and assume that $m_u = m_d$, then we get $m_u = m_d = 462$ MeV, $m_s = 584$ MeV, $m_c = 1702$ MeV, and $B = 154$ MeV. We can predict the masses of other mesons and baryons, then, with this formula. Some results are shown in Table 1. Clearly these are truly spectactular success for such a trivial model!

There are also some spectacular failures of the IQM. Most obvious is that it predicts the same mass independent of the quark spin, i.e. $M_\pi = M_\rho$ and $M_p = M_\Delta$. Also apparent from Table 1 is that the agreement is poorer for the heavy quark systems, but that is not unexpected. We assumed a single binding energy, but there is likely some potential well, into which the quarks are more tightly bound when the quarks are heavier and the system is smaller. Putting all this together will lead to mixings and splittings and this is where all the effort goes when formulating a "real" Quark Model.

There has been a good deal of progress on the static $q\bar{q}$ potential, based on lattice QCD. Figure 10 shows the result of one such calculation[24]. The fit is to, essentially, the phenomenological form

$$V(r) = ar - \frac{b}{r}$$

which embodies a short range attractive potential as well as a long range confining potential. This is generally the starting point for a dynamical model of quarks bound into hadrons.

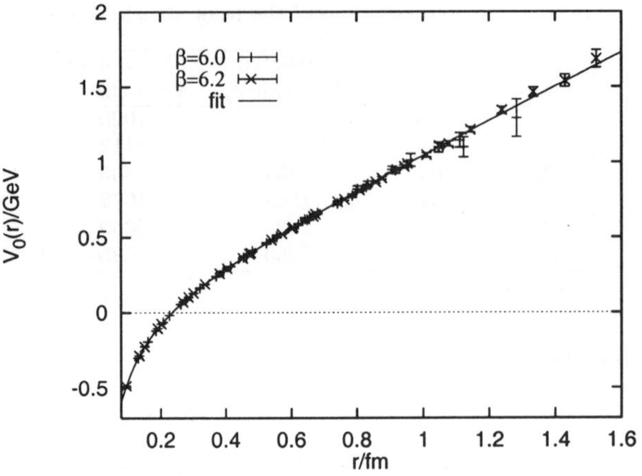

Figure 10: The static $q\bar{q}$ potential calculated in lattice QCD.

Of the two main possibilities for spin dependent potentials, the spin-orbit and spin-spin forms, it seems clear that the dominant effect is from the spin-spin interaction, also known as the "hyperfine" interaction thanks to its analogy in atomic physics, that is $V_{HF} = \vec{\mu}_1 \cdot \vec{\mu}_2$ where the magnetic moment $\vec{\mu}$ is inversely proportional to the mass. Following this analogy, a "one gluon exchange" potential should have the form

$$\begin{aligned} V_{HF}(q_1\bar{q}_2) &= \frac{k}{m_1 m_2}\vec{s}_1 \cdot \vec{s}_2 \\ &= \frac{k}{m_1 m_2}\left(-\frac{3}{4}\right) \quad \text{for } S=0 \\ &= \frac{k}{m_1 m_2}\left(\frac{1}{4}\right) \quad \text{for } S=1 \end{aligned} \tag{6}$$

This behavior is shown[18] very clearly in Fig.11 where the splitting of the pseudoscalar mesons systematically increases as the mass of the constituent quarks decreases. Furthermore, the parameters one derives from the ρ/π splitting for the hyperfine potential do a very good job of predicting[21] the splitting between the Δ and proton. On the other hand, the orbitally excited mesons in Fig.11 do not show any strong mass dependence, suggesting another origin. In addition, if we consider the masses of the $L=1$, $S=1$ isovector mesons (the $a_0(1450)$, $a_1(1260)$, and $a_2(1320)$), isoscalar mesons (the $f_0(1300)$, $f_1(1285)$, and $f_2(1270)$), and the strange mesons (the $K_0(1430)$, $K_1(1270)/K_1(1400)$,

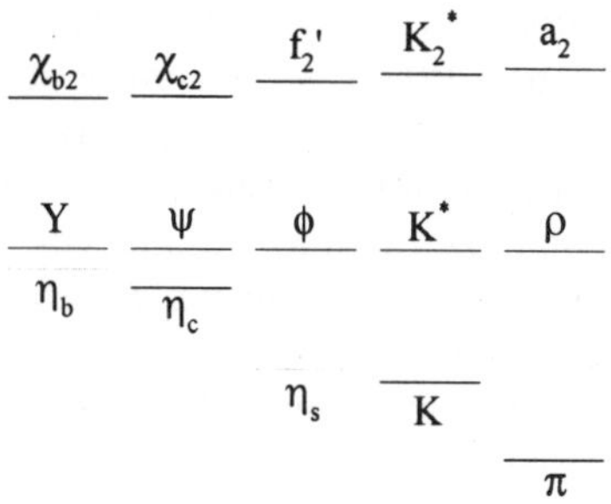

Figure 11: Mass splittings of the pseudoscalar, vector, and tensor mesons.

and $K_2(1430)$), there is no clear dependence on $\vec{L} \cdot \vec{S}$.

Some homework problems were assigned at this point, mostly considering how one can calculate the hyperfine parameters and modify the Idiot Quark Model. The bottom line is that this very simple addition does a pretty good job of fixing many of the problems.

Putting all these ingredients together, along with a long list of other subtle effects and lots of hard work, gives us the Quark Model we use today to predict the observed states and their properties, and to test for new states that are beyond the Quark Model.

5 Interlude: What We've Done and Where We're Going

We're about to get into the real stuff for which I was invited to give these lectures, namely the photo and electroproduction of baryon resonances. First, let me take a moment to review what we have so far.

My first goal was to have it clear in your mind what we mean by "resonance". I used a classical oscillator as an example, and then showed you what happens in nonrelativistic and then relativistic quantum mechanics. The main lesson is all the same. A resonance shows a peak in the magnitude of the amplitude, as the phase of the amplitude increases through 90°. This is the same as saying that the imaginary part of the amplitude shows a peak at the same place that the real part decreases through zero.

Next, we examined how this worked with some real data. We studied the various meson resonances that show up in the reaction $e^+e^- \to \pi^+\pi^-$. One *big* advantage here is that this reaction, proceeding through one photon exchange, can only produce states with $J^{PC} = 1^{--}$. That is, there is only one "partial wave", and we were able to study what happens when these resonances with the same quantum numbers interfere with each other.

This point is important. When we generally study the spectroscopy of mesons or baryons, we produce all sorts of quantum numbers and they can all interfere with each other (so long as the various conservation laws are respected). Before we can analyze these for resonances, we need to separate out the contributions from the different quantum numbers. This is a significant task for meson resonances, but it is even harder for baryon resonances. We won't be discussing meson partial wave analysis here[e] but I will show you the basics of how it works for baryons.

Baryons are harder than mesons for two reasons. First, there are more ways to combine three quarks than two. That leads to more states and more ways for them today, so overlaps are much more serious. Secondly, baryon decays always end up with a nucleon in the final state and that extra spin degree of freedom complicates the analysis and wreaks havoc on your common sense. For example, we will see that the angular distribution for decay of an $\ell = 1$ $N\pi$ resonance has very little to do with $P_1(\cos\theta)$.

We then diverted for one lecture and reviewed the quark model. I mostly did this to set the stage for the physics some of us are very interested in, namely the notion that *violations* of the quark model can be more interesting than the excellent agreement it generally enjoys with data.

6 Lecture Three: Partial Wave Analysis of Baryon Resonances

Now, finally, back to baryons. These are half-integer spin hadrons which, in the Quark Model, are composed of three quarks. Learning the properties of these states teaches us about the fundamental dynamics in the Quark Model, but a careful study of these states may also take us *beyond* the Quark Model to reveal features that are directly due to QCD.

Unlike meson states, baryon resonances are broader and more numerous. Furthermore, the various ways to combine the final spin-$\frac{1}{2}$ proton or neutron in the decay will give us some problems when we try to describe the angular distribution in an experiment. That is, the partial wave analysis we have to employ in order to separate out the different quantum numbers in a particular reaction, will be a lot tougher. We cannot rely on bump hunting alone, and if we are going to look for physics beyond the Quark Model, we need to be as model independent as possible in the analysis.

We'll begin with a discussion of the $\pi^{\pm}p$ cross section and the clear evidence for baryon resonances. Then, we'll undertake a detailed discussion of the

[e] If you want to see more on meson resonances, you can look at the review paper[15] I wrote with Steve Godfrey.

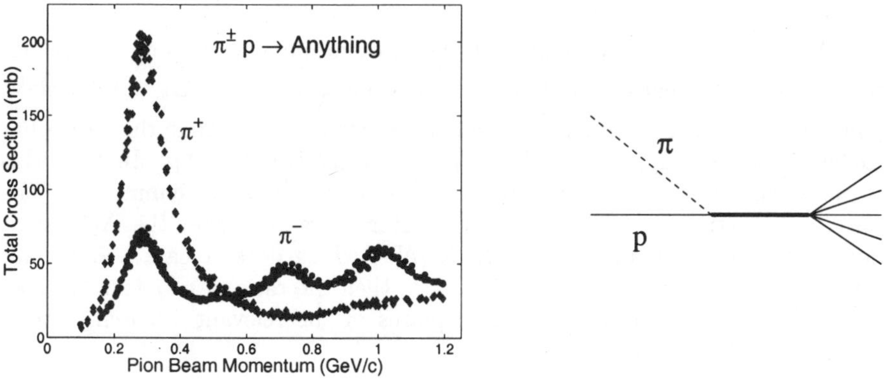

Figure 12: Total cross section fo π^+p and π^-p scattering. The diagram on the right shows how we picture the process proceeding through and intermediate baryon resonance.

angular distribution for πN elastic scattering for an intermediate state of a particular total spin J, and we'll compare this to some data for the $\Delta(1232)$. We'll then extend this to a discussion of photoproduction amplitudes and introduce the language of the "multipole expansion" used extensively in the literature. Next we'll talk a little about the practical difficulties in these types of analyses, and then there will be an example problem on fitting the $\gamma p \rightarrow \pi^+ n$ angular distribution in the "first" and "second" resonance regions. We'll finish up with a brief discussion of some of the current research problems that people are working on.

6.1 The $\Delta(1232)$ in πp Scattering

Figure 12 shows data[1] on the total cross section for π^+ and π^- beams on a hydrogen target, that is, the reactions $\pi^{\pm}p \rightarrow$ *Anything*. The reaction, shown schematically in the figure, is pictured as the pion and proton coming together to form a baryon resonance which subsequently decays any way it wants. This picture is clearly supported in the figure. There are many "bumps" which indicated peaks in resonant amplitudes, and these are the excited baryons which, presumably, are predicted by the Quark Model. Of course, a lot more work needs to be done in order to establish that these are indeed resonances, and that these resonances have the quantum numbers, mass, widths, and other properties that can establish them as Quark Model states, or perhaps as manifestations of physics beyond the Quark Model.

Our job now is to start looking at this work that needs to be done to go beyond the "bump hunting" stage.

We'll do this first by looking at the most prominent "bump" in Fig. 12, namely the peak near 300 MeV/c, or $M = \sqrt{s} = \sqrt{m_p^2 + 2m_p E_\pi} \approx 1220$ MeV. This is the $\Delta(1232)$ resonance. We can immediately tell that this state has isospin $\frac{3}{2}$ since it appears in the $\pi^+ p$ channel as well as $\pi^- p$. In fact, this is one of the classic demonstrations of isospin as a dynamic symmetry of the strong interactions[11]. If the elastic scattering cross section at the $\Delta(1232)$ is dominated by a matrix element $\langle N\pi|\mathcal{O}|N\pi\rangle$ which doesn't care about what kind of pion or nucleon we're dealing with, then the ratio of $\pi^+ p$ to $\pi^- p$ cross sections should be the ratio of the squares of the relevant Clebsch-Gordan coefficients. That is, we predict

$$\frac{\sigma(\pi^+ p \to \Delta^{++})}{\sigma(\pi^- p \to \Delta^0)} = \frac{\left|\langle 1, +1, \frac{1}{2}, +\frac{1}{2}|\frac{3}{2}, +\frac{3}{2}\rangle\right|^2}{\left|\langle 1, -1, \frac{1}{2}, +\frac{1}{2}|\frac{3}{2}, -\frac{1}{2}\rangle\right|^2} = \frac{1}{\frac{1}{3}} = 3$$

This agrees quite well with the cross sections at the peak of the $\Delta(1232)$ as shown in Fig. 12.

Now, let's study the angular distribution of elastic (or charge exchange) scattering to get a handle on its spin J.

I will follow the discussion in Gasciorowicz[11]. Since the pion is spinless, we need to have $M = \pm\frac{1}{2}$, regardless of J, when we quantize along the direction of the spin of the target proton. The angular momentum J of the πp system will be constructed from the relative πp orbital angular momentum ℓ and the proton spin $s = \frac{1}{2}$. We therefore choose $\ell = J \pm \frac{1}{2}$, depending on the parity P of the resonance, i.e. $P = (-1)^{\ell+1}$. (Note that the pion has negative intrinsic parity, while the proton has positive parity.) We can then make a table of the eight different possibilities for getting J, four for each parity P:

ℓ	M	m_ℓ	m_s	ℓ	M	m_ℓ	m_s
$J-\frac{1}{2}$	$+\frac{1}{2}$	0	$+\frac{1}{2}$	$J+\frac{1}{2}$	$+\frac{1}{2}$	0	$+\frac{1}{2}$
		$+1$	$-\frac{1}{2}$			$+1$	$-\frac{1}{2}$
	$-\frac{1}{2}$	-1	$+\frac{1}{2}$		$-\frac{1}{2}$	-1	$+\frac{1}{2}$
		0	$-\frac{1}{2}$			0	$-\frac{1}{2}$

There is some obvious redundancy in this table, but I want to make the point that there are separate contributions to the wave function, depending on the spin and parity of the resonance.

Now let's construct the wave function. This wave function is specified by the two quantum numbers J and $\ell = J \pm \frac{1}{2}$, and for a given $M = \pm\frac{1}{2}$ it has two orthogonal components, each consisting of a spacial part given by the appropriate $Y_\ell^{m_\ell}$ and a spin part given by $\chi_s^{m_s}$. We read these two components off from the table we've constructed above. For example, for $M = +\frac{1}{2}$ and $\ell = J - \frac{1}{2}$, the wave function is

$$\Psi_J^{\frac{1}{2}} = \langle J\frac{1}{2}|J-\frac{1}{2},0,\frac{1}{2},\frac{1}{2}\rangle Y_{J-\frac{1}{2}}^0(\theta,\phi)\chi_{\frac{1}{2}}^{\frac{1}{2}} + \langle J\frac{1}{2}|J-\frac{1}{2},1,\frac{1}{2},-\frac{1}{2}\rangle Y_{J-\frac{1}{2}}^1(\theta,\phi)\chi_{\frac{1}{2}}^{-\frac{1}{2}}$$

where we used the two values of m_ℓ from our table. The Clebsch-Gordan coefficients can be found in any book on angular momentum, or even in the PDG[1], and you find

$$\Psi_J^{\frac{1}{2}} = \sqrt{\frac{J+\frac{1}{2}}{2J}}Y_{J-\frac{1}{2}}^0(\theta,\phi)\chi_{\frac{1}{2}}^{\frac{1}{2}} + \sqrt{\frac{J-\frac{1}{2}}{2J}}Y_{J-\frac{1}{2}}^1(\theta,\phi)\chi_{\frac{1}{2}}^{-\frac{1}{2}}$$

The other three possibilities of M and ℓ would be done the same way. I didn't include any indication of the parity of the resonances in the notation for Ψ_J^M because parity is conserved in the strong interactions, so wave functions of different parities will not mix.

Now, if the target is unpolarized and there is no sensitivity to the spin of the outgoing proton (or neutron), then (aside from phase space and kinematic factors) the intensity distribution is

$$I_J(\theta) = \sum_{m_s}\left[|\Psi_J^{\frac{1}{2}}|^2 + |\Psi_J^{-\frac{1}{2}}|^2\right]$$

which is straightforward, but tedious[f] to evaluate. You find that, for $J = \frac{1}{2}$,

$$I_{\frac{1}{2}}(\theta) = \text{Constant}$$

and for $J = \frac{3}{2}$, for example the $\Delta(1232)$, you find

$$I_{\frac{3}{2}}(\theta) = \text{Constant} \times \left(1 + 3\cos^2\theta\right)$$

Note that the angular distribution does not depend on the orbital angular momentum ℓ of the decay, and therefore on the parity of the resonance. This is in stark contrast to the case for mesons, and has to do with averaging over the spin of the nucleon.

Figure 13 shows data[g] for $\pi^+ p \to \pi^+ p$ for pion momentum at the peak of the $\Delta(1232)$ resonance. The angular distribution we derived above, for a

[f] Of course, it is best to do this sort of algebra with a symbolic math program like MAPLE or MATHEMATICA.

[g] Data obtained from the VPI SAID website at http://said.phys.vt.edu/.

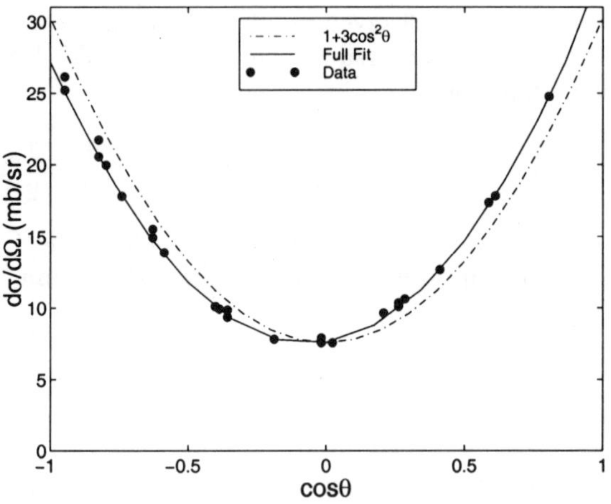

Figure 13: Differential cross section for π^+p elastic scattering at the peak of the $\Delta(1232)$ resonance. The dot-dash line is the prediction based on the simple quantum mechanics for an isolated $J = \frac{3}{2}$ resonances, as described in the text. The solid line is the result of a more complete analysis.

pure, isolated, $J = 3/2$ state, does a very good job of fitting the data. It's not perfect, of course, because this resonance is not isolated, although it does stand out very predominantly, as shown in Fig. 12. Furthermore, the simple schematic reaction also shown in Fig. 12 is not the whole story, and other mechanisms of π^+p elastic scattering should also be included. The line labeled "Full Fit" in Fig. 13 shows the result of a more complete analysis[25] in which amplitudes from different resonances as well as other processes are included and allowed to interfere.

This interference makes it possible to begin to unravel the different resonances which populate the cross section in Fig. 12, and therefore to start to uncover the many overlapping baryon resonance states. This is particularly true when data on polarization observables are included. The wave functions are constructed in the same way we showed above, and each is given a complex amplitude which is varied to obtain a good fit to the data. The result of such a fit is compared to the data in Fig. 13. However, these fits are done in individual pion momentum bins (i.e. different values of $M = \sqrt{s}$), and you can study the real and imaginary parts of these amplitudes as a function of M.

This is nicely shown[25] in Fig. 14 for two of the amplitudes included in the analysis. In the notation of baryon resonance spectroscopy, these are the $P33$ (i.e. $\ell = 1$, $J = \frac{3}{2}$, so positive parity, and $I = \frac{3}{2}$), and $D13$ amplitudes. Com-

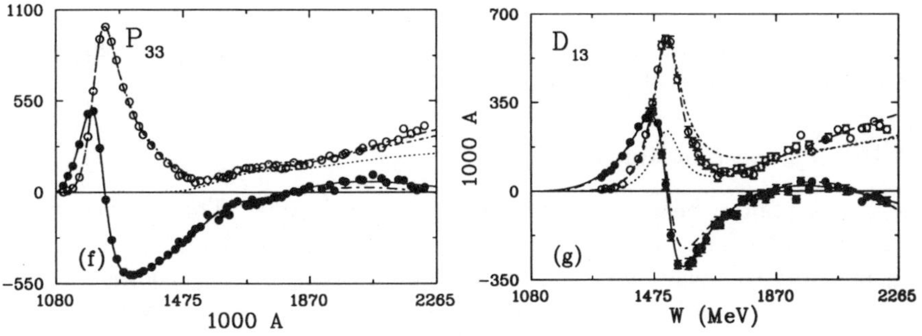

Figure 14: Results of a fit to data for overlapping baryon resonances in πN scattering. Each figure shows the result of a partial wave analysis for the real and imaginary parts of an amplitude (i.e. coefficient of a wave function) for a specific ℓ and J for the πN system. The amplitudes are plotted as a function of center of mass energy W (which we call M or $\sqrt{s}$ in the text). These amplitudes clearly show resonance behavior, and strongly suggest two baryon states, each overlapping with each other but isolated from other resonances with the same quantum numbers.

pare the shapes of the real and imaginary parts to the right side of Fig. 6. In particular, note that the real part passes through zero, and the imaginary part peaks with the shape of a Breit-Wigner resonance. These are clearly consistent with two resonant states, each relatively isolated from other states with the same quantum numbers, but very much overlapping with each other. A total of sixteen amplitudes have been analyzed[25] this way. The $P33$ amplitude is dominated by the $\Delta(1232)$, probably the best known baryon resonance, but clearly not the only thing buried in the total cross section.

I want to emphasize that only after a partial wave analysis is completed, and the separate amplitudes extracted, can we fit Breit-Wigner line shapes through the amplitudes. That is, we can only do the analogy of what we carried out in Section 3.3 for baryons after we've done the hard work of isolating "data" corresponding to the same quantum numbers. We got this for free because of the reaction we used when studying the pion form factor.

This should make it clear how the baryon spectrum has traditionally been studied. Pion beams are incident on proton (or, for neutrons, deuterium) targets, and careful measurements are made of the differential cross sections as a function of beam momentum. By combining these angular distributions with other measurements of polarization observables in the target and outgoing nucleon, the πN data can be decomposed into strengths of the various quantum numbers. By analyzing the real and imaginary parts of these amplitudes, or alternatively the magnitude of the amplitude and the phase, one can identify different states.

Now let's move on and see what can be learned in addition to this using photon beams instead of pion beams.

6.2 *Photoproduction Amplitudes*

One treats photon beams the same way as pion beams. That is, you write down the wave function for a γp state with specific spin and parity. You then multiply this wave function by an "amplitude", i.e. a variable complex number, which is adjusted so that the predicted angular distributions (and other observables) match the data. Finally, you examine the behavior of the real and complex parts of this amplitude, as a function of photon beam energy (and therefore as a function of $\sqrt{s}$) for evidence of resonance behavior.

The one major complication of photons, however, is that photons have spin 1 whereas pions have spin 0. This means that for a particular J, one can have $M = \pm\frac{3}{2}$ as well as $M = \pm\frac{1}{2}$, depending on the orientation of the proton spin with respect to the photon spin. The formalism, although straightforward in principle, becomes rather complicated in practice. A particular nomenclature and methodology has been developed to determine these partial wave amplitudes for photon beams.

This was originally laid down, for the most part, by Walker[26] for analysis of the reaction $\gamma p \to \pi N$, and that terminology has pretty much stuck to this day. (An earlier analysis was carried out by Peierls[27], and in fact is somewhat more explicit with regard to the predicted angular distributions.)

Just about all quantum mechanics textbooks discuss the general problem of scattering a plane wave of spinless particles from a spherically symmetric central potential. The incident wave function is e^{ikz} and some of it is scattered so that the outgoing wave function is written

$$\psi(r \to \infty) = e^{ikz} + f_k(\theta)\frac{e^{ikr}}{r}$$

where the scattering amplitude is $f_k(\theta)$ is then expanded in "partial waves" according to the relative orbital angular momentum ℓ between the scattered particle and the target. That is

$$f_k(\theta) = \frac{1}{k}\sum_{\ell=0}^{\infty}(2\ell+1)e^{i\delta_\ell(k)}\sin\delta_\ell(k)P_\ell(\cos\theta)$$

The job here is to generalize this for collisions of spin-1 photons with spin-$\frac{1}{2}$ protons. The general phenomenology of particles with spin was worked out by Jacob and Wick[28]. For γp collisions, the scattering amplitude works out to be

$$A_{\mu,\lambda} = \sum_J A^J_{\mu,\lambda}(2J+1)d^J_{\mu,\lambda}(\theta)e^{i(\lambda-\mu)\phi} \tag{7}$$

for $\mu = \pm\frac{1}{2}$ and $\lambda = \frac{1}{2}$ or $\lambda = \frac{3}{2}$, and J is the spin of the intermediate state, a positive half-integer number. Each term with a given J is analagous to the two possibilities for $M = \pm\frac{1}{2}$ that we had for πN scattering, but here we write μ instead of M. The difference is that, because of the photon spin, there are two new possibilities, corresponding to $\lambda = \frac{1}{2}$ or $\frac{3}{2}$. These two possibilities are sometimes called $A_{\frac{1}{2}}$ and $A_{\frac{3}{2}}$. In the case of electroproduction, the photon can have a longtidudinal component as well as a transverse component, and the new amplitude is generally called $C_{\frac{1}{2}}$.

Each of the four helicity amplitudes $A_{\mu,\lambda} = H_i$, $i = 1,4$, are orthogonal to each other, so the intensity is $I(\theta) = \sum_i |H_i|^2$. We can therefore, in principle, analyze the angular distribution to determine the individual amplitudes coefficiencts $A^J_{\mu,\lambda}$ in Eq. 7 and ask if they form resonances. First, however, we need to do some rearranging and redefinitions.

The helicity eigenstates are actually of mixed parity. Since both the electromagnetic and strong interactions conserve parity, it is better to rewrite the amplitudes in terms of parity eigenstates. These are conventionally written

$$\begin{aligned}
A_{n+} &= -\frac{1}{\sqrt{2}}\left(A^J_{\frac{1}{2},\frac{1}{2}} + A^J_{-\frac{1}{2},\frac{1}{2}}\right) \\
A_{(n+1)-} &= \frac{1}{\sqrt{2}}\left(A^J_{\frac{1}{2},\frac{1}{2}} - A^J_{-\frac{1}{2},\frac{1}{2}}\right) \\
B_{n+} &= \left[\frac{2}{n(n+2)}\right]^{1/2}\left(A^J_{\frac{1}{2},\frac{3}{2}} + A^J_{-\frac{1}{2},\frac{3}{2}}\right) \\
B_{(n+1)-} &= -\left[\frac{2}{n(n+2)}\right]^{1/2}\left(A^J_{\frac{1}{2},\frac{3}{2}} - A^J_{-\frac{1}{2},\frac{3}{2}}\right)
\end{aligned}$$

where $n = J - \frac{1}{2}$ and $n \geq 1$ for the B's.

Next, for reasons that are partly historical but also useful when discussing "electric" or "magnetic" multipoles, these amplitudes are rearranged once again. These multipoles reduce, nonrelativistically, to familiar operators in electromagnetic transition operators, very common in low energy nuclear physics. (This notation was originally developed[29] by Chew, Goldberger, Low, and Nambu.) We write

$$E_{\ell+} = \frac{1}{\ell+1}\left[A_{\ell+} + \frac{\ell}{2}B_{\ell+}\right]$$

Table 2: Angular distributions for some pure baryon resonance states, produced in photoproduction with no polarization observables.

State	Multipole	$I(\theta)$
$P33$	M_{1+}	$\frac{1}{2}\left(5-3\cos^2\theta\right)$
	E_{1+}	$\frac{1}{2}\left(1+\cos^2\theta\right)$
$S11$	E_{0+}	1
$D13$	M_{2-}	$\frac{1}{2}\left(1+\cos^2\theta\right)$
	E_{2-}	$\frac{1}{2}\left(5-3\cos^2\theta\right)$

$$\begin{aligned} M_{\ell+} &= \frac{1}{\ell+1}\left[A_{\ell+} - \frac{\ell+2}{2}B_{\ell+}\right] \\ E_{(\ell+1)-} &= -\frac{1}{\ell+1}\left[A_{(\ell+1)-} + \frac{\ell+2}{2}B_{(\ell+1)-}\right] \\ M_{(\ell+1)-} &= \frac{1}{\ell+1}\left[A_{(\ell+1)-} + \frac{\ell}{2}B_{(\ell+1)-}\right] \end{aligned}$$

for $\ell \geq 1$. For $\ell = 0$, $E_{0+} = A_{0+}$ and $M_{1-} = A_{1-}$. In this notation, ℓ is the orbital angular momentum of the outgoing pion, and the intermediate state has spin $J = \ell \pm \frac{1}{2}$. These are the amplitudes that are conventionally extracted from data.

As we did with $\pi^+ p$ scattering, let's just take a look at what we predict for a "pure" resonance and compare it to data. (We'll leave the extraction of the amplitudes to you and the experts.) Again, because the photon has spin-1, there will generally be two amplitudes for each J, namely $E_{\ell\pm}$ and $M_{\ell\pm}$, where $J = \ell \pm \frac{1}{2}$. This is just a matter of algebra, and looking up things like $d^J_{\mu,\lambda}(\theta)$ functions, starting with Eq. 7. Some results are shown in Table 2. (Note that these are tabulated by Peierls[27], although with a different notation.) Similar to πp scattering, the angular distribution seems to depend mainly on the spin of the intermediate baryon state, getting the same result for the two spin-$\frac{3}{2}$ resonances, although the angular distribution is switched between multipoles.

Figure 15 compares these amplitudes with the data, similar to the way we did for $\pi^+ p$ elastic scattering in Fig. 13. In this case, in addition to switching to the reaction $\gamma p \to \pi^+ n$, we plot the angular distribution for two photon energies. These two photon energies correspond to the first two "bumps" in Fig. 4.

The figure on the left is for $E_\gamma \approx 340$ MeV, that is, the region of the $\Delta(1232)$. Remember, now, that unlike the case for the spinless pion beam, there are two amplitudes for photoproduction. We show the amplitudes for a

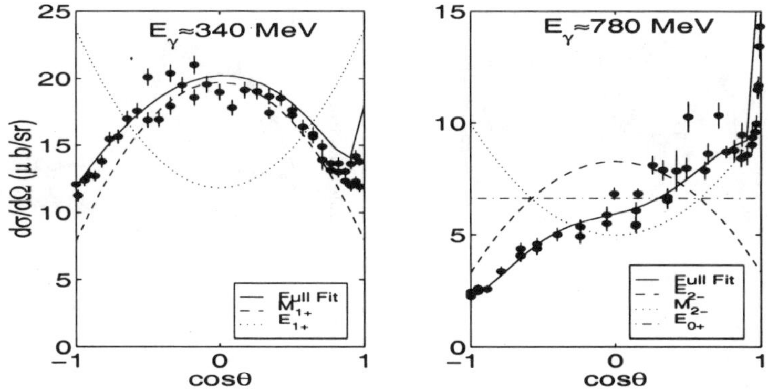

Figure 15: Angular distributions for the reaction $\gamma p \to \pi^+ n$, in the regions corresponding to the first two "bumps" in the γp total cross section. The data are compared to multipoles corresponding to selected pure states, as well as a "full fit" based on a sophisticated partial wave analysis.

pure $J^P = \frac{3}{2}^+$ resonance, consistent with the $\Delta(1232)$. The data is certainly more consistent with the M_{1+} amplitude than the E_{1+}, and therein lies a nice piece of physics. In the quark model, the $p \to \Delta(1232)$ transition is just a quark spin-flip, analagous to the magnetic dipole transition which is very familiar in nuclear physics. The M_{1+} amplitude is, nonrelativistically, just the magnetic dipole transition operator, so it is somewhat of a triumph for the simple quark model that this amplitude is dominant here. There is, however, some controversey[30] on this subject.

The figure on the right is for $E_\gamma \approx 780$ MeV, or $W = \sqrt{s} = 1530$ MeV, in the "second" resonance region of Fig. 4. Here the angular distribution is obvsiously more complicated. It is known, certainly from πp scattering, that there are two prominent states here. These are the $D13(1520)$, i.e. $J^P = \frac{3}{2}^-$ with possible multipoles E_{2-} and M_{2-}, and the $S11(1535)$, with just E_{0+}. Clearly, the data are inconsistent with any of these individually, and it would seem that the waves are interfering with each other. In fact, the angular distribution is decidedly antisymmetric in $\cos\theta$ and therefore must be the result of amplitude interference.

Each of the two data sets in Fig. 15 shows a "full fit" based on extracted amplitudes[31] from a full partial wave analysis. Figure 16 plots the isospin-$\frac{3}{2}$ E_{1+} (k) and M_{1+} (l) amplitudes on the left, and isospin-$\frac{1}{2}$ E_{2-} (m) and M_{2-} (n) amplitudes on the right, for the reaction $\gamma p \to \pi N$. That is, the left column should include the $\Delta(1232)$ baryon resonance, while the right shows

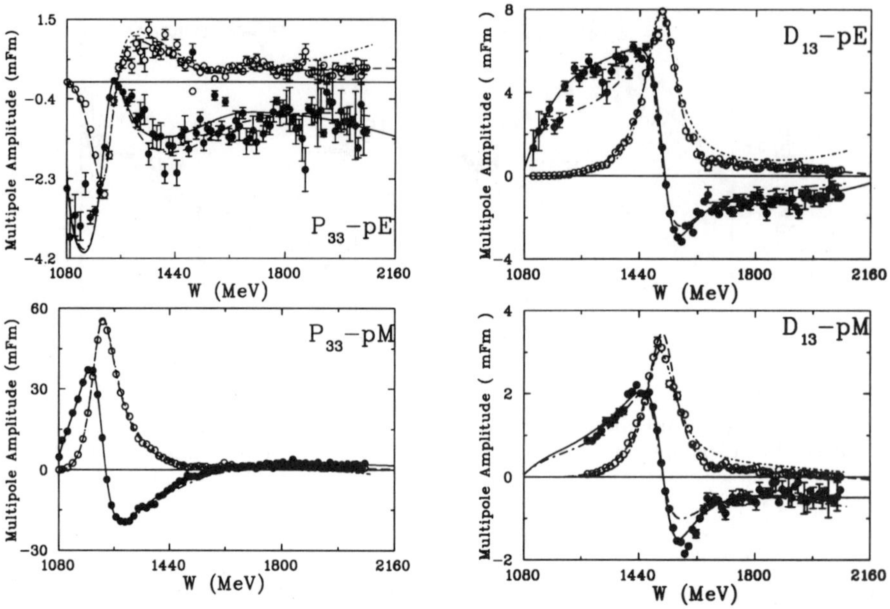

Figure 16: Results of a fit to data on single pion photoproduction from nucleons. The left side shows extracted amplitudes relevant to $\Delta(1232)$ production, while the right is for the $D13(1520)$. Details in the text.

data relevant to the $D13(1520)$ baryon resonance.

In the region near $W \approx 1232$, the left hand side of Fig. 16 confirms what we saw in the left side of Fig. 15. That is, the reaction $\gamma p \to \Delta(1232) \to \pi N$ is dominated by the M_{1+} amplitude, but at the peak of the resonance there is a small but clearly resonant contribution from E_{1+} as well, with $E_{1+}/M_1+ \approx -2.3/60$ based on the peak values of the imaginary parts. There seems to be some evidence of higher mass resonance behavior in the E_{1+} as well.

The $D13$ amplitudes on the right side of Fig. 16 show clear and comparably strong resonant behavior for both available amplitudes. In this case, however, unlike the simple spin-flip transition for the $\Delta(1232)$, one cannot draw any immediate conclusions from the relative size of these amplitudes.

Masses and widths of various Breit-Wigner type fits to these data have been carried out[31] and the results are tabulated. The size of the transition amplitudes can be compared to various model calculations[32,33].

Of course, once the intermediate baryon state is produced, it can decay any way it wants. If it wants to decay in some unique way, in fact, that can also help isolate it from nearby, overlapping states. For example, the $S11(1535)$ has a strong branch to ηN. Consequently, the reaction $\gamma p \to \eta p$ distinctly shows[34] the flat angular distribution of the lone E_{0+} multipole. As a flat angular distribution is expected near threshold in any case (because of barrier penetration factors), one can do a good job in extracting the strong couplings of this excited baryon [35].

7 Current Research Directions

I've tried to take a lot of the mystery out the quantum mechanics and the language that is used to analyze data on the formation of baryons in the fusion of photons and protons. However, this is just a start, the bare basics. I hope that it is enough for you to go ahead and read the papers on the calculations and experiments that are currently hot topics in this field.

One glance at Figures 13 and 15 make it clear that the quality of data in pion scattering is far superior to that in photoproduction. Indeed, this has always been one of the primary reasons for building the CEBAF facility at Jefferson Lab. The research program in baryon resonances, however, is much more extensive than simply expanding the data set on single pion photoproduction. The list[h] of new experiments on the spectroscopy and dynamics of baryon and meson resonances is quite long.

I would like to finish up with a quick look at a couple of recent and current experiments that extend our knowledge of baryon resonances.

7.1 The E_{1+}/M_{1+} Ratio in $\gamma^\star p \to \Delta(1232)$ at High Q^2

Nearly all the discussion has been on photoproduction of baryon amplitudes, as opposed to electroproduction. Formally, there are very few changes, the most important of which is the addition of one more multipole amplitude because of the longitudinal polarizability of the virtual photon. Physically, however, this leads to a very important test of the degrees of freedom relevant for describing the strong interactions.

It has been pointed out[3] that the cross section for baryon formation in $\gamma^\star p$ reactions tends to fall with Q^2 in a way suggestive of perturbative QCD. (Here, $Q^2 \equiv -q^2$ and the four momentum $q \equiv k - k'$ where k and k' are the four momenta of the beam and scattered electron, respectively.) This would mean that at short enough distances, which are still accessible experimentally, the appropriate degrees of freedom for describing this reaction are the *current* quarks of QCD, not the constituent quarks of the Quark Model.

---

[h]For a complete listing of approved Jefferson Laboratory experiments, including sorting by field, see http://www.jlab.org/exp_prog/experiments/.

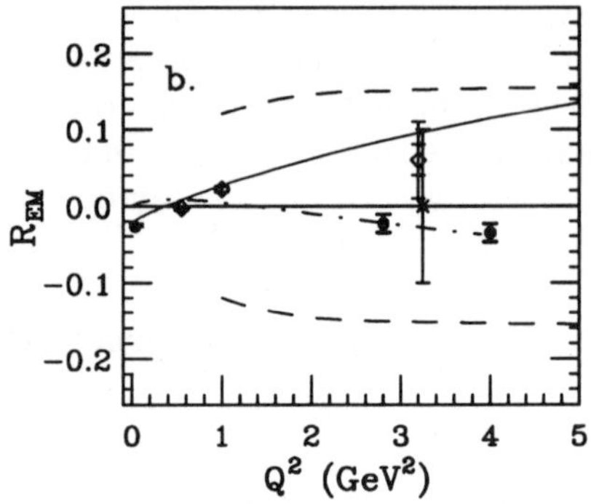

Figure 17: Measurement of the ratio $R_{EM} \equiv E_{1+}/M_{1+}$, as a function of Q^2, for $\gamma^* p \rightarrow \Delta(1232) \rightarrow \pi^0 p$. If this process were describable in terms of perturbative QCD, the ratio should tend towards +1 at the highest momentum transfer.

A precise test of this idea, however, can be carried out by examining the behavior of the individual multipole amplitudes instead of simply the total cross section. The perturbative QCD prediction, in fact, is that E_{1+}/M_{1+} should approach +1 at high Q^2, and not be the very small number that is evident in Figures 15 and 16.

This measurement was carried out[36] using the spectrometer facilities in Hall C at Jefferson Lab. Figure 17 plots values of $R_{EM} \equiv E_{1+}/M_{1+}$ as a function of Q^2 from this experiment (solid black points at $Q^2 = 2.8$ GeV^2 and 4.0 GeV^2) and others. Clearly, the E_{1+} amplitude remains quite small, even at these relatively large Q^2, more or less in line with Quark Model calculations[32] and making it difficult to accomodate perturbative QCD. There are still caveats, of course, and more experimental and theoretical work is in progress.

7.2 A Search for Missing Baryons in $\gamma p \rightarrow p\pi^+\pi^-$

It has been known for quite some time that there are too few baryon resonances known to fully account for the $N = 2$ band of the Quark Model. One suggestion[16] has been that the Quark Model does not accurately reflect the appropriate degrees of freedom for baryons, which should instead be described by a model based on "diquarks". If this were true, it would point towards an unexpected conclusion based on our current understanding of QCD and how it describes baryons.

However, most of the data used to study the baryon spectrum comes from reactions initiated with pion beams. There is a decent data set, which we have studied above, for $\gamma p \rightarrow \pi N$, but there is still the requirement that the intermediate state couple of the πN final state. If it happens that these missing baryon states have fortuitously small couplings to πN, then they would have

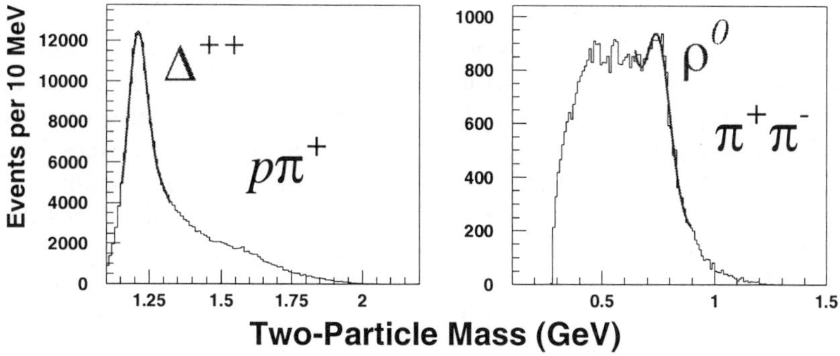

Figure 18: Preliminary data from the CLAS detector on the reaction $\gamma p \to p\pi^+\pi^-$. It is clear that at least a subset of the data is dominated by $\Delta^{++}\pi^-$ and $p\rho^0$ quasi-two body production.

been very difficult to identify in the available measurements.

In fact, state-of-the-art Quark Model calculations[37] do indeed show a preponderance of small πN couplings for these missing states. However, the predicted couplings[38] to quasi-twobody final states, such as $\Delta\pi$ or $N\rho$, are not small in general. Furthermore, these states have predicted γN couplings[32] that are in line with the well established states.

Consequently, a measurement of $\gamma p \to p\pi^+\pi^-$ is in progress using the CLAS detector at Jefferson Lab[17]. This would allow us to study intermediate states that decay to $\Delta^{++}\pi^-$, $\Delta^0\pi^+$, $p\rho^0$, and a number of other combinations. Data was taken with a broad band tagged photon facility in Hall B and the CLAS detector. An event sample was isolated with three charged particles, consistent with the reaction kinematics of $\gamma p \to p\pi^+\pi^-$.

Figure 18 shows some preliminary[39] results from a small subset of this data. We plot the two-body invariant mass combinations for $p\pi^+$ and for $\pi^+\pi^-$. There are clear peaks showing the presence of $\gamma p \to \Delta^{++}\pi^-$ and $\gamma p \to p\rho^0$. These data are integrated over all photon energies, and deeper analysis shows that the relative amounts of these two quasi-two body reactions are very dependent on photon energy. In fact, one of the challenges of this analysis is to separate the diffractive production of $\gamma p \to \rho^0 p$ from the production of an intermediate baryon which then decays to $\rho^0 p$.

The partial wave analysis of this data is also considerably more difficult.

One makes use of the quasi-twobody nature of the reaction and builds an "isobar model" of the decay, preserving angular momentum and isospin at each step, and using this to construct a "decay amplitude". This can be used to compare predicted angular distributions to data, and also to fit in search of underlying baryon resonances. Procedures along these lines have been carried out for[40] $\pi N \to \pi\pi N$, and also[6] for the very sparse amount of data in $\gamma p \to p\pi^+\pi^-$. The new data from Jefferson Lab will make it possible to do a much more detailed study of the photoproduction reaction using a sophisticated partial wave analysis program[41], and a search for the missing baryon states will finally be possible in reactions which do not involve πN in either the initial or final state.

8 Finale

As I stated in the beginning, the purpose of these lectures was not to bring you up to date on the current research in the field. Instead, I wanted to make it possible for you to understand the underlying quantum mechanics and phenomenology which people actually use in their research. This way, you can read those papers, understand what people are doing, and perhaps find new and better ways to pursue these problems in your own research.

Good luck in your careers!

Acknowledgments

I would very much like to thank the organizers, and José Goity in particular, for their kind invitation to tell you about this nice physics. The hospitality of the laboratory is kindly acknowledged, in particular Sue Ewing for her help with so many of the logistics. I owe much of my knowledge and appreciation of this subject to a number of people, but in particular Nathan Isgur, Simon Capstick, Carl Carlson, Nimai Mukhopadhyay, Paul Stoler, and Dennis Weygand. Thanks also to Igor Strakovsky for giving me access to some of the original plots of the VPI analysis.

References

1. C. Caso, *et al*, *Eur. Phys. J* C **3**, 1 (1998).
2. M. MacCormick, *et al*, *Phys. Rev.* C **53**, 41 (1996).
3. P. Stoler, *Phys. Rev.* D **44**, 73 (1991).
4. Cambridge BC group, *Phys. Rev.* **155**, 1477 (1967).
5. Cambridge BC group, *Phys. Rev.* **146**, 994 (1966).
6. Cambridge BC group, *Phys. Rev.* **163**, 1510 (1967).

7. AABHHM group, *Phys. Rev.* **175**, 1669 (1968).
8. L.M. Barkov, *et al*, *Nucl. Phys.* B **256**, 365 (1985).
9. D. Bisello, *et al*, *Phys. Lett.* B **220**, 321 (1989).
10. M. Peskin and D. Schroeder, "An Introduction to Quantum Field Theory", Addison-Wesley (1995).
11. S. Gasciorowicz, "Elementary Particle Physics", Wiley (1966).
12. J.D. Jackson, *Nuovo Cimento* **34**, 1644 (1964).
13. A. Bernicha, *et al.*, *Phys. Rev.* D **50**, 4454 (1994).
14. S. Gardner and H.B. O'Connell, *Phys. Rev.* D **59**, 076002 (1999).
15. S. Godfrey and J. Napolitano, *Rev. Mod. Phys.* **71**, October (1999).
16. D.B. Lichtenberg, *Phys. Rev.* **178**, 2197 (1969).
17. Jefferson Laboratory experiment E93-033, J. Napolitano and D. Weygand, spokespersons.
18. S. Godfrey and N. Isgur, *Phys. Rev.* D **32**, 189 (1985).
19. S.S.M. Wong, "Introductory Nuclear Physics", Wiley (1998).
20. F. Halzen and A. Martin, "Quarks and Leptons", Wiley (1984).
21. S. Capstick and N. Isgur, *Phys. Rev.* D **34**, 2809 (1986).
22. D. Faiman and A. Hendry, *Phys. Rev.* **173**, 1720 (1968).
23. H. Frauenfelder and E. Henley, "Subatomic Physics", Prentice Hall (1991).
24. G. Bali, K. Schilling, and A. Wachter, *Phys. Rev.* D **56**, 2566 (1997).
25. R. Arndt, *et al.*, *Phys. Rev.* C **52**, 2120 (1995).
26. R.L. Walker, *Phys. Rev.* **182**, 1729 (1969).
27. R.F. Peierls, *Phys. Rev.* **118**, 325 (1960).
28. M. Jacob and G.C. Wick, *Annals of Physics* **7**, 404 (1959).
29. G.F. Chew, *et al.*, *Phys. Rev.* **106**, 1345 (1957).
30. R.M. Davison, *et al.*, *Phys. Rev.* C **59**, 1059 (1999).
31. R. Arndt, *et al.*, *Phys. Rev.* C **53**, 430 (1996).
32. S. Capstick, *Phys. Rev.* D **46**, 2864 (1992).
33. R. Bijker, F. Iachello, A. Leviatan, *Phys. Rev.* C **54**, 1935 (1996).
34. B. Krusche, *et al.*, *Phys. Rev. Lett.* **74**, 3736 (1995).
35. M. Benmerrouche, N.C. Mukhopadhyay, and J.-F. Zhang, *Phys. Rev. Lett.* **77**, 4716 (1996)
36. V.V. Frolov, *et al.*, *Phys. Rev. Lett.* **82**, 45 (1999).
37. S. Capstick and W. Roberts, *Phys. Rev.* D **47**, 1994 (1993).
38. S. Capstick and W. Roberts, *Phys. Rev.* D **49**, 4570 (1994).
39. M. Klusman, Ph.D. Thesis, Rensselaer Polytechnic Institute (in progress)
40. D.M. Manley, *et al.*, *Phys. Rev.* D **30**, 904 (1984).
41. J.P. Cummings and D.P. Weygand, "PWA 2000" (in progress)

STRANGE QUARKS AND PARITY VIOLATION

P. A. SOUDER
Department of Physics, Syracuse University,
Syracuse, NY 13244, USA,
E-mail souder@suhep.phy.syr.edu

These lectures discuss how it is possible to measure the strangeness content of form factors in nuclei and nucleons by measuring the tiny parity-violating asymmetry in the scattering of polarized electrons. First, the theory is developed for the simplest case of potential scattering from nuclei. Here all the relevant ideas are developed with a minimum of formalism. Next, the theory is extended to the case of the nucleon. The experimental techniques unique to this field are discussed. Finally, recent experimental results are described.

1 Introduction

The proton is a state of three quarks bound by the strong force of QCD through the exchange of gluons. Quark-antiquark pairs are also likely to be abundant given the small masses of $u\bar{u}$ and $d\bar{d}$ pairs and the strength of the coupling. When QCD was first developed in the 1970's, it was thought that gluons and quark-antiquark pairs were part of an inert "sea" that played little role in properties of the proton such as its charge radius or magnetic moment. Therefore, the details of the properties of antiquarks and gluons in the proton were not considered important.

Data on spin-structure functions published in 1988 [1], which led to the famous "spin crisis," changed this perception of the inertness of the sea. Specifically, the data could be interpreted to mean that gluons, antiquarks, and strange quarks all contribute to the spin of the nucleon and therefore represent relevant degrees of freedom for describing the proton at low energy.

Considerable data on spin-structure functions have been obtained more recently, confirming that relatively little, only 30-40%, of the spin of the proton comes from the quarks. The data, together with information on hyperon decays, also suggest the striking conclusion that ~10% of the spin comes from strange quarks. However, by invoking flavor SU(3) breaking in the hyperon decay, this effect of strange quarks might be explained away.

There is additional evidence that strangeness is an important degree of freedom in the nucleon, but such evidence is also subject to interpretation. One example is the pion Σ term from an analysis of low energy pion-nucleon scattering which suggests that $s\bar{s}$ pairs contribute significantly to the mass of the nucleon. [2] Certain results in $p\bar{p}$ annihilations can also be explained by $s\bar{s}$

pairs in the nucleon.[3] Unfortunately, these effects might also be explained in other ways. The role of strange quarks in the proton is not firmly established.

The subject of these lectures is an unambiguous method to establish the importance of $s\bar{s}$ quarks to the structure of the proton.[4] By measuring the parity-violating asymmetry in the elastic scattering of polarized electrons from hydrogen, we can observe an electroweak interference that depends on the contribution of strange quarks to the elastic form factors.[5,6] The only other input is the assumption of isospin symmetry and data from ordinary elastic scattering of electrons from protons and neutrons. If $s\bar{s}$ quarks do contribute to the relevant matrix elements, this contribution can be measured cleanly.

The subject has been covered in detail in a review article.[7] In these lectures, I will give a description of how the method works for the simple case of potential scattering and how to measure the contribution of strange quarks to the charge radius of a nucleus. This example contains the essential ideas. Next, extensions to the realistic case for nucleons are given. An important characteristic of parity-violation experiments is the small size of the effects measured, typically on the order of 1-100 ppm (parts per million). I will describe some of the general experimental techniques required to measure such small effects. Finally, I will survey the experiments in the field that have been done or are underway.

2 Theory: How to Measure a Strange Radius

2.1 Potential Scattering

Both weak and electromagnetic scattering from a point nucleus can be described by simple quantum mechanical scattering of electrons from the potential

$$V(r) = ke^2 g^b g^t \frac{\exp[-Mr]}{4\pi r} \tag{1}$$

Here g^b is the charge of the beam projectile, g^t is the charge of the target, both in units of the electron charge e. Also, k is the strength of the coupling, and M is a parameter that controls the range of the potential. Even though it is zero for electromagnetic scattering, the limit $M \to 0$ must be taken at the end of the computation. For the neutral weak interaction, $M = M_Z$, the mass of the Z-boson.

The electromagnetic charges q_{em} of the relevant particles, u, d, and s quarks and the electron are familiar and are given in Table 1. For the weak interactions, particles of different helicity are treated differently, and they have different charges g^R and g^L. The spin-averaged charge is given by $g^V = (g^R + g^L)/2$ and the difference in the charges is given by $g^A = (g^R - g^L)/2$. These

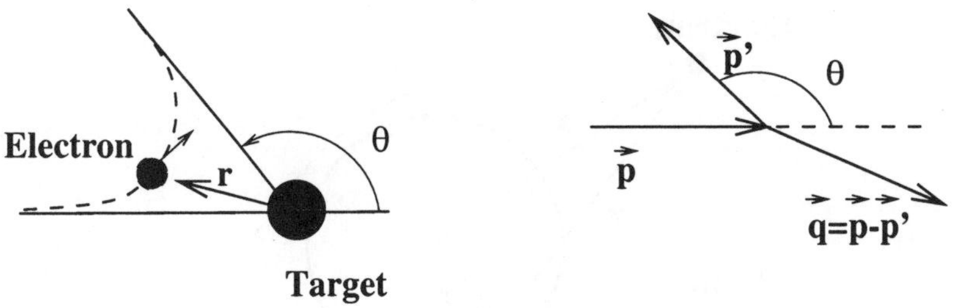

Figure 1: Kinematics for scattering from a point potential.

Table 1: Weak and electromagnetic charges for electrons and light quarks in units of the electron charge e.

Particle	q^{em}	g^V	g^A
e^-	-1	$-\frac{1}{4}+\sin^2\theta_W$	$\frac{1}{4}$
u	$\frac{2}{3}$	$\frac{1}{4}-\frac{2}{3}\sin^2\theta_W$	$-\frac{1}{4}$
d, s	$-\frac{1}{3}$	$-\frac{1}{4}+\frac{1}{3}\sin^2\theta_W$	$\frac{1}{4}$

charges are also given in Table 1 with $\sin^2\theta_W \sim 0.23$ being the weak mixing angle. For electromagnetic scattering, k=1, and for neutral weak scattering $k = k_Z = (\sin\theta_W \cos\theta_W)^{-2}$. In these units, the fine structure constant $\alpha = e^2/4\pi$.

The differential cross section for scattering by an angle θ is given in terms of the scattering amplitude f by:

$$\frac{d\sigma}{d\Omega} = |f(\theta)|^2 \tag{2}$$

where f is given in the Born approximation for a potential $V(r)$ by:

$$f(q) = -\frac{m}{2\pi}\int V(\mathbf{r})\exp[i\mathbf{q}\cdot\mathbf{r}]d^3r = \frac{(2km\alpha g^b g^t)}{(q^2+M^2)}. \tag{3}$$

and where $q = 2p\sin(\theta/2)$. The relevant variables are shown in Fig. 1. Since f depends on θ only through q, we often write $f(\theta)$ as $f(q)$. The result is:

$$\left(\frac{d\sigma}{d\Omega}\right)_{Point} = \frac{(2km\alpha g^b g^t)^2}{(q^2+M^2)^2} \tag{4}$$

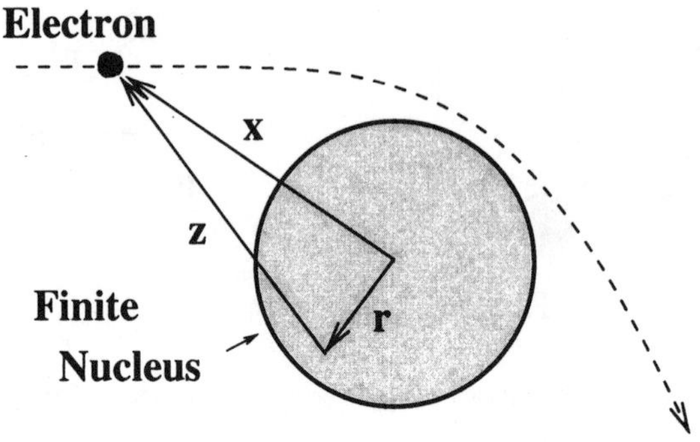

Figure 2: Kinematics for scattering from a potential due to a distributed charge.

For electromagnetic scattering, $\sigma \sim 1/q^4$ and the cross section is forward peaked. For weak scattering $\sigma \sim 1/M_Z^4$. The cross section is isotropic but very small since M_Z is large. Both cases arise from the same calculation.

We will apply the above result to the case of ultra-relativistic electrons. The correct calculation, which uses the Dirac equation, gives the same integral for the scattering amplitude. However, the m in the numerator is replaced by the relativistic total energy E and an additional factor of $\cos(\theta/2)^2$ in the cross section. These factors are not relevant for the issues under discussion, and we choose to use the non-relativistic derivation for simplicity.

The above describes scattering from a point source. For a nucleus of finite size, the scattering amplitude f is computed from a smeared distribution of point charges [8]. We start with a toy model in which there is only one kind of charge. Later, we will do the realistic case of three kinds of charge corresponding to $u, /d,,$ and s quarks. If we define $\rho(r)$ as the density of point weak or electromagnetic charges, the differential potential is given by

$$dV(r) = ke^2 g^b g^t \frac{\exp[-Mz]}{4\pi z} \rho(r) d^3r \qquad \rho(r) = \#/\text{unit Volume} \tag{5}$$

where the variables are given in Fig. 2

The scattering amplitude is then:

$$f(\theta) = -\frac{m}{2\pi} \int d^3r \int d^3x dV \exp[i\mathbf{q} \cdot \mathbf{r}] \tag{6}$$

The integral factors into two parts. One part is defined to be the form factor:

$$F(q^2) = \int d^3 r \rho(r) \exp[i\mathbf{q}\cdot\mathbf{r}]. \tag{7}$$

The other factor is just Eq. 3, and the scattering amplitude is:

$$F(\theta) = \frac{(2km\alpha g^b g^t)}{(q^2 + M^2)} F(q^2). \tag{8}$$

In terms of the form factor, the cross section is

$$\left(\frac{d\sigma}{d\Omega}\right) = \left(\frac{d\sigma}{d\Omega}\right)_{Point} |F(q^2)|^2. \tag{9}$$

The cross section in this case is just the point-like cross section from Eq. 4 times the square of the form factor. The key point is that if the nucleus had only one type of charge, the ***same form factor appears for electromagnetic and weak scattering.*** This is somewhat remarkable given that the electromagnetic interaction is long range and the weak interaction is very short range.

Finally, to measure the shape of an extended potential distribution, one measures the cross section, extracts $F(q^2)$, and then determines the density $\rho(r)$ by taking the Fourier transform:

$$\rho(r) = \frac{1}{(2\pi)^3} \int d^3 q F(q^2) \exp[-i\mathbf{q}\cdot\mathbf{r}] \tag{10}$$

Some charge distributions and form factors for some simple cases are given in Table 2. The first example is the point charge, yielding a form factor of unity, consistent with Eq. 4. The second is an exponential charge distribution which yields a 'dipole" form factor characteristic of nucleons. The last example, a uniform spherical distribution, is a crude approximation to a large nucleus. The form factor has oscillations, as do form factors for actual nuclei.

By convention, the total density is normalized to unity. For small q, the form factor may be expanded as

$$F(q^2) \sim 1 - \frac{1}{6} q^2 \langle r^2 \rangle \tag{11}$$

where

$$\langle r^2 \rangle = \int d^3 r \; r^2 \rho(r) \tag{12}$$

defines the mean square radius of the distribution. Hence $F(0)$ gives the total charge and $\langle r^2 \rangle = -6(\partial F/\partial q^2)$ gives the radius.

Table 2: Simple charge distributions and their corresponding form factors

Probability Density $\rho(r)$	Form factor $F(q^2)$
$\delta(r)$	1
$\rho_0 \exp[-r/a]$	$(1+q^2a^2)^{-2}$
$\rho_0, r \leq R; 0, r > R$	$3[\sin(qR) - qR\cos(qR)]/(qR)^3$

2.2 Scattering from Quarks in a Nucleus

The above discussion applied to the case of a single interaction from a single kind of charge. We now consider realistic nuclei which have u, d, and possibly s quarks, each with their own density[a] ρ_j where $j = u, d, s$. In analogy with Eq. 7, we define quark density form factors for each quark flavor:

$$F_j(q^2) = \int d^3r \rho_j(r) \exp[i\mathbf{q} \cdot \mathbf{r}] \tag{13}$$

The normalization is such that in a nucleon, $F_d(0) = 1,\ F_u(0) = 2$ (since there are two up quarks) and $F_s(0) = 0$ (since there is no net strangeness). If the s and $\bar{s}$ quarks have different spatial distributions, the mean squared strange radius $\langle r_s^2 \rangle$ is finite and to leading order

$$F_s(q) = -\langle r_s^2 \rangle q^2/6 \tag{14}$$

In addition, the scattering involves different interactions i, where i spans both electromagnetic and weak interactions. Thus we must add many terms. Do we add amplitudes or cross sections? When we observe an elastically scattered electron, is there any way to tell which quark flavor caused the scattering or whether a photon or Z-boson was exchanges? The answer is no. Therefore scattering is coherent, and we ***add amplitudes.***

$$f(q) = \sum_{i,j} f_{ij}(q), \tag{15}$$

where i lists all forces and j lists the three quark flavors.

The individual amplitudes are just the amplitudes f computed above in Eq. 8:

$$f_{ij}(q) = \frac{2k_i m \alpha g_i^b g_{ij}^t}{q^2 + M_i^2} F_j(q^2) \tag{16}$$

[a]Formally, one can write, for example, $\rho_u(r) \sim \int d^3p \langle N|\overline{u}\gamma_0 u|N\rangle \exp[ip \cdot r]$.

Thus the only properties of the nucleus needed to compute any type of elastic scattering are the three form factors F_u, F_d, and F_s. The same form factors also describe neutrino scattering.

The most precise data on scattering from nuclei is confined to electron scattering. The weak contribution is suppressed by a factor of $(q^2/M_Z^2)^2$ and is totally negligible. In this case, only one combination of form factors contributes: $F_\gamma = \frac{2}{3}F_u - \frac{1}{3}(F_d + F_s)$. Here F_γ is the charge form factor. Decomposing F_γ into quark flavors adds nothing to the interpretation of that data. There is no reason to go beyond the analysis that led to Eq. 9. In addition, there is no way to determine if the strange quarks contribute, i.e., whether $F_s = 0$ or not.

2.3 Parity Violation

Additional information about the form factors becomes available when the helicity of the electron beam can be controlled. One can measure the parity-violating asymmetry

$$A^{PV} = \frac{d\sigma_R - d\sigma_L}{d\sigma_R + d\sigma_L} \tag{17}$$

This asymmetry has the following features:

1. $A^{PV} = 0$ for electromagnetic scattering because the interaction is invariant under parity inversion.

2. Weak interaction potentials with one axial-vector coupling violate parity and contribute to A^{PV}.

3. Most systematic errors that arise in cross section measurements, such as the solid angle, target thickness, beam energy, etc, cancel in the ratio.

4. Sensitivities as low as 10^{-8} are feasible, so weak effects can be measured.

To compute A^{PV}, we use Eq. 15. The relevant potentials are V_γ for the electromagnetic and V_Z for the weak interaction:

$$V_\gamma(r) = -e^2 \frac{q_j^{em}}{4\pi r} \qquad\qquad V_Z^{L(R)}(r) = k_Z e^2 g_e^{L(R)} g_i^V \frac{\exp[-M_Z r]}{4\pi r}. \tag{18}$$

For spinless targets, the quarks have no net spin and the spin-averaged weak charge g_i^V for the quarks is appropriate.

The corresponding amplitudes are:

$$f_\gamma(\theta) = -\frac{2m\alpha}{q^2}\sum q_j^{em} F_j(q^2); \quad f_Z^{L(R)}(\theta) = \frac{2m\alpha k_Z g^{L(R)}}{M_Z^2}\sum g_j^V F_j(q^2) \quad (19)$$

Next we add amplitudes for each helicity: $f^R = f_\gamma + f_Z^R$, and $f^L = f_\gamma - f_Z^L$. The asymmetry is, neglecting terms in f_Z^2:

$$A^{PV} = \frac{|f^R|^2 - |f^L|^2}{|f^R|^2 + |f^L|^2} = \frac{f_\gamma(f_Z^R - f_Z^L)}{f_\gamma^2} = \frac{f_Z^R - f_Z^L}{f_\gamma} \quad (20)$$

Also, since $g^R - g^L = 2g^A = 1/2$, we have

$$f_Z^R - f_Z^L = \frac{mk_Z}{2M_Z^2}\sum g_j^V F_j(q^2). \quad (21)$$

Then substituting the weak and electromagnetic charges from Table 1 yields:

$$\begin{aligned} A^{PV} &= -\frac{k_Z q^2}{2M_Z^2}\frac{\sum g_j^V F_j}{\sum q_j F_j} \\ &= -\frac{k_Z q^2}{2M_Z^2}\frac{(\frac{1}{4} - \frac{2}{3}\sin^2\theta_W)F_u + (-\frac{1}{4} + \frac{1}{3}\sin^2\theta_W)(F_d + F_s)}{\frac{2}{3}F_u - \frac{1}{3}(F_d + F_s)} \end{aligned} \quad (22)$$

For a nucleus with no spin or isospin, such as ^{4}He or ^{12}C, the distribution of up and down quarks must be the same under the assumption of isospin invariance. This implies

$$F_u = F_d \equiv F \quad (23)$$

When substituted into Eq. 22, the result is:

$$\begin{aligned} A^{PV} &= -\frac{k_Z q^2}{2M_Z^2}\left(\frac{-\frac{1}{3}\sin^2\theta_W(F - F_s) - \frac{1}{4}F_s}{\frac{1}{3}(F - F_s)}\right) \\ &= \frac{k_Z q^2}{2M_Z^2}\left(\sin^2\theta_W + \frac{F_s}{4F_\gamma}\right) \end{aligned} \quad (24)$$

where we used for the last step

$$F_\gamma = \frac{2}{3}F_u - \frac{1}{3}(F_d + F_s) = \frac{1}{3}(F - F_s). \quad (25)$$

The implications of Eq. 24 are as follows:

1. The first term is independent of nuclear structure! It can be used for test of electroweak theory for kinematics where $F_s \sim 0$.

2. The only effect of nuclear shape on A_{PV} is due to F_s. This is a very clean way to measure the effect of strange quarks.

The only assumptions are the standard model and isospin as given in Eq. 23.

2.4 Relativistic Scattering

The above discussion gives the essence of how to cleanly measure the effects of strange quarks in a spinless nucleus. Complications due to relativity were avoided. However, to describe scattering from a proton, relativistic theory must be used. Moreover, relativistic theory is needed to clean up a few points glossed over in the above discussion.

This section covers material that is in textbooks [9]. However, the perspective is that spin is the most important degree of freedom. This is a purely personal bias arising from the fact that most of the experiments I have done in my life have involved spin. I hope that the reader will find that this perspective is of some pedagogical value.

Electrons follow the Dirac equation $(i\gamma^\mu \partial_\mu - m)\psi \equiv (i\partial\!\!\!/ - m)\psi = 0$. For this discussion, we will use the representation

$$\gamma^0 = \begin{pmatrix} 1 & 0 \\ 0 & -1 \end{pmatrix}; \quad \vec{\gamma} = \begin{pmatrix} 0 & \vec{\sigma} \\ \vec{\sigma} & 0 \end{pmatrix}; \quad \gamma^5 = \begin{pmatrix} 0 & 1 \\ 1 & 0 \end{pmatrix} = i\gamma^0\gamma^1\gamma^2\gamma^3 \tag{26}$$

with the Pauli matrices

$$\sigma^1 = \begin{pmatrix} 0 & 1 \\ 1 & 0 \end{pmatrix}; \quad \sigma^2 = \begin{pmatrix} 0 & -i \\ i & 0 \end{pmatrix}; \quad \sigma^3 = \begin{pmatrix} 1 & 0 \\ 0 & -1 \end{pmatrix} \tag{27}$$

The plane wave solutions are $\psi = u(p)e^{-ip\cdot x} \Rightarrow (p\!\!\!/ - m)u = 0$ where the Dirac spinor is given by

$$u = N \begin{pmatrix} \xi \\ \frac{\sigma\cdot p}{E+m}\xi \end{pmatrix}: \tag{28}$$

Here ξ is a two component spinor that describes the spin degree of freedom of the electron and N is a normalization constant.

The helicity of the electron is given by the operator

$$\frac{1}{2}\Sigma\cdot\hat{p} = \frac{1}{2}\begin{pmatrix} \sigma\cdot\hat{p} & \\ & \sigma\cdot\hat{p} \end{pmatrix} \tag{29}$$

which commutes with the Hamiltonian and has eigenvalues $\pm\frac{1}{2}$. We stated above that the weak charge of the electron depends on its helicity, which is impossible because helicity is not Lorentz invariant. The helicity of an electron depends upon the velocity of the observer.

The solution of this problem is to use the operator γ^5, which is Lorentz invariant. For ultra-relativistic particles, γ_5 is approximately the helicity operator of Eq. 29:

$$\gamma^5 u = \begin{pmatrix} \frac{\sigma\cdot p}{E+m} & \\ & \frac{\sigma\cdot p}{E-m} \end{pmatrix} u \Rightarrow \gamma^5 \approx \Sigma\cdot\hat{p}\ (m\to 0) \tag{30}$$

Using γ^5, we can define projection operators $P_L = (1-\gamma^5)/2$ and $P_R = (1+\gamma^5)/2$ which follow the rules $P_L + P_R = 0$, $P_L^2 = P_L$, $P_R^2 = P_R$, and $P_L P_R = 0$. We then define

$$u_L \equiv P_L u; \quad u_R \equiv P_R u \tag{31}$$

Then u_L and u_R can have different charges. For weak interactions, they indeed do have different charges. To a good approximation, u_L and u_R are helicity eigenstates as indicated above.

The scattering amplitude T_{fi} for an electron interacting with a spin 1 boson such as the photon or Z is given by

$$T_{fi} = -i\int j_\mu^{fi}(x) A^\mu d^4x \tag{32}$$

where A_μ is the boson field and the electron current $j_\mu^{fi}(x)$ is given by

$$j_\mu^{fi}(x) = Q\overline{\psi}_f\gamma_\mu\psi_i = Q\overline{u}_f\gamma_\mu u_i e^{i(p_f-p_i)\cdot x} = j_\mu e^{iq\cdot x}. \tag{33}$$

For the charged (neutral) weak interaction A_μ is replaced by $W_\mu(Z_\mu)$.

In Eq. 33, there are some implicit definitions. First $q^\mu = p_f^\mu - p_i^\mu$ is the four momentum transfer. The dot product $q\cdot x$ means $q^\mu x_\mu$, and $|q\cdot q| \equiv Q^2$. For potential scattering, energy is conserved, and $Q^2 = q^2$, the three momentum transfer used in the sections above. Indeed, this is the reason why the same result is obtained for potential scattering for both relativistic and non-relativistic treatments. For scattering from a nucleon, there is significant recoil. Also $Q \neq q$, and a relativistic treatment is essential. Finally, the current without the plane-wave x-dependence is denoted j_μ, whereas with the x-dependence is denoted $j_\mu(x)$.

For electrons, there are two currents for each particle a exchanged:

$$Q_L^a j_L^\mu = Q_L^a \overline{u}_L\gamma^\mu u_L; \quad Q_R^a j_R^\mu = Q_R^a \overline{u}_R\gamma^\mu u_R \tag{34}$$

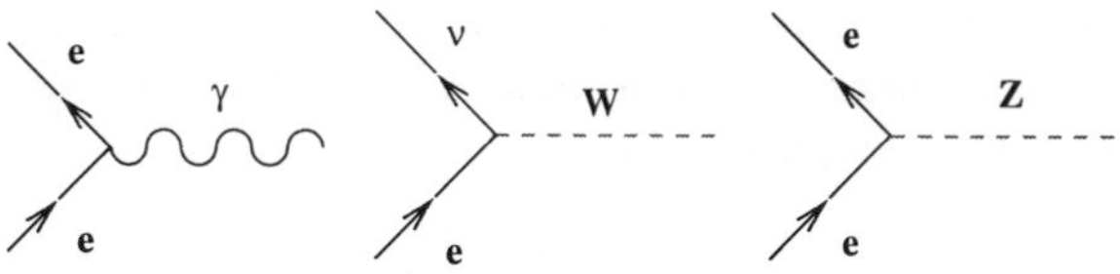

Figure 3: Interactions of electrons with the photon, W and Z.

one for left chirality and one for right chirality. The units are such that $e = 1$. There are several cases of interest.

1. Charged weak current: $Q_R = 0$. Since $\overline{u}_L = u^\dagger\gamma^0 = \overline{u}\frac{1}{2}(1+\gamma^5)$ because $\{\gamma^5, \gamma^\mu\} = 0$, we have also $Q_L j_L^\mu = Q_L\overline{u}\gamma^\mu(1-\gamma^5)u/2$. Thus j_L is the V-A current of the charged weak interactions. We note that axial current is $j_A^\mu \equiv \overline{u}\gamma^\mu\gamma_5 u = \overline{u}_R\gamma^\mu u_R - \overline{u}_L\gamma^\mu u_L$.

2. Electromagnetic current: $Q_R = Q_L = -1$. Then $-(j_L^\mu + j_R^\mu) = -\overline{u}\gamma^\mu u$ is proportional to the vector current $j_V^\mu \equiv j_l^\mu + j_R^\mu$.

3. Neutral weak current: $Q_L \neq Q_R$.

There are two choices of basis currents, j_L^μ and j_L^μ or j_V^μ and j_A^μ. Both have their advantages in different contexts.

In the standard model, only two currents (and two charges) are needed. In this section, we will define currents with unit charges and add the charge as part of the interaction. One is a "hypercharge" current. The other is a left-handed weak isospin current. The isospin current is made from left-handed isodoublets, eg.

$$\chi_L = \begin{pmatrix} \nu_L \\ e_L \end{pmatrix} \tag{35}$$

where ν is the neutrino spinor and e the electron spinor. We use the isospin operators:

$$\tau^1 = \begin{pmatrix} 0 & 1 \\ 1 & 0 \end{pmatrix}; \quad \tau^2 = \begin{pmatrix} 0 & -i \\ i & 0 \end{pmatrix}; \quad \tau^3 = \begin{pmatrix} 1 & 0 \\ 0 & -1 \end{pmatrix} \tag{36}$$

with $\tau_\pm \equiv (\tau_1 \pm i\tau_2)/2$. The charged currents are

$$J_\mu^\pm = \overline{\chi}_L\gamma_\mu\tau^\pm\chi_L. \tag{37}$$

Since τ is a vector, there is a third neutral component

$$J_\mu^3 = \overline{\chi}_L\gamma_\mu\frac{1}{2}\tau^3\chi_L. \tag{38}$$

The charge corresponding to the weak isospin current of Eq. 37 is $T^i = \int J_0^i d^3x$. The hypercharge Y then is defined by $Q^{em} = T_3 + \frac{Y}{2}$ and the hypercharge current is given implicitly by

$$j_\mu^{em} = J_\mu^3 + \frac{1}{2} j_\mu^Y \tag{39}$$

Since $[Y, \vec{T}] = 0$, both Y and $\vec{T}$ can be used to label states.

Thus in the standard model, there are only two currents, a left-handed isospin current j^i with charge g and a helicity-independent current j^Y with charge g'. The interaction is

$$-ig(J^i)^\mu W_\mu^i - i\frac{g'}{2}(j^Y)^\mu B_\mu \tag{40}$$

There is a complication. The masses are generated by symmetry breaking that mixes W^3 and B:

$$\begin{aligned} A_\mu &= B_\mu \cos\theta_W + W_\mu^3 \sin\theta_W \\ Z_\mu &= -B_\mu \sin\theta_W + W_\mu^3 \cos\theta_W \end{aligned} \tag{41}$$

where θ_W is the weak mixing angle which is determined from experiment. Thus the neutral current is:

$$\begin{aligned} &-i(g \sin\theta_W J_\mu^3 + g' \cos\theta_W \frac{1}{2} j_\mu^Y) A^\mu \\ &-i(g \cos\theta_W J_\mu^3 - g' \sin\theta_W \frac{1}{2} j_\mu^Y) Z^\mu \end{aligned} \tag{42}$$

The top part is the electromagnetic current. To be consistent with Eq. 39, we must have

$$g \sin\theta_W = g' \cos\theta_W = e \Rightarrow -iej_\mu^{em} A^\mu \text{ interaction.} \tag{43}$$

The result is the desired current. The bottom part is the weak neutral current interaction:

$$j_\mu^Z = J_\mu^3 - \sin^2\theta_W j_\mu^{em} \Rightarrow -i\frac{g}{\cos\theta_W} j_\mu^Z Z^\mu \text{ interaction} \tag{44}$$

For the processes considered here, only quarks and leptons appear in the initial and final states. Thus the boson propagator

$$A^{\mu\nu} = \frac{g_{\mu\nu} - q_\mu q_\nu / M_B^2}{q^2 - M_B^2} \rightarrow \frac{g_{\mu\nu}}{q^2 - M_B^2} \tag{45}$$

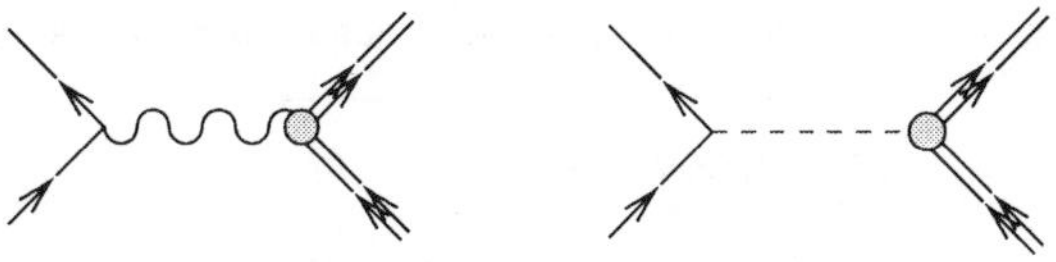

Figure 4: Tree level diagrams for $\vec{e}p$ scattering.

is needed and the scattering amplitude becomes

$$T_{fi} = -i \int g^1 j^1_\mu(x) \left(\frac{-1}{q^2 - M_B^2} \right) g^2 j^{2\mu}(x) d^4x \tag{46}$$

This leads to the current-current interaction $T_{fi} = -i(2\pi)^4 \delta^4(\Sigma p_k) \mathcal{M}$ where

$$\mathcal{M} = g^1 j^1_\mu \left(\frac{-1}{q^2 - M_B^2} \right) g^2 j^{2\mu} \tag{47}$$

This corresponds to diagrams shown in Fig. 4

The standard model currents are thus related to the phenomenological Fermi interactions as follows:

$$\mathcal{M}^W = \frac{4G_F}{\sqrt{2}} J^{+\mu} J^{+\dagger}_\mu \Rightarrow \frac{4G_F}{\sqrt{2}} = \frac{g^2}{2M_W^2} \tag{48}$$

$$\mathcal{M}^Z = \frac{4G_F}{\sqrt{2}} 2\rho j^{Z\mu} j^Z_\mu \Rightarrow \rho \frac{4G_F}{\sqrt{2}} = \frac{g^2}{2M_Z^2 \cos^2\theta_W} = \frac{e^2}{2M_Z^2 \cos^2\theta_W \sin^2\theta_W} \tag{49}$$

where $\rho = M_W^2 / M_Z^2 \cos^2\theta_W = 1$ from the Higgs sector. There are 6 parameters e, g, g', θ_w, M_W, M_Z. Eq. 43 gives three relations, so there are three independent constants to be determined from experiment. The most precisely measured quantities are $\alpha = e^2/4\pi$, G_F, and M_Z. We can use G_F as input by using Eq. 48. The three constants are sufficiently well known so that if the standard model is correct, they introduce negligible error in the interpretation of any anticipated low energy data. If there are extensions to the standard model, such as the existence of extra $Z's$, or if quarks or leptons are actually composite, additional corrections are necessary.

To compute the current for a specific particle, we use the quantum numbers listed in Table 3 and Eq. 44. As an example, the electron currents are done below. The electromagnetic current is $j^{em}_\mu = (\bar{e}_R \gamma_\mu e_R + \bar{e}_L \gamma_\mu e_L) = \bar{e}\gamma_\mu e$. The couplings are $g^R = g^L = g^V = 1$; $g^A = 0$. For the charged weak interaction $j^+_\mu = \chi_L \gamma_\mu \tau_+ \chi_L = \chi \gamma_\mu \frac{1}{2}(1+\gamma^5)\tau_+ \chi$. Thus $g^L = 1$ and $g^R = 0$. Since

$$g^V + g^A = g^R, \; g^V - g^A = g^L$$

Table 3: Weak isospin and hypercharge quantum numbers for leptons

Lepton	T	T^3	Q	Y
ν_e	$\frac{1}{2}$	$\frac{1}{2}$	0	-1
e_L^-	$\frac{1}{2}$	$-\frac{1}{2}$	-1	-1
e_R^-	0	0	-1	-1

$$g^V = \frac{1}{2}(g^R + g^L),\ g^A = \frac{1}{2}(g^R - g^L) \tag{50}$$

$g^V = \frac{1}{2}$ and $g^A = -\frac{1}{2}$. Finally, for the neutral weak interaction, we have:

$$j_\mu^Z = J_\mu^3 - \sin^2\theta_W j_\mu^{em} = -\frac{1}{2}\bar{e}_L\gamma_\mu e_L - \sin^2\theta_W[-(\bar{e}_R\gamma_\mu e_R + \bar{e}_L\gamma_\mu e_L)] \tag{51}$$

Thus $g^R = -\sin^2\theta_W,\ g^L = \frac{1}{2} - \sin^2\theta_W,\ g^V = \frac{1}{4} - \sin^2\theta_W$, and $g^A = -\frac{1}{4}$. In general,

$$g^R = -Q\sin^2\theta_W;\ \ g^L = T_3 - Q\sin^2\theta_W. \tag{52}$$

2.5 Scattering of Electrons from Protons

We are now ready to compute the weak and electromagnetic scattering of electrons from protons. We will proceed in stages, starting with relativistic potential scattering, then scattering from a Dirac particle (we will choose the muon), and finally scattering from a real proton.

For potential scattering, we start with Eq. 32 which gives an interaction $-iej_\mu^{em}A^\mu \to -ie\bar{u}\gamma_0\phi u \to u^\dagger uV$. The result is essentially the same as Eq. 3 except that $m \to E$ and there is a factor $u_f^\dagger u_i$ (ignoring normalization). Normally computations involving Dirac spinors are done with the help of trace theorems and projection operators

$$\Lambda_\pm(s_\mu) = \frac{\not{p}+m}{2m}\cdot\frac{1\pm\gamma_5\not{s}}{2} \to \frac{\not{p}}{2m}\cdot\frac{1\pm\gamma_5}{2} \tag{53}$$

where s_μ is the spin four-vector and the second formula is the ultra-relativistic limit. For summing over spins, the second factor is omitted. This is indeed the easiest method for working out most of the examples that follow. However, performing the present calculation without traces provides more physical insight. The relevant spinors, using Eq. 28, are

$$u_i = \begin{pmatrix} 1 \\ 0 \\ 1 \\ 0 \end{pmatrix}; \quad u_f^+ = \begin{pmatrix} 1 \\ 0 \\ \cos\theta \\ \sin\theta \end{pmatrix}, \quad u_f^- = \begin{pmatrix} 0 \\ 1 \\ \sin\theta \\ \cos\theta \end{pmatrix},$$

The spin of the incident particle u_i was chosen to be in the $+z$ direction. The result is independent of this choice. The final particle $u_f^\pm$ has scattered an angle θ and may have spin in either the + or $-z$-direction. Thus there are two amplitudes, $u_f^{+\dagger} u_i = 1 + \cos\theta = 2\cos^2(\theta/2)$ and $u_f^{-\dagger} u_i = \sin\theta = 2(\cos\theta/2)\sin(\theta/2)$. In contrast to the other cases where we have added amplitude, one can tell the spin of the final state, so this sum must be *incoherent* and probabilities added: $|u_f^{+\dagger} u_i|^2 + |u_f^{-\dagger} u_i|^2 = 4\cos^2(\theta/2)$. This results in the extra factor of $\cos^2(\theta/2)$ that multiplies Eq. 4 mentioned above. The cross section is

$$\left.\frac{d\sigma}{d\Omega}\right|_{lab} = \left(\frac{\alpha^2}{4E^2 \sin^4 \frac{\theta}{2}}\right)\{\cos^2 \frac{\theta}{2}\} \tag{54}$$

From the relative probabilities for having $\sigma_z = \pm 1$, we can compute $\langle \vec{\sigma} \rangle = \hat{p}$, which implies that the interaction has not changed the helicity of the electron. This is an important general feature of all the interactions that we are studying. Moreover, it justifies the idea that particles of different chiralities have different charges.

The next example is the electromagnetic scattering of muons from electrons. The spin-dependence of this process is a useful way to measure the polarization of a high energy muon beam. The process

$$e^-(k) + \mu^-(p) \to e^-(k') + \mu^-(p')$$

is governed by the matrix element

$$\mathcal{M} = -\frac{e^2}{q^2}[\overline{u}(k')\gamma^\mu u(k)][\overline{u}(p')\gamma_\mu u(p)].$$

For the impractical case where the muon is at rest, the kinematics are

$$Q^2 = 4EE' \sin^2 \frac{\theta}{2}; \; \nu \equiv E - E' = \frac{Q^2}{2M}, \; \frac{E'}{E} = (1 + 2\frac{E}{M}\sin^2 \frac{\theta}{2})^{-1}$$

Here M is the muon mass, E is the beam energy and E' is the energy of the scattered electron. The mass of the electron is neglected. These kinematic equations also apply for the practical case of scattering from a proton at rest, where the proton takes the role of the muon. The spin-averaged scattering cross section is

$$\left.\frac{d\sigma}{d\Omega}\right|_{lab} = \left(\frac{\alpha^2}{4E^2\sin^4\frac{\theta}{2}}\right)\frac{E'}{E}\{\cos^2\frac{\theta}{2} + \frac{Q^2}{2M^2}\sin^2\frac{\theta}{2}\} \tag{55}$$

In the limit $M \to \infty \Rightarrow E = E'$, the result is the same as Eq. 54. The additional term is due to the magnetic moment of the muon.

It is useful for the student to do this calculation. A more interesting case is to compute the spin dependence of a beam of longitudinally polarized muons from electrons in a magnetized foil. The calculation, like most calculations involving spin, is most easily done by using the projection operators of Eq. 53. The asymmetry for $\theta = \pi$ is 100%. The reason is that at high energies, helicity is conserved for vector interactions. This implies that the spin must flip for both particles in the center of mass frame. Due to angular momentum conservation, this is impossible for the case where beam and target spins are parallel. Similarly, large asymmetries also occur in deep inelastic scattering of polarized electrons from polarized protons. In this case, the struck quark must have its spin antiparallel to the electron for backscattering to occur. This is the basis for spin-structure function measurements.

Elastic scattering from the proton is similar to that from the muon except that the nucleon is a composite particle. The effect of this compositeness cannot be computed presently. However, we can write the proton current in the general form

$$j_\mu^a(proton) = \langle p(k', s')|j_\mu^a|p(k, s)\rangle =$$

$$\overline{u}(k', s')\big[F_1^a(Q^2)\gamma_\mu + \frac{i}{2M}F_2^a(Q^2)\sigma_{\mu\nu}q^\nu + F_A^a(Q^2)\gamma_\mu\gamma^5 + F_P^a(Q^2)\gamma^5 q_\mu\big]u(k, s) \tag{56}$$

where the $F^{a'}s$ are form factors that depend only upon Q^2. The index a refers to the particle exchanged. The above form is the most general assuming

1. Lorentz invariance.

2. Current conservation ($q_\mu j^\mu = 0$).

3. Time reversal invariance$\Rightarrow$ F's are real.

Finally, we note that induced pseudoscaoar form factor F_P is negligible except for muon capture and will not be considered further. Notice that we are now using vector and axial vector currents instead of chirality currents.

The form factors are the information that we can obtain from experiments about the proton from elastic scattering. The cross section is computed in the same way as was done for muon-electron scattering. The result is

$$\left.\frac{d\sigma}{d\Omega}\right|_{lab} = \left(\frac{\alpha^2}{4E^2\sin^4\frac{\theta}{2}}\right)\frac{E'}{E}\{\left((F_1^\gamma)^2+\tau(F_2^\gamma)^2\right)\cos^2\frac{\theta}{2}+2\tau(F_1^\gamma+F_2^\gamma)^2\sin^2\frac{\theta}{2}\} \tag{57}$$

with $\tau = Q^2/4M^2$. The cross section is simpler in terms of the Sachs form factors:

$$G_E^a \equiv F_1^a - \tau F_2^a; \quad G_M^a \equiv F_1^a + F_2^a. \tag{58}$$

The result is

$$\left.\frac{d\sigma}{d\Omega}\right|_{lab} = \left(\frac{\alpha^2}{4E^2\sin^4\frac{\theta}{2}}\right)\frac{E'}{E}\left\{\frac{(G_E^\gamma)^2+\tau(G_M^\gamma)^2}{1+\tau}\cos^2\frac{\theta}{2}+2\tau(G_M^\gamma)^2\sin^2\frac{\theta}{2}\right\} \tag{59}$$

By measuring the cross section at different angles, G_E and G_M can be independently measured. $G_E^\gamma(0)=1$ and $G_M^\gamma(0)=\mu$=2.79 is the anomalous magnetic moment of the proton. It is tempting but dangerous to interpret $G_E^\gamma(Q^2)$ as the charge distribution of the proton in analogy to the nuclear form factors. The problem is that the mass in natural units of the proton is comparable to its radius. This fact implies that for Q^2 values sufficient to observe the effect of the charge radius, the nucleon has significant recoil. The effects of the recoil on the form factor cannot be determined without an accurate microscopic theory of the proton [10].

One problem is that G_E^γ contributes little to the cross section at large values of Q^2 at any angle. The reason is that the contributions of the form factors add in quadrature. If instead one uses polarized electrons and measures the polarization of the recoil proton, terms proportional to $G_E G_M$ appear that are easy to measure. Computing this spin-dependence is a useful exercise for the student.

2.6 Parity Violation for Proton Targets

Finally, we compute the parity-violating asymmetry for $\vec{e}p$ scattering. We follow the procedure of Sec. 2.3 and in particular Eqs. 17-24. The main difference is that we use the general currents of Eq. 56. The result in terms of weak and electromagnetic form factors is

$$A^{PV} = \frac{\sigma_R - \sigma_L}{\sigma_R + \sigma_L} = \left[\frac{-G_F Q^2}{\pi\alpha\sqrt{2}}\right] \tag{60}$$

$$\times \frac{\varepsilon G_E^{p\gamma} G_E^{pZ} + \tau G_M^{p\gamma} G_M^{pZ} - \frac{1}{2}(1 - 4\sin^2\theta_W)\varepsilon' G_M^{p\gamma} G_A^{pZ}}{\varepsilon (G_E^{p\gamma})^2 + \tau (G_M^{p\gamma})^2}$$

where $\varepsilon = [1+2(1+\tau)\tan^2(\theta/2)]^{-1}$ is the transverse polarization of the virtual photon exchanged and $\varepsilon' = \sqrt{\tau(1+\tau)(1-\varepsilon^2)}$.

One surprising feature of this result is that even though the target is unpolarized, it involves the axial form factor G_A^{pZ}. The reason for this is easiest to see for backward scattering in the ultra-relativistic limit. In that case, as in the argument in the paragraph following Eq. 55, only particles of the same helicity can interact. Then the cross section is proportional to $g_b^R g_t^R - g_b^L g_t^L = -2(g_b^V g_t^A + g_b^A g_t^V)$ and an axial term is induced. Here $b(t)$ refers to the beam(target) particle.

The result of Eq. 60 involves more form factors. This raises the crucial question as to what are the relationships among the form factors. The weak and electromagnetic interactions of the proton arise from interactions with the quarks that compose the proton. Thus, just as we did for Eq. 13, we can decompose the form factors in terms of flavors:

$$\langle p|j_\mu^a|p\rangle = \langle p|\sum g_i^{Va}\overline{u}_i\gamma_\mu u_i|p\rangle + \langle p|\sum g_i^{Aa}\overline{u}_i\gamma_\mu\gamma_5 u_i|p\rangle$$

$$= \overline{u}_p\Big[\sum g_i^{Va}\big(F_1^i\gamma_\mu + \frac{i}{2M}F_2^i\sigma_{\mu\nu}q^\nu\big) + g_i^{Aa}F_A^i\gamma_\mu\gamma_5\Big]u_p \tag{61}$$

where the g_i^a is the coupling of the current to boson a by quark i. The proton spinor is denoted u_p and the quark spinors are denoted u_i. Thus the flavor form factors are defined:

$$\langle p|\overline{u}_i\gamma_\mu u_i|p\rangle = \overline{u}_p\big(F_1^i\gamma_\mu + \frac{i}{2M}F_2^i\sigma_{\mu\nu}q^\nu\big)u_p$$

$$\langle p|\overline{u}_i\gamma_\mu\gamma^5 u_i|p\rangle = \overline{u}_p F_A^i\gamma_\mu\gamma^5 u_p \tag{62}$$

Thus we have a total of nine flavor form factors for the proton, three flavors times three Lorentz invariants:

$$F_1^u,\ F_2^u,\ F_A^u$$

$$F_1^d,\ F_2^d,\ F_A^d$$

$$F_1^s,\ F_2^s,\ F_A^s \tag{63}$$

By using isospin symmetry, we can also use the above quark form factors to compute form factors for the neutron. Under isospin:

$$p \to n \quad \Rightarrow \quad u \to d,\ d \to u,\ s \to s. \tag{64}$$

This implies the analog of Eq. 23:

$$F_1^u \equiv F_1^{pu} = F_1^{nd}; \quad F_1^d \equiv F_1^{pd} = F_1^{nu}; \quad F_1^s \equiv F_1^{ps} = F_1^{ns} \tag{65}$$

This is a complete set from which we can compute all the following precesses:

1. $n \to p\overline{\nu}e$; β Decay
2. $\overline{\nu} \to e^+n$; Charge Current Scattering
3. $\nu p \to \nu p$; Neutral Current Scattering
4. $\nu n \to \nu n$
5. $\overline{\nu}n \to \overline{\nu}n$
6. $ep \to ep$; Electromagnetic Scattering
7. $\vec{e}p \to \vec{e}p$; Parity Violation

All that needs to be done is to take the form factors of Eq. 63, Apply Eq. 65, and multiply by the appropriate charges.

For example, the electromagnetic form factor is computed from the quark form factors by multiplying by the electric charges:

$$j_\mu^\gamma = \frac{2}{3}\overline{u}\gamma_\mu u - \frac{1}{3}\overline{d}\gamma_\mu d + \frac{2}{3}\overline{c}\gamma_\mu c - \frac{1}{3}\overline{s}\gamma_\mu s + \cdots.$$

Then the electromagnetic form factors for protons and neutrons may be written:

$$F_{ip}^\gamma = \frac{2}{3}F_i^u - \frac{1}{3}F_i^d - \frac{1}{3}F_i^s; \quad F_{in}^\gamma = \frac{2}{3}F_i^d - \frac{1}{3}F_i^u - \frac{1}{3}F_i^s. \tag{66}$$

The neutral weak current coupling to the Z can be determined from Eq. 44 or Eq. 52 and the quantum numbers for quarks in Table 4 and Eq. 50 as was done for the electron in Eq. 51. Left-handed quarks form weak isodoublets just as the electron and neutrino do:

$$\chi_L^q = \begin{pmatrix} u_L \\ d_L \end{pmatrix}; \quad \begin{pmatrix} c_L \\ s_L \end{pmatrix} \tag{67}$$

The partner of the s-quark is the c-quark, which is expected to have negligible matrix elements in the nucleon due to its large mass. The result is

$$\begin{aligned} j_\mu^Z &= (\frac{1}{4} - \frac{2}{3}\sin^2\theta_W)\overline{u}\gamma_\mu u - (\frac{1}{4} - \frac{1}{3}\sin^2\theta_W)\overline{d}\gamma_\mu d \\ &+ (\frac{1}{4} - \frac{2}{3}\sin^2\theta_W)\overline{c}\gamma_\mu c - (\frac{1}{4} - \frac{1}{3}\sin^2\theta_W)\overline{s}\gamma_\mu s \\ &- \frac{1}{4}\overline{u}\gamma_\mu\gamma_5 u + \frac{1}{4}\overline{d}\gamma_\mu\gamma_5 d - \frac{1}{4}\overline{c}\gamma_\mu\gamma_5 c + \frac{1}{4}\overline{s}\gamma_\mu\gamma_5 s \end{aligned} \tag{68}$$

Table 4: Weak isospin and hypercharge quantum numbers for quarks

Quark	T	T^3	Q	Y
u_L, c_L	$\frac{1}{2}$	$\frac{1}{2}$	$\frac{2}{3}$	$\frac{1}{3}$
d_L, s_L	$\frac{1}{2}$	$-\frac{1}{2}$	$-\frac{1}{3}$	$\frac{1}{3}$
u_R, c_R	0	0	$\frac{2}{3}$	$\frac{4}{3}$
d_R, s_R	0	0	$-\frac{1}{3}$	$-\frac{2}{3}$

Consequently:

$$F_i^{pZ} = (\frac{1}{4} - \frac{2}{3}\sin^2\theta_W)F_i^u - (\frac{1}{4} - \frac{1}{3}\sin^2\theta_W)(F_i^d + F_i^s) \tag{69}$$

Both G_E^γ and G_M^γ are known for both the proton and neutron. Thus instead of just quark form factors, it is convenient to use the electromagnetic form factors and G_E^s, G_M^s as the complete set. Then G_E^s and G_M^s are the only unknowns. This procedure is the analog of Eq. 25. First, Eq. 69 can be written:

$$F_{ip}^Z = \frac{1}{4}(F_i^u - F_i^d - F_i^s) - \sin^2\theta_W\left(\frac{2}{3}F_i^u - \frac{1}{3}(F_i^d + F_i^s)\right)$$

By using Eq. 66, this becomes

$$F_i^{pZ} = \frac{1}{4}(F_i^{p\gamma} - F_i^{n\gamma}) - \sin^2\theta_W F_i^{p\gamma} - \frac{1}{4}F_i^s$$

Finally, converting to Sachs form factors, the result is:

$$G_{E,M}^{pZ} = \frac{1}{4}(G_{E,M}^{p\gamma} - G_{E,M}^{n\gamma}) - \sin^2\theta_W G_{E,M}^{p\gamma} - \frac{1}{4}G_{E,M}^s \tag{70}$$

Also

$$G_A^{pZ} = -\frac{1}{2}G_A^\beta + \frac{1}{4}F_{As} \tag{71}$$

where $G_A^\beta = \frac{1}{2}(F_{Au} - F_{Ad})$ is the axial form factor that appears in beta decay of the neutron. $G_A^\beta(0) = 1.26$.

Substituting this into Eq. 60 gives an expression for A^{PV} that involves only known quantities, including proton and neutron electromagnetic form factors, and the strange form factors. If the strange form factors are zero, we have a

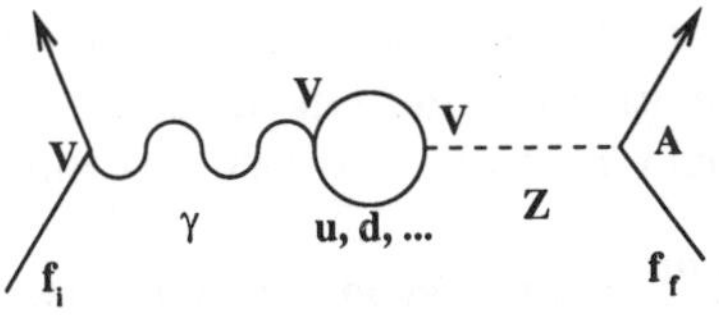

Figure 5: One diagram needed for computing radiative corrections

firm prediction for A^{PV}. If a measured value for A^{PV} is different from this, there must be non-zero strange form factors.

This method of measuring the strange form factors is based on using both strong and weak isopsin. The key to the separation of the strange quarks from the other flavors is that strong (Eq.64) and weak (Eq. 67) isospin are different. In the former, $s \rightarrow s$ whereas in the latter $s \rightarrow c$.

2.7 Radiative Corrections

The above analysis treats the electroweak theory to lowest order, or tree level. The strength of the theory is that it is renormalizable, and higher order contributions can be computed giving a result to high precision. The theory of these corrections is a vast and interesting subject in itself. However, here we will focus briefly on what the corrections are, what the uncertainties are, and how these uncertainties affect the interpretation of the data in terms of strange form factors.

In the theory, the coupling constant $\alpha/4\pi \sim 0.001$ is small and the corrections should be small. Indeed, in some cases the corrections are fairly small, but in others they are large and even uncertain for the following reasons:

1. The radiative corrections involve large factors like $\ln(M_w/m_e)$.

2. The tree level interaction may involve small charges whereas the corrections involve large charges.

3. The radiative corrections may involve hadronic structure that is difficult to calculate.

The diagram in Fig. 5 illustrates the second point. If two fermions of unit charge scatter, the parity-violating part must have one axial coupling and one vector coupling. The vector coupling of the Z has a suppression factor of $1 - 4\sin^2\theta_W \sim 0.08$, so the parity violation is small. However, the radiative correction diagram has a quark loop bubble that allows the Z to couple to fractional charges and avoid the $1 - \sin^2\theta_W$ factor. Now the photon provides a large vector coupling to the second fermion, and the contribution of the radiative correction is relatively large. On the other hand, corrections to large parity-violating diagrams tend to have the suppression factor of $1 - 4\sin^2\theta_W$ and are quite small.

For $\vec{e}p$ scattering, A^{PV} can be decomposed into two parts. The first, which is the result of Eq. 60 if G_A^{pZ} is set to zero, has large couplings and corresponds to graphs with axial electron couplings. The radiative corrections [11] can be included if the result is multiplied by a factor $\hat{\rho}'_{eq} = 0.9897$ and if $\sin^2\theta_W$ is replaced by $\hat{\kappa}_{eq}\hat{s}^2_Z$. Here $\hat{\kappa}_{eq} = 1.0029$ and $\hat{s}^2_Z$, which comes from LEP data, is 0.23124 ± 0.00017. If the mass of the Higgs is 300 GeV instead of the value favored by fits to the LEP data, $\hat{s}^2_Z$ increases by 0.00024. These corrections are small, well known, and to a good approximation independent of Q^2.

The other term, proportional to ε', involves a vector coupling to the electron with its suppression factor of $1 - 4\sin^2\theta_W$. The radiative correction may be included by adding a factor $(1 + R_A)$ where $R_A = -0.34 \pm 0.34$ [13]. R_A is large because it has no suppression and is uncertain because it involves hadronic excitations.

Thus for $\vec{e}p$ scattering at small angles such as for the HAPPEX experiment [12], the radiative corrections are small and known. For the SAMPLE experiment [13,14,7], the uncertainty on R_A is one of the limiting errors. In an upcoming run, the SAMPLE collaboration will measure the asymmetry for quasi-elastic scattering from the deuteron in order to determine R_A empirically.

2.8 Measuring Strange Form Factors

Combining the results from Eqs. 60, 70, 71, and adding in the radiative corrections, we obtain the final expression for A^{PV}:

$$A^{PV} = \left[\frac{-G_F M_p^2 \tau}{\pi\alpha\sqrt{2}}\right]\left\{\hat{\rho}'_{eq}(1 - 4\hat{\kappa}_{eq}\hat{s}^2_Z) - \right. \tag{72}$$

$$\frac{\hat{\rho}'_{eq}[\varepsilon G_E^{p\gamma}(G_E^{n\gamma} + G_E^s) + \tau G_M^{p\gamma}(G_M^{n\gamma} + G_M^s)]}{\varepsilon(G_E^{p\gamma})^2 + \tau(G_M^{p\gamma})^2} -$$

$$\left.\frac{(1 - 4\sin^2\theta_W)(1 + R_A)\sqrt{\tau(1+\tau)}\sqrt{1-\varepsilon^2}G_M^{p\gamma}(G_A^\beta - \frac{1}{2}F_A^s)}{\varepsilon(G_E^{p\gamma})^2 + \tau(G_M^{p\gamma})^2}\right\}$$

Since the electromagnetic form factors have been measured, the above equation has three unknown form factors that are functions of Q^2 or equivalently τ, $G_E^S(Q^2)$, $G_M^S(Q^2)$, and $F_A^s(Q^2)$. For fixed τ, there is still freedom to vary the scattering angle θ between small angles and 180°, corresponding to $1 < \varepsilon < 0$. The G_E terms are multiplied by ε, the G_A term by $\sqrt{1-\varepsilon^2}$, and the G_M terms are independent of ε. By choosing three different angles, we can determine three different linear combinations of the strange form factors, invert the matrix, and determine each form factor. Forward angles are insensitive to G_A^{pZ} and backward angles are insensitive to G_E^{pZ}.

Unfortunately, there are a number of issues that complicate this simple picture. First is the knowledge of the electromagetic form factors $G_{E,M}^{p\gamma}$ and $G_{E,M}^{n\gamma}$. At Q^2=0, the form factors are constrained to be:

$$G_E^p = 1, \quad G_E^n = 0, \quad G_M^p = \mu_p = 2.79, \quad G_M^n = -1.91.$$

The form factor $G_M^p(Q^2)$ has been measured to fit the "dipole" function

$$G_M^p(Q^2) = \mu_p G_D : \quad G_D = (1 - Q^2/M_V^2)^{-2} = (1 + 4.97\tau)^{-2}$$

where $M_V = 0.843$ GeV. It is also a good approximation that

$$G_E^p = G_D, \quad G_M^n = -\mu_n G_D$$

which is valid to several percent for the low Q^2 values of interest here. Data presently exists that is more accurate than the dipole approximation and can be used if needed. Recent data from JLab on G_E^p indicate that the dipole approximation fails severely for $Q^2 \gg 1(\text{GeV/c})^2$. Information on G_E^n is more of a problem. Present data is described by the "Galster fit" [15]

$$G_E^n(Q^2) = -\tau\mu_n G_D G_n : \quad G_n = (1 + 5.6\tau)^{-1}. \tag{73}$$

The error in this form factor is hard to estimate, but probably lies between 20-50% of the Galster fit. This places a limit on the interpretability of precise experiments at forward angles where the contribution of G_E^n to A^{PV} is large. Fortunately, experiments are planned that will greatly improve our knowledge of G_E^n.

A second problem is the uncertainty in the radiative correction for the axial form factor R_A. This error is much larger than the contribution of F_A^s, so it is probably impossible to measure F_A^s by electron scattering. We note that $F_A^s(0) = 2\Delta s$, the contribution of strange quarks to the nucleon made famous by the "spin crisis." Rather, elastic neutrino scattering is the best way to measure F_A^s [16]. Instead, $(1 - R_A)G_ApZ$ becomes an unknown effective form factor that must be measured to interpret A^{PV} except at the most forward angles.

A third problem is that measuring three angles doesn't work well. Forward angles provide one combination and backward angles measure another, but intermediate angles measure approximately a linear combination of the forward and backward results. Extracting three quantities greatly inflates the errors, and the third measurement is of little use.

There are two other strategies for separating the strange form factors. One, used by the SAMPLE collaboration, is to measuring quasi-elastic scattering from deuterium [17]. Another approach is to scatter from a nucleus with spin and isospin of 0. The asymmetry is [18,19]

$$A^{PV} = \frac{G_F Q^2}{\pi\alpha\sqrt{2}}\left[\sin^2\theta_W + \frac{G_E^s}{2(G_E^p + G_E^n)}\right].$$

This expression follows from Eq. 24 with the additional assumptions that the nuclear form factors are a product of the nucleon form factor times another form factor arising from the shape of the nucleus.

3 Experiments

3.1 Introduction

To date, five experiments have been published that use polarized electron beams to study parity violation. Several new parity experiments are being planned or are at various stages of completion. These experiments are listed in Table 5. The first such experiment, SLAC E122 [20], studied deep inelastic scattering from deuterium. It observed a convincing non-zero asymmetry that led to the general acceptance that indeed the Z couplings are parity-violating and that the standard model is the correct theory from among a number of models viable at that time. The next two experiments published, one performed at Mainz[21] and the other at Bates [22], were also planned as tests of the standard model and gave results consistent with theory.

These experiments are relevant for the present discussion in that they developed many of experimental techniques that are used for parity experiments today. Presently, most of the activity in the field, as seen in Table 5, is devoted to searching for strange form factors in the nucleon. The SAMPLE and HAPPEX experiments, which will be the main examples discussed below, have recently published data that are sensitive to strange quarks in the nucleon.

The striking feature of all of these experiments is the small size of the asymmetry measured, ranging from $\sim 10^{-4} - 10^{-7}$. They are done at accelerator facilities that are designed to measure various form factors with errors typically at the percent level or bigger. Hence measuring these small asymmetries present two unique problems:

Table 5: History of parity experiments.

Experiment	Reaction	Physics	A^{PV}
	Published Experiments		
SLAC E122 [20]	$\vec{e}D$ (DIS)	PV of Z	10^{-4}
Mainz [21]	$\vec{e}\,{}^9$Be QE	New Physics	10^{-5}
Bates [22]	$\vec{e}\,{}^{12}$C Elastic	New Physics	10^{-6}
SAMPLE(Bates) [13]	$\vec{e}P$ Elastic	$G_M^s(0) = \mu_s$	10^{-5}
HAPPEX(TJNAF) [12]	$\vec{e}P$ Elastic	$G_M^s + 0.39G_E^s$	10^{-5}
	Approved Experiments		
Mainz [23]	$\vec{e}P$ Elastic	G_M^s, G_E^s	10^{-5}
G_0(JLab) [24]	$\vec{e}P$ Elastic	G_M^s, G_E^s	10^{-5}
^{4}He(JLab) [25]	$\vec{e}\,{}^4$He Elastic	G_E^s	10^{-5}
Moller(SLAC) [26]	$\vec{e}e$	New Physics	10^{-7}
	Possible new experiments		
(JLab)	$\vec{e}P$	New Physics	10^{-7}
(JLab) [27]	$\vec{e}$ Pb	Neutron Radius	10^{-6}

1. Collection of a sufficient number N of events so that the statistical error $\Delta A^{PV} \sim 1/\sqrt{N}$. For the experiments in the Table, $N \sim 10^{10} - 10^{16}$.

2. Control of systematic errors at an unprecedented level.

Accelerator facilities do measure cross sections corresponding to form factors squared at the $10^{-10} - 10^{-12}$ level. Thus they automatically have sufficient luminosity to generate the required number of events at kinematics with Q^2 values where the form factors are near unity and the event rates are correspondingly higher. It turns out that many cases of high physics interest involve large form factors. For example, strange form factors are probably most interesting at low Q^2.

By their nature, asymmetries can be measured with much smaller systematic errors than cross sections. The reason is that an asymmetry is a *ratio* in which many effects, such as target thickness, solid angle acceptance, and electromagnetic radiative corrections, cancel. Moreover, the helicity of the beam is reversed rapidly, typically between 15 and 300 Hz. An asymmetry is measured

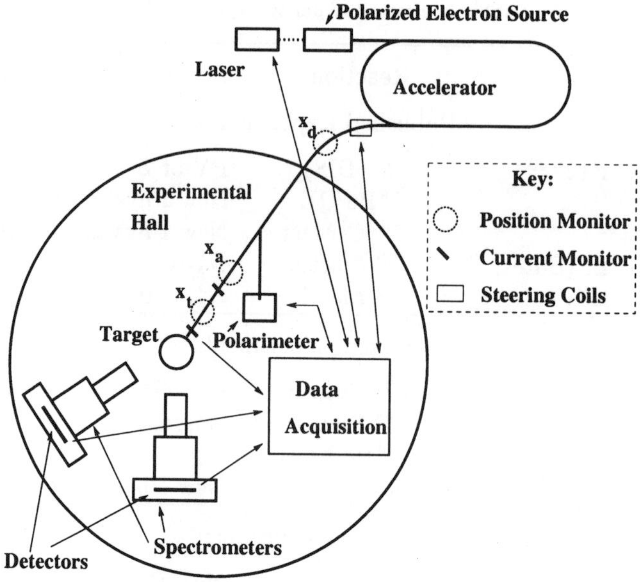

Figure 6: A generic parity experiment. The entire facility: electron source, accelerator, beam line, target, spectrometers and detectors all must work in concert to measure small asymmetries

each cycle

$$A_i = \frac{\sigma_i^R - \sigma_i^L}{\sigma_i^R + \sigma_i^L} \tag{74}$$

and the A_i's are averaged over the run. By using this procedure, slow drifts in any parameter, such as gains in amplifiers, also cancel.

One important effect that does not cancel in the asymmetry, however, is any correlation between the helicity of the beam and any other beam parameters such as intensity, position, angle, or energy. A big part of the experimental challenge of this field is ensuring that these correlations are small and that their effects are well understood.

A generic parity experiment is shown schematically in Fig. 6. Maintaining small helicity correlations and measuring the effects of the remaining correlations requires treating the entire facility from polarized source, through the accelerator, and into the experimental hall as a single entity. The experimental data-acquisition system must communicate with and in some cases control the accelerator in ways unique to parity experiments. Parity experiments are more like table-top experiments in atomic physics than traditional high-energy experiments.

3.2 Polarized Electrons

The heart of parity-violation experiments in electron scattering is the polarized electron source. The requirements of an ideal source are high intensity, high polarization, rapid reversal of helicity, and insensitivity of any beam parameters to helicity reversal. The technique of optical pumping of GaAs, which has been developed over many years, is approaching the ideal.

A schematic diagram of a typical source is given in Fig. 7. Electrons in the crystal are excited into the conduction band by light from a laser. Due to the band structure of GaAs, the polarization of the excited electrons depends on the circular polarization of the laser. The surface is carefully prepared to have a negative electron affinity which implies that the polarization electrons in the conduction band escape into the vacuum when they reach the surface. They are then accelerated electrostatically and injected into the main accelerator.

The polarization of the electrons is then determined by the polarization of the laser, which in turn is controlled by a Pockels cell. A Pockels cell is a birefringent crystal whose birefringence is controlled by an applied high voltage. When inserted into a linearly polarized light beam and set for ±quarter-wave retardation, the result is light of ±circular polarization. The high voltage can be switched quickly, even between beam bursts on a pulsed accelerator. At a CW facility, a time "window" is chosen and the Pockels cell voltage is changed between windows.

Various patterns of reversal are possible. If there is a source of noise at a specific frequency, a pattern can be chosen so that A_i is insensitive to that frequency. For example, the 60 Hz line frequency often causes 60 Hz noise to occur in the beam. Choosing 33 ms windows locked to the line frequency eliminates this noise and its harmonics. The helicity for each window may be chosen randomly. This procedure ensures that noise at any frequency is uncorrelated with helicity. At JLab, we use a slight variation in which windows are grouped in pairs. The helicity of the first is chosen randomly and the second helicity is the complement of the first. Then an asymmetry A_i may be computed for each pair. This procedure simplifies the statistical analysis.

Ideally, changing the voltage on the Pockels cell changes no laser beam parameters except for the helicity. In practice, the intensity and to some extent the position of the laser on the cathode is correlated with helicity. Careful alignment of the Pockels cell is important, but even the best procedures found to date are imperfect.

The biggest problem, but one that can be fixed, is intensity. Without

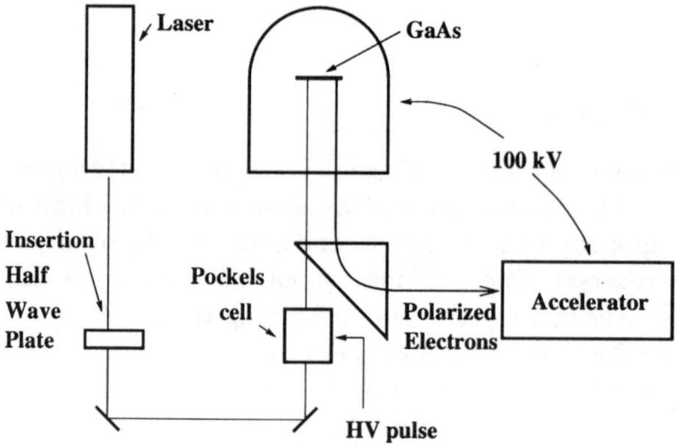

Figure 7: Diagram of a typical polarized electron source.

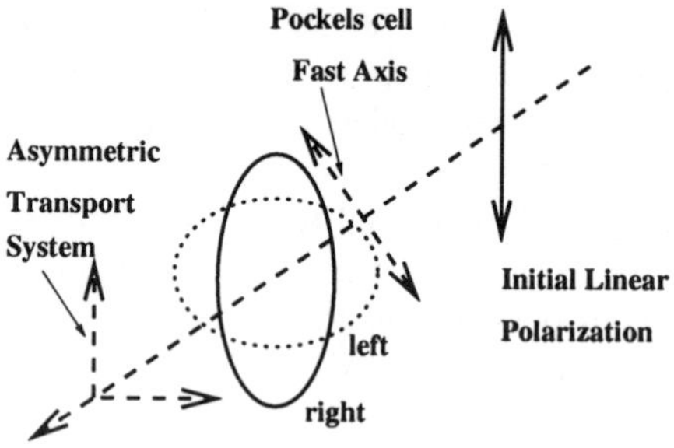

Figure 8: Polarized Induced Transport Asymmetry.

intervention, intensity correlation with helicity is typically several hundred ppm or more. It is desirable to have the correlation smaller than A^{PV} or a few ppm.

One mechanism for intensity correlations is shown in Fig. 8. The Pockels cell is not a perfect quarter wave plate, and there is residual linear polarization that changes with helicity. The transmission efficiency of light usually depends on its linear polarization, so the difference in linear polarization results in a difference in intensity. We call this the Polarization Induced Transport Asymmetry (PITA) effect. The PITA effect tends to vary in time over periods of hours, perhaps due to temperature variations.

The intensity correlation ΔI is linear in the difference ΔV in the magnitudes of the HV applied to the Pockels cell for each helicity. Thus $\Delta I = \alpha \Delta V$

where α is called the PITA slope. The PITA slope is reasonably stable. Consequently, the PITA effect may be exploited in a feedback loop. Data are obtained over a convenient period of time Δt. Then ΔI is computed and the HV control on the Pockels cell is modified so that $\Delta V_{new} = \Delta V_{old} - \Delta I/\alpha$. This procedure works well and net helicity correlations of less that 1 ppm are easy to obtain.

The Pockels cell also steers the position of the laser on the cathode in a helicity-correlated manner. This deflection effect can be ameliorated by using a lens chosen so that the beam angle instead of position changes at the surface of the cathode or by using mirrors mounted on piezo-electric crystals that are part of a feedback system.

This position correlation presently appears to be the main limitation to ultra-high precision parity experiments. It would be useful to better understand the physics behind the problem and develop a better solution.

3.3 Corrections for Nonzero Helicity Correlations

The helicity correlated differences in beam parameters presumably originate with differences in laser intensity and position at the cathode. These differences propagate through the accelerator and result in helicity correlations with intensity, position angle, and energy at the target. The function of the beam monitors is to measure these differences and make corrections to the asymmetry if necessary.

Typical beamline instrumentation, illustrated in Fig. 6, includes intensity monitors I and horizontal(vertical) position monitors $x_t(y_t)$, $x_a(y_a)$, and x_d. The position at the target is given by $x_t(y_t)$. The difference between $x_t(y_t)$ and $x_a(y_a)$ gives the beam angle. The energy is monitored by the position of the beam x_d located at a point of high dispersion such as in the middle of an arc transporting the beam into the experimental hall. The five monitors are alternatively denoted x_i. The helicity-correlated differences in these quantities are denoted by Δx_i, etc.

Devices for measuring intensity are toroids for pulsed machines and resonant microwave cavities in CW machines. Position is measured by cavity or stripline monitors. These devices give a signal proportional to the beam intensity times the position relative to the axis of the device.

The cross section is then defined to be (to within normalization factors that cancel in the asymmetry)

$$\sigma^R = \frac{D^R}{I^R}; \quad \sigma^L = \frac{D^L}{I^L} \tag{75}$$

where D is the detector signal, and the superscript refers to the beam helicity. By this means, the direct effect of any helicity correlated intensity difference is eliminated from the asymmetry.

Similarly, the effect of position differences can be corrected by using the equation:

$$\Delta\sigma = \Delta\sigma_{raw} - \sum_i \frac{\partial\sigma}{\partial x_i}\Delta x_i \tag{76}$$

Typical values for $\partial\sigma/\partial x_i$ are on the order of a few ppm/μm. Eq. 76 assumes that the measured value of σ depends linearly on five beam parmameters, the position, angle (in both the x and y directions) and energy, and that the positions x_i are linear combinations of the five beam parameters.

The correction coefficients $\partial\sigma/\partial x_i$ can be measured by modulating beam steering coils C_j in the beam transport line and by modulating the energy of the beam. Then the correlations $\partial\sigma/\partial C_j$ and $\partial x_i/\partial C_j$ are measured. Then

$$\frac{\partial\sigma}{\partial x_i} = \frac{\partial\sigma}{\partial C_j}\frac{\partial C_j}{\partial x_i}; \quad \frac{\partial C_j}{\partial x_i} = \left(\frac{\partial x_i}{\partial C_j}\right)^{-1} \tag{77}$$

where the inverse refers to matrix inversion. This modulation can be done while the primary experiment is running. The effect of the modulation is small compared to the statistical noise in σ for a given window and the random helicity reversal ensures that the effect averages to zero. However, the correlation between the coils and σ can clearly be seen when averaged over the entire run. This procedure has two advantages:

1. The correction coefficients are measured under the exact conditions under which the data are obtained. For example, if the correction coefficients depend on the average beam position, it does not matter.

2. No extra beam time is required to measure the corrections. This permits parity experiments to be performed limited only by statistics.

It is desirable to have the helicity correlations as small as possible for the following reasons:

1. Small corrections make the result convincing.

2. If the corrections are large, higher order terms, which are very difficult to correct for, are likely to become significant.

Another problem is that differences in intensity often result in differences in energy due to beam loading effects in the accelerating cavities. The electrons are accelerated by the RF voltage. However accelerating the electrons requires energy which comes from decreasing the RF power in the cavities and consequently decreasing the RF voltage applied to following electrons. The changed RF fields can directly cause position differences or induce them because of dispersion in the beam line.

Differences in position can cause differences in intensity if beam is lost by scraping on some slit. Beam lost at a slit can also cause helicity-dependent backgrounds. The result is additional complicated helicity correlations in the parameters of the beam delivered to the target. Hence it is desirable to have all correlations as small as possible.

3.4 Beam Noise and Electronics

The ability to perform a parity experiment efficiently depends in part on minimizing the jitter in the accelerated beam. The issues are illustrated in Fig. 9. The upper left panel shows the beam intensity as a function of time. The jitter in the data is at the percent level, which is good for an accelerator beam. The upper right panel gives the distribution of current asymmetries for pulse pairs:

$$A(I) = \frac{I^R - I^L}{I^R + I^L}. \tag{78}$$

The RMS width $\sigma[x]$ of any quantity x is defined by,[b]

$$\sigma[x] \equiv \sqrt{\langle (x^R - x^L)^2 \rangle} \tag{79}$$

which we denote $\sigma[A(I)]$ is only 150 ppm. This width is small because the drifts in intensity occur slowly compared to the flipping frequency.

The width $\sigma[a(I)]$ is the essential parameter that determines how precisely the helicity correlation can be measured:

$$\frac{\Delta I}{I} = \frac{\sigma[A(I)]}{\sqrt{N}} \tag{80}$$

where N is the total number of pulse pairs. Thus the current asymmetry at JLab can be measured to 1.5 ppm in 10,000 pulse pairs or about 10 min. This low noise allows us to quickly evaluate intensity correlations at the desired level of sensitivity.

The bottom left plot shows the correlation between two identical current monitors in the beam. They are highly correlated because both monitors are

[b] Here we have assumed that the noise per window is much greater than the average: $\Delta x \ll \sigma[x]$.

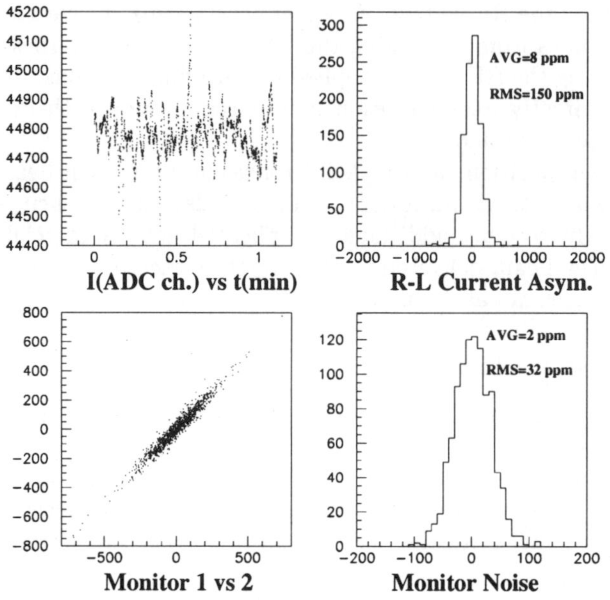

Figure 9: Data on beam current monitors at JLab.

measuring the same intensity jitter. The distribution of differences in the asymmetries of the two monitors is graphed in the bottom right panel. This provides a measure of the noise in the electronics. It is only this electronic noise and not the beam jitter that contributes noise to A_i as defined in Eqs. 75 and 74. Therefore the electronic noise must be smaller than the noise due to counting statistics. For studying the beam, it is also useful to have the electronics quieter than the beam jitter as is the case in Fig. 9.

The ~ 200 ppm beam intensity jitter at JLab as shown in Fig. 9 is unusually small. Normally, accelerators run satisfactorily with noise in the current asymmetry 100 times greater. Unfortunatetly, with 1% beam noise, it is more difficult to control the helicity correlations to the desired level.

Just like the case for intensity, the spread $\sigma[\Delta x_i]$ in the measured beam position differences for each window also has contributions both from real jitter in the beam position and from electronic noise. By analogy, small position jitter facilitates identifying and reducing helicity correlations. Small electronic noise reduces the error in the corrections of Eq. 76. At JLab, the position noise of the beam per window is on the order of 10μ and the monitors have electronic noise at the 2μ level, providing for excellent monitoring of these systematic effects. The average correlations Δx_i can be as small as about 3 nm.

3.5 Target

A target for a parity experiment has special requirements:

1. The target must be able to withstand the high beam currents required to produce sufficient statistics. For SAMPLE and HAPPEX, on the order of 0.5 kW of power is deposited in the target due to beam heating, yet the temperature must be maintained at $\sim$ 19 K.

2. The thickness should be independent of the position of the beam to minimize sensitivity to helicity-position correlations.

3. The thickness must be constant in time. For a cryotarget such as liquid hydrogen in an intense beam, this implies that any density fluctuations due to boiling must be negligible.

The effect of target boiling on the width of the asymmetry distribution $\sigma[A_i]$ is

$$\sigma[A_i] = \sqrt{\frac{1}{N_i} + \left(\frac{\sigma[l]}{l}\right)^2}$$

where N_i is the number of events integrated during the window, l is the effective target length and $\sigma[l]$ is the RMS fluctuation in l for an average window. Thus we require $\sigma[l]/l \ll 1/N$.

There are two ways to determine the effects of boiling. The first is to measure the width of the asymmetry $\sigma[A_i]$ as a function of intensity I as shown in Fig. 10. If there is no boiling and $\sigma[l]/l$ is negligible,

$$\sigma[A_i] \propto \sqrt{I^{-1}} \tag{81}$$

A plot of $\sigma[A_I)]$ vs I from HAPPEX is shown in Fig. 10. For this test, extremely high counting rates were obtained by running at a low beam energy. Here we see that Eq. 81 holds down to quite small values of $\sigma[A_i]$. This test proved that the boiling is negligible at the lower rates corresponding to the 3.3 GeV energy of the HAPPEX experiment.

The second test can be performed simultaneously while taking data. It requires that there be two (or more) independent detectors measuring asymmetries A_1 and A_2.. Then a correlation between the asymmetries may be

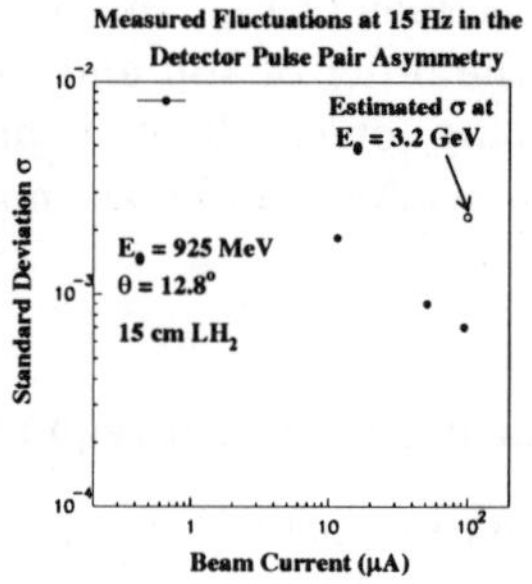

Figure 10: Noise in the standard deviation of the asymmetry $\sigma[A_i]$ as a function of beam current.

measured:

$$C_B \equiv \frac{\langle A_1 A_2 \rangle}{\sigma[A_1]\sigma[A_2]} = \frac{(\sigma[l]/l)^2}{1/N + (\sigma[l]/l)^2}.$$

If C_B is small, boiling is negligible.

3.6 Spectrometers and Detectors

Spectrometers for parity experiments have two main requirements:

1. Have sufficient solid angle to collect the necessary events.

2. Isolate events from the desired reaction.

We will start with a discussion of solid angle. The running time for an experiment is inversely proportional to the "statistical figure of merit" M_s defined by

$$M_s = \frac{d\sigma}{d\Omega} \Delta\Omega P_e^2 (A^{PV})^2 \sim \frac{1}{T} \qquad (82)$$

where P_e is the beam polarization and T is the total running time. The solid angle $\Delta\Omega \sim \Delta \cos\theta \Delta\phi$ where the variables are shown schematically in Fig. 11. For scattering at small angles, Q^2 and hence the asymmetry varies rapidly with angle θ. Thus the spectrometer must have a finite acceptance $\Delta\theta$ corresponding

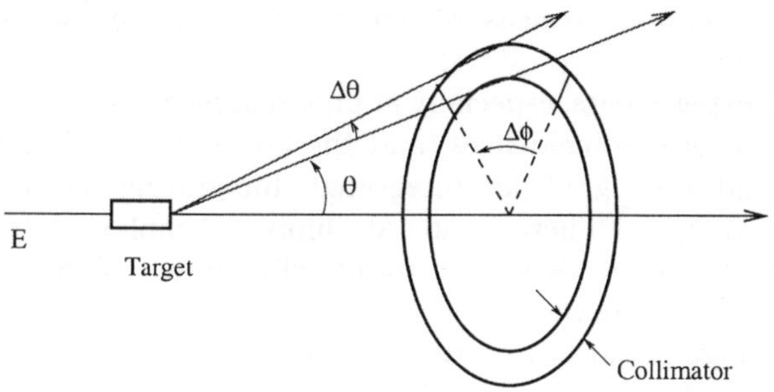

Figure 11: Geometry for computing the figure of merit for forward scattering

to a finite $\Delta Q^2/Q^2 \equiv b \leq 0.3$. Larger values of b correspond to measuring several kinematic points at once, an approach to be used by G_0. Assuming further $A^{PV}/Q^2 =$Const, $d\sigma/d\Omega = \sigma_{Mott}|F|^2$, and $\sigma_{Mott} = E^2\cos^2(\theta/2)/Q^4$, M_s is

$$M_s \sim Q^2|FF|^2 b\cos^2(\theta/2)\Delta\phi. \tag{83}$$

The issues for the design of a spectrometer in a paritiy experiment are

1. Maximize $Q^2|FF|^2$. For elastic nuclear (nucleon) scattering, this occurs at $Q^2 \sim 0.1(1.0)(\mathrm{GeV/c})^2$
2. Maximize $\Delta\phi$
3. Choose a beam energy E large enough so $\cos^2(\theta/2) \sim 1$
4. Achieve energy resolution ΔE sufficiently good to reject backgrounds. For elastic scattering from nuclei, resolution on the order of several MeV is appropriate since nuclear inelastic levels for favorable cases are several MeV below the elastic peak. For elastic nucleon scattering, π^0 production is the first inelastic process, and resolution of only about 50 MeV is ample.

Meeting these requirements for the various experiments in Table 5 has led to quite a variety of spectrometers. Examples include traditional magnetic spectrometers (SLAC E122 and HAPPEX), a large solid angle air Cerenkov counter (Mainz and SAMPLE), and a toroidal spectrometer (G_0).

In traditional experiments with magnetic spectrometers, an electron is identified by a coincidence between a tracking detector, a gas Cerenkov counter, and a large pulse in a shower counter. The momentum of the electron is determined from the details of the track and a knowledge of the magnetic fields in the spectrometer. These techniques are very powerful for rejecting background and permit the measurement of very small cross sections and hence small form factors.

For parity experiments, especially at pulsed accelerators, the counting rates are too high to form coincidences, and the above powerful technique is not possible. Instead, the signals are integrated. Integrating the signal does have important advantages. There is no rate limit. Complex corrections due to dead time and rate dependence are totally eliminated. The one drawback is poor rejection of background.

One advantage of parity experiments measuring elastic scattering at kinematics where the form factors are large is that the rates are high, so other processes are swamped by comparison. Also for elastic scattering, the electron has the highest possible energy of any particle scattered at that angle. There are no radiative tails to contaminate the data sample. Given these features, it is possible to integrate the signals and still achieve small backgrounds.

The detectors for parity experiments have special requirements. When a signal is integrated, the counting statistics are:

$$\sigma = \frac{1}{\sqrt{N}}\sqrt{1+\left(\frac{\Delta E}{E}\right)^2} \tag{84}$$

where ΔE is the energy (E) resolution. Once the energy resolution is $< 15\%$, there is negligible degradation of counting statistics. One implication of this formula is that it is essential that response of the detector not have high energy tails. For example, thin scintillators with their large Landau tails with $\Delta E_i \gg E$ have large values for ΔE and result in significant loss of statistics.

Now that CW facilities are available at JLab and Mainz, some parity experiments with great segmentation are counting events and even using coincidence methods. Examples include the G_0 experiment [24] at JLab and the $\vec{e}p$ experiment [23] at Mainz.

We will now illustrate the above principles with two examples, HAPPEX and SAMPLE. The design of the HAPPEX experiment is based on two high-resolution spectrometers each with fixed solid angle of 5.5 mr. When the spectrometers are set at their most forward angle of 12.5°, the 11 msr total solid angle is a respectable fraction of the maximum possible solid angle of ~50 msr for $b = 0.3$. For much larger scattering angles, however, the solid angle is poor. For fixed angle, $M_s \sim E^2$. Thus, decreasing the beam energy is also costly in statistics. For optimum statistics, the spectrometers must be run at the most forward angles and at the highest possible energy.

The maximum energy of 3.2 GeV corresponds to $Q^2 \sim 0.47$ (GeV)^2. The energy resolution of the spectrometer is ample to reject electrons from any inelastic process, even single pion production. Thus, there is no drawback to

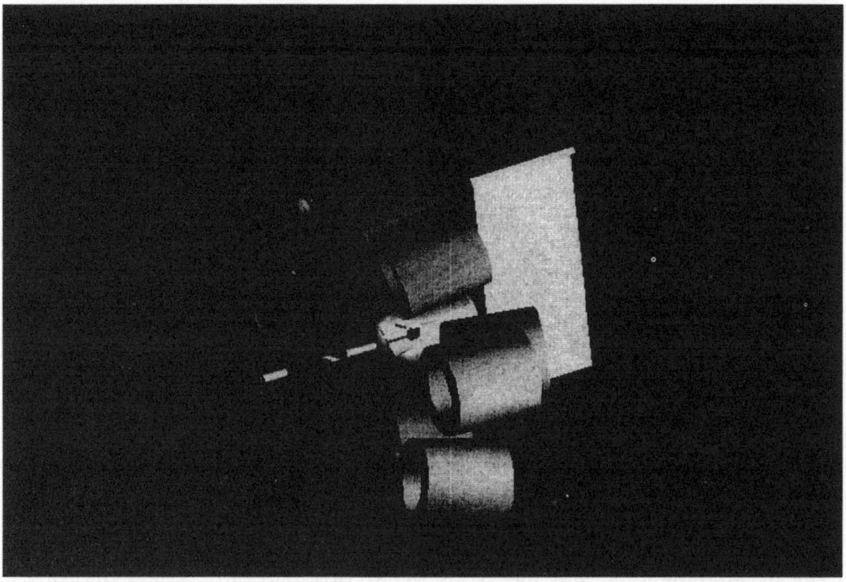

Figure 12: The SAMPLE apparatus. The beam impinges from the left as shown. The curved surfaces on the left are the Cerenkov mirrors. The heavy cylinders are the phototube shields. The domed structure in the middle is the high-power hydrogen target, and the box behind encloses the heat exchangers for the target.

integrating the signal. The detector chosen was a sandwich of five layers each of lead and lucite. The total thickness was about 12 radiation lengths, so most of the energy of the scattered electrons is deposited in the detector. Since the light is produced by Cerenkov radiation, low energy particles cannot produce background. The detector is thinner and less segmented than traditional designs, but the achieved resolution of 15% is appropriate in light of Eq. 84. The detector also produces many photoelectrons, swamping single photoelectrons from the phototube arising from dark current or due to interactions of low energy particles with the photocathode.

For backward angle scattering as in the SAMPLE experiment, Q^2 varies slowly with angle. The arguments of Eq. 83 are irrelevant. Rather, a very large solid angle is desirable. SAMPLE has achieved a remarkable ~ 2 sr by using an array of mirrors to collect Cerenkov light from the scattered particles passing through air. The apparatus is shown in Fig. 12. Since the beam energy is chosen to be 200 MeV, pion production is almost forbidden by kinematics, so there is little background from scattered high-energy particles. One drawback

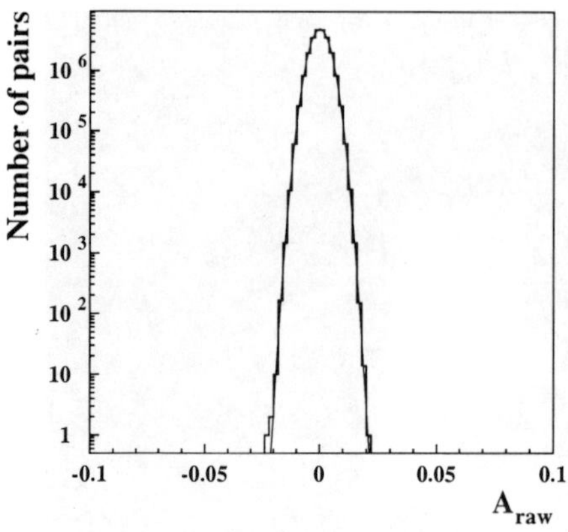

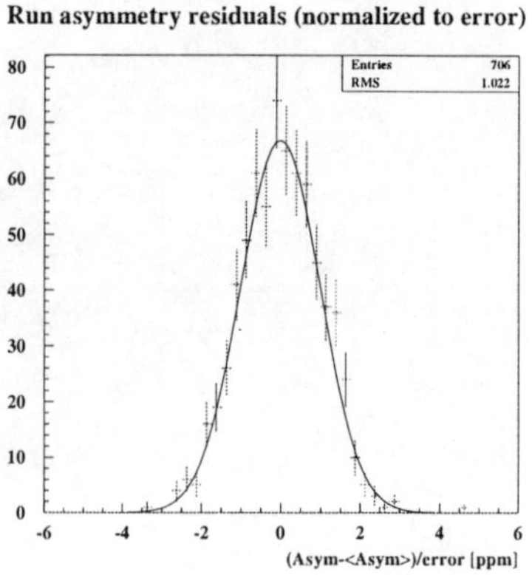

Figure 13: Left: Distribution of A_i for the HAPPEX run. There are 23 million entries. Right: Distribution of average asymmetries for ~1 hr HAPPEX runs normalized to the predicted error. There are 706 runs.

to the arrangement is that electrons in air produce only a few photons and the phototubes detecting these photons are difficult to shield. Thus there is significant background from low energy particles produced in the target. Fortunately, these particles are produced independently of helicity, and the correction is straightforward.

3.7 Statistical and Systematic Tests

During a typical parity experiment, the asymmetry A_i is measured frequently, often more that 10^7 times. This vast amount of data on the same quantity allows for some interesting and useful statistical tests. We will illustrate the basic ideas with data from the HAPPEX experiment.

The distribution of asymmetries A_i for individual pairs of windows is shown in a log plot in the left panel of Fig. 13. The shape is Gaussian over seven orders of magnitude. This ideal behavior, which is due to the excellent stability of the JLab facility, simplifies the analysis by minimizing the need for cuts. etc.

The statistical error in A_i is given by the width of the distribution $\sigma[A_i]$. We note that this procedure is equivalent to choosing σ so that the χ^2/degree of freedom (dof) is one.

The data are divided into $\sim$ 700 runs of about one hour duration that have $N_j^{run} \sim$ 50,000 values of A_i. For each run j, σ_j^{run} is computed. With this many events, σ_j^{run} is determined to better than 1%. The error in σ_j^{run} can be computed by noting that for a large number of degrees of freedom n:

$$\frac{\delta\chi^2}{\chi^2} = \frac{1}{\sqrt{2n-1}}$$

and the error in the χ^2 governs the error in σ when the χ^2/dof is set to one.

The average asymmetries over the N_j pairs in run j are computed:

$$A_j^{run} = \langle A_i \rangle \pm \sigma^{run}; \quad \sigma_j^{run} = \sigma[A_i]/\sqrt{N_j}.$$

The plot the distribution of A_j^{run}/σ_j^{run} for all runs in the experiment is shown in the right panel of Fig. 13. The result is another perfect Gaussian. The width is unity as expected to within a few percent, showing ideal statistical behavior. The error in this width is again related to $\delta\chi^2$, which is small due to the large number (700) of points. This convincing test requires $> 10^6$ events since it takes >1000 events to get a good measure of $\sigma[A_i]$ for each run and $\sim$ 1000 runs are needed for the A_j^{run}/σ_j^{run} plot.

This result of the error decreasing as the square root of the number of measurements by more than a factor of 1000 assumes that the measurements are uncorrelated and that all relevant parameters are stable. For many laboratory experiments, these assumptions are violated and simply repeating measurements does not yield stable results. However, for parity experiments where we measure an asymmetry that cancels many slow drifts and flip the helicity randomly to eliminate correlated noise, this ideal statistical behavior is natural.

One important systematic check for electron parity experiments is the insertion of a half-wave plate into the laser beam as shown in fig 7. This procedure reverses the helicity of the electrons relative to normal operation and reverses the apparent sign of A^{PV}. Averaging the asymmetry with this sign included cancels any systematic errors that do not depend on the status of the half wave plate.

Data from HAPPEX as a function of half-wave plate setting are shown in Fig. 14. The status of the half-wave plate was reversed after obtaining data sets of approximately 40 runs. With 40 runs, the errors are small enough so that a nonzero asymmetry can be observed. The asymmetry indeed reverses sign as expected. In addition, the asymmetry averaged by ignoring the half-wave plate is zero as expected. This demonstrates that a large class of systematic errors are small.

For the data in Fig. 14, the χ^2 is only 6.7 for 13 degrees of freedom, rather far from the expected χ^2 of 13. However, with so few degrees of freedom, the statistical uncertainty in the χ^2 is high, and the plot does not provide a precise measure of the statistical quality of the data.

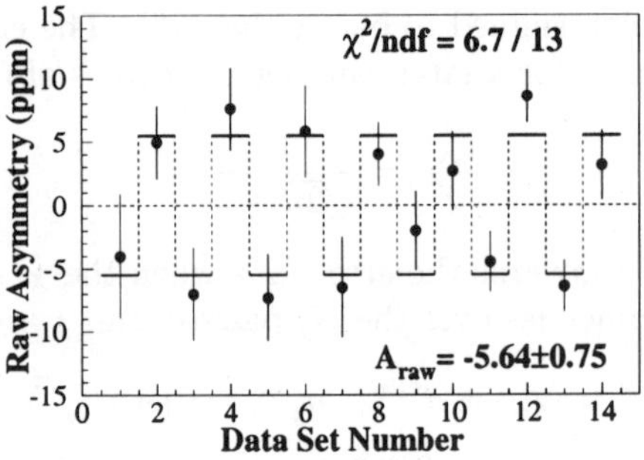

Figure 14: Average values of the asymmetry for each data set. Odd-numbered data sets have the half-wave plate inserted in the laser beam.

Some helicity correlations, such as the Δx_i, sometimes change sign when the half wave plate is inserted. Thus the half-wave plate insertion does not guarantee the absence of systematic errors. Rather, they must be evaluated by using Eq. 76. For HAPPEX, small values of Δx_i, typically < 3 nm, resulted in totally negligible corrections, providing strong evidence for the lack of spurious sources of asymmetry.

3.8 Polarimetry and other Corrections

In a parity measurement, the physics asymmetry A^{PV} is related to the raw experimental asymmetry $\langle A_i \rangle$

$$A^{PV} = \langle A_i \rangle \frac{1}{P_e} \left(1 + r \left[1 - \frac{A_B}{\langle A_i \rangle} \right] \right) \tag{85}$$

where P_e is the polarization of the beam, r is the ratio of signal to background and A_B is the background asymmetry.

Measuring P_e precisely is one of the major challenges of the experiments. The most common method is Moller scattering from polarized iron in a magnetized foil. The experimental asymmetry is large, typically several percent from practical targets. Statistics are easy to obtain. The problem is that there are many possible systematic errors, including measuring the magnetization of the foil, knowing the g-factor in the foil, and carefully controlling the kinematics. Achieving better than 5% systematic error is a challenge. With careful work, 1% errors might be achieved.

Another approach being developed at JLab is the Compton Polarimeter. For this device, the polarized electrons are scattered from polarized photons in a laser beam. In principle, this method is quite accurate in that the underlying physics of the scattering is well understood. A major advantage of the Compton Polarimeter is that it can be run simultaneously with data-taking with the parity experiment. Possible time variations of the polarization are easily studied. In contrast, the experiment must be stopped and the foil must be inserted for the Moller polarimeter. Polarization measurements are then performed every day or two at a cost of $\sim$ 2 hrs per measurement.

Mott scattering [28] of transversely polarized electrons from an atom is a third approach. This method works for energies as high as 5 MeV, and is suitable for measuring the polarization of the beam near the injector of the main accelerator. Since depolarization of the beam due to acceleration is expected to be small, this method is suitable. Ultimate precisions at the few percent level or so may be possible. Given the difficulty in assessing systematic errors in polarimetry, having different kinds of polarimeters is valuable.

Another correction indicated by Eq. 85 arises from backgrounds. Both the background fraction r and its asymmetry A_B must be determined. These measurements must be performed with auxiliary measurements. Designing an experiment with small or negligible backgrounds greatly increases the efficiency by minimizing the time required for these measurements.

Finally, A^{PV} depends at least linearly in Q^2. Thus the energy of the beam and average scattering angle must be carefully measured.

4 Strange Quarks: Theoretical Predictions and Experimental Results

Considerable theoretical effort has been devoted to estimating the size of the strange form factors [29]. Numerous approaches have been used, including lattice calculations [30], Skyrme models [32,33], chiral symmetry [34], pole fits [35,36,38], meson loops, NJL model [31], etc. None of these approaches can yet produce reliable estimations of the strange form factors. The main significance of this literature in my opinion is the demonstration that no matter how one tries to deal with the nucleon quantitatively, it is relevant to know whether or not the presence of strange quarks is important.

The estimates of strange form factors that appear in the above literature do serve as an interesting guide to the experimenter. Experiments that rule out models, even those with the largest strangeness contributions, provide useful information about strangeness in the nucleon.

Most of the calculations focus on the properties of the strange form factors

Table 6: Some published theoretical estimates for ρ_s and μ_s.

Method	ρ_s	μ_s	Author
Pole fits	-2.1 ± 1.0	-0.31 ± 0.09	Jaffe [35]
	-2.9 ± 0.5	-0.24 ± 0.03	Hammer [36]
Kaon Loops	0.2	-0.03	Koepf [37]
	0.5 ± 0.1	-0.35 ± 0.05	Ramsey-Mulolf [39]
	0.3	-0.12	Ito [40]
Unquenched quarks	0.6	0.04	Geiger [41]
Meson Exchange	0.03	0.002	Meissner [38]
NJL	3.0 ± 0.08	-0.15 ± 0.10	Weigel [31]
Skyrme	1.6	-0.13	Park [32]
	-0.7	-0.05	Park [33]
Lattice QCD	1.7 ± 0.7	-0.36 ± 0.20	Dong [30]

near $\tau = 0$ where

$$\mu_s = G_M^s(0); \quad \rho_s \equiv \left.\frac{dG_E^s}{d\tau}\right|_\tau = 0.$$

Another parameter is the strange charge radius

$$\langle r_s^2 \rangle = -6\frac{dF_1^s}{dQ^2}; \quad \rho_s = -\frac{2}{3}M_p^2\langle r_s^2 \rangle - \mu_s.$$

Predictions from representative models in terms of these parameters are given in Table 6.

A number of references [36,31,33,30] also give the strange form factors as a function of Q^2. One paper [7] suggests the following convenient form inspired by the Galster fit (Eq. 73) of G_E^n:

$$G_E^s(Q^2) = \rho_s \tau G_D (1 + \xi_s \tau)^{-1} : \quad G_M^s = \mu_s G_D \tag{86}$$

where ξ_s is an unknown parameter that might be +5 or even larger. We have chosen to use this form to extrapolate predictions for which no Q^2-dependence is given. This procedure is questionable; it is conceivable that there is a zero crossing in the strange form factors below $Q^2 \sim 1(\text{GeV/c})^2$.

4.1 Results from SAMPLE and HAPPEX

The SAMPLE experiment measured A^{PV} for a beam energy of 0.20 GeV and a scattering angle $130° < \theta < 170°$. The kinematics were chosen to provide

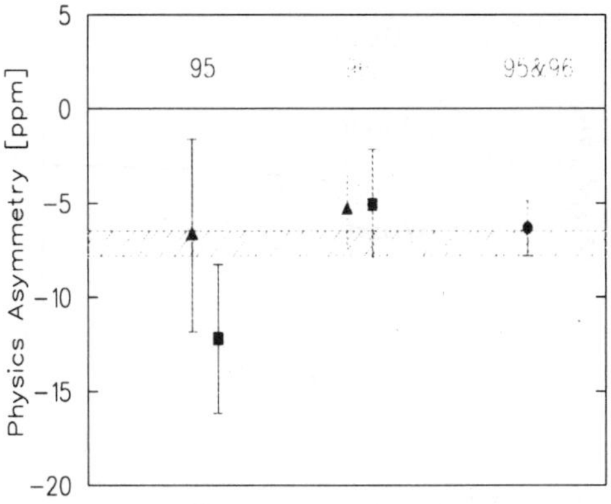

Figure 15: Results of the SAMPLE experiment. The shaded band is the theoretical prediction in the absence of strange quarks. The width of the band is given by the error in the radiative correction R_A

good sensitivity to μ_s. The published result is [13]

$$A^{PV} = -6.34 \pm 1.45 \pm 0.53\text{ppm}$$

where the first error is statistical, and the second is the estimated systematic error. From this number the authors obtain

$$\mu_s = 0.23 \pm 0.37 \pm 0.15 \pm 0.19$$

where the third error arises from the uncertainty in the radiative correction R_A discussed in Section 2.8.

The SAMPLE collaboration has recently obtained significantly more data and is planning to also measure the asymmetry for quasi-elastic scattering from deuterium to measure R_A. The anticipated final error on μ_s is ± 0.27.

The HAPPEX experiment performed a measurement at forward angles, where the asymmetry is very sensitive to predicted values of G_E^s if Eq. 86 is valid with a small value of ξ_s. Moreover, the contribution from R_A is negligible. Thus, it is an excellent place to search for strange form factors.

The HAPPEX experiment measured the asymmetry

$$A_{exp} = -14.5 \pm 2.0(stat) \pm 1.1(syst)\text{ppm}$$

at $Q^2 = 0.48$ (GeV/c)^2 and an angle of 12.5°. This corresponds to a measurement of strange form factors as follows:

$$G_E^s + 0.39 G_M^s = 0.023 \pm 0.034\ (stat) \pm\ 0.022\ (syst) \pm\ 0.026\ (\delta G_E^n).$$

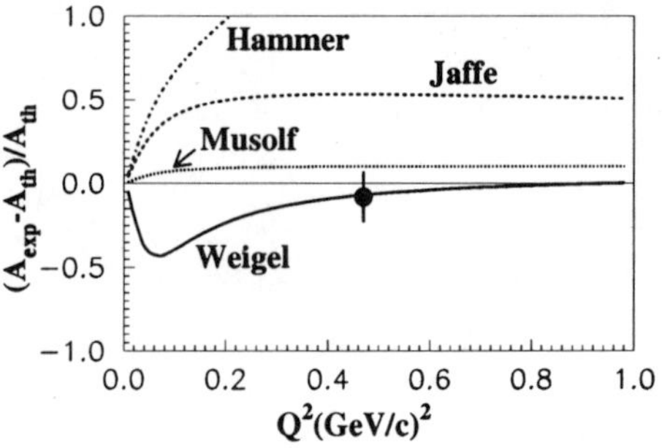

Figure 16: Results of the HAPPEX experiment relative to A_{th}, which is the theoretical prediction in the absence of strange quarks. Also shown are predictions for the additional contributions due to the presence of strange form factors predicted by selected models.

Again, the result is consistent with negligible contribution from strange quarks. The result is shown together with predictions from representative models including Hammer et al. [36], Jaffe [35], Musolf and Burkardt [39], and Weigel et al. [31] For the models labeled Musolf and Jaffe, we assume a Q^2 dependence as given by Eq. 86. The uncertainty due to G_E^n is also neglected in the figure. HAPPEX is expected to ultimately halve these errors. Future experiments at JLab and elsewhere should greatly reduce the error in G_E^n.

It is interesting to speculate on what the HAPPEX result tells us about ρ_s. If one assumes:

1. $\mu_s \sim 0$.
2. Eq. 86 is valid.
3. $\xi_s = 0$.
4. G_E^n is given by Eq. 73.

Then a value for ρ_s may be obtained. The result is $\rho_s = 0.4 \pm 0.8$. Thus if there are large strange form factors, one of the above assumptions must be wrong.

4.2 Future Directions

A number of parity experiments, as listed in Table 5, are planned for the future to continue the search for strange form factors. The G_0 collaboration at JLab plans to measure A^{PV} to high precision over the Q^2 range 0.15-1 $(\text{GeV}/c)^2$. The HAPPEX collaboration is proposing experiments at $Q^2 \sim 0.1$ $(\text{GeV}/c)^2$ for both hydrogen and ^{4}He. An experiment at Mainz is also measuring A^{PV} at $\theta \sim 35°$. It is reasonable to anticipate that these experiments will in the near future either establish the presence of strange form factors or set impressive limits.

In addition, the technique of parity violation in electron scattering is being used for physics other than strange quarks. The E158 experiment [26] at SLAC is planning to test the standard model by studying Moller scattering [42]. The HAPPEX collaboration has proposed [27] to measure A^{PV} for ^{208}Pb in order to measure the radius of the neutron distribution in a complex nucleus [43,44].

In conclusion, parity violation in electron scattering is a practical and important tool in sub-atomic physics. It is the method of choice for measuring the strangeness content of the vector form factors.

Acknowledgments

This work is supported by the U.S. Department of Energy under Grant DE-FG02-84ER40146.

1. J. Ashman , *et al., Phys. Lett.* B **206**, 364 (1988), *Nucl. Phys.* B **328**, 1 (1989).
2. J. Gasser, H. Leutwyler, and M. E. Sanio, *Phys. Lett.* B **253**, 163 (1991).
3. J. Ellis, *et al., Phys. Lett.* B **353**, 319 (1995).
4. D. B. Kaplan and A. Manohar, *Nucl. Phys.* B **310**, 527 (1988).
5. R. D. McKeown, *Phys. Lett.* B **219**, 140 (1989).
6. E. J. Beise and R. D. McKeown, *Comments in Nucl. and Part. Phys* **20**, 105 (1991).
7. M. J. Musolf, *et al., Phys. Rep.* **239**, 1 (1994) and references therein.
8. For a textbook treatment, see H. Frauenfelder and E. M. Henley *Subatomic Physics*, Prentice-Hall, Englewood Cliffs, NJ, (1974).
9. See, for example, F. Halzen and A. D. Martin, *Quarks and Leptons: An Introductory Course in Modern Particle Physics*, John Wiley and Sons, NY, 1984.
10. For a recent discussion of this point, see N. Isgur, JLab preprint JLAB-THY-98-48.
11. Particle Data Group, C. Caso, *et al.*, Eur. Phys. J. C **3**, 1 (1998).
12. K. Aniol, *et al., Phys. Rev. Lett.* **82**, 1096 (1999).
13. B. Mueller, *et al., Phys. Rev. Lett.* **78**, 3824 (1997).
14. M. J. Musolf and B. R. Holstein, *Phys. Lett.* B **242**, 461 (1990).
15. S. Galster, *et al., Nucl. Phys.* B **32**, 221 (1971).

16. G. T. Garvey, *et al.*, *Phys. Rev.* C **48**, 1919 (1993).
17. E. Hadjimichael, G. I. Poulis, and T. W. Donnelly, *Phys. Rev.* C **45**, 2666 (1992).
18. D. H. Beck, *Phys. Rev.* D **39**, 3248 (1989).
19. M. J. Musolf, R. Schiavilla, and T. W. Donnelley *Phys. Rev.* C **50**, 2173 (1994).
20. C. Y. Prescott, *et al.*, *Phys. Lett.* B **84**, 524 (1979)).
21. W. Heil, *et al.*, *Nucl. Phys.* B **327**, 1 (1989).
22. P. A. Souder, *et al.*, *Phys. Rev. Lett.* **65**, 694 (1990).
23. Mainz propoosal A4/1-93 (D. von Harrach, spokesperson).
24. JLab experiment 91-017 (D. Beck, spokesperson).
25. JLab experiment 91-004 (E. J. Beise, spokesperson).
26. SLAC experiment E158 (K. S. Kumar, spokesperson, E. W. Hughes and P. A. Souder, deputy spokespersons).
27. JLab experiment 99-012 (R. Michaels and P. A. Souder, spokespersons).
28. N. F. Mott, *Proc. Roy. Soc. (London)* **A135**, 429 (1932); ibid. **A124** 425 (1929).
29. C. V. Christov *et al.*, *Prog. Part. Nucl. Phys.* **37**, 1 (1996).
30. S. J. Dong, K. F. Liu, and A. G. Williams, *Phys. Rev.* D **58**, 074504 (1998).
31. H. Weigel, *et al.*, *Phys. Lett.* B **353**, 20 (1995).
32. N. W. Park, J; Schecter, and H. Weigel, *Phys. Rev.* D **43**, 869 (1991).
33. N. W. Park and H. Weigel, *Nucl. Phys.* A **541**, 453 (1992).
34. T. R. Hemmart, Ulf-G. Meissner, and S. Steinenger, *Phys. Lett.* B **437**, 184 (1998).
35. R. L. Jaffe, *Phys. Lett.* B **229**, 275 (1989).
36. H. -W. Hammer, Ulf-G. Meissner, and D. Drechsel, *Phys. Lett.* B **367**, 323 (1996).
37. W. Koepf, E. M. Henley, and J. S. Pollock, *Phys. Lett.* B **288**, 11 (1992).
38. Ulf-G. Meissner, V. Mull, J. Speth, J. W. Van Orden, *Phys. Lett.* B **408**, 381 (1997).
39. M. J. Musolf and M. Burkhardt, *Z. Phys.* C **61**, 433 (1994).
40. H. Ito, *Phys. Rev.* C **52**, R1750 (1995).
41. P. Geiger and N. Isgur, *Phys. Rev.* D **55**, 299 (1997).
42. K. S. Kumar, E. W. Hughes, R. Holmes, and P. A. Souder, *Mod. Phys. Lett.* **A10**, 2979 (1995).
43. T. W. Donnelly, T. W. Dubach, and Ingo Sick, NPA **503**, 589 (1989).
44. C. J. Horowitz, *Phys. Rev.* C **57**, 3430 (1998).